Cancer Treatment and Research

Volume 177

This book series provides detailed updates on the state of the art in the treatment of different forms of cancer and also covers a wide spectrum of topics of current research interest. Clinicians will benefit from expert analysis of both standard treatment options and the latest therapeutic innovations and from provision of clear guidance on the management of clinical challenges in daily practice. The research-oriented volumes focus on aspects ranging from advances in basic science through to new treatment tools and evaluation of treatment safety and efficacy. Each volume is edited and authored by leading authorities in the topic under consideration. In providing cutting-edge information on cancer treatment and research, the series will appeal to a wide and interdisciplinary readership. The series is listed in PubMed/Index Medicus.

More information about this series at http://www.springer.com/series/5808

Craig Meyers
Editor

HIV/AIDS-Associated Viral Oncogenesis

Second Edition

Springer

Editor
Craig Meyers
Penn State Cancer Institute
Penn State University
Hershey, PA, USA

ISSN 0927-3042 ISSN 2509-8497 (electronic)
Cancer Treatment and Research
ISBN 978-3-030-03501-3 ISBN 978-3-030-03502-0 (eBook)
https://doi.org/10.1007/978-3-030-03502-0

Library of Congress Control Number: 2018960201

This Springer imprint is published by the registered company Springer Nature Switzerland AG
The registered company address is: Gewerbestrasse 11, 6330 Cham, Switzerland

Contents

Overview

According to the WHO, there are currently over 36 million people living with HIV globally, but with the successes of the antiretroviral therapy (ART), deaths due to AIDS have continued to decline, and people living with HIV (PLWH) are now living a longer and normal life span. However, non-AIDS-associated diseases are now increasing in PLWH, and cancer has now become a leading cause of morbidity and mortality. It has been estimated that cancer is responsible for over one-third of all deaths in HIV-infected individuals. The majority of cancers in AIDS patients are known to associate with co-infection with known oncogenic viruses such as human papilloma virus (HPV), Epstein Barr virus (EBV), the Kaposi's sarcoma associated herpesvirus (KSHV) or human herpesvirus-8 (HHV-8), as well as hepatitis B (HBV) and hepatitis (HCV), and more recently the Merkel cell polyoma virus (MCPyV). With the successes of ART, a number of the AIDS-associated malignancies such as Kaposi's sarcoma (KS) and some lymphomas have declined in the developed countries, but the KS disease burden remains high in Africa; the incidence of KS has reported to be as high as prostate cancer in the US. In addition, several additional non-AIDS defining malignancies (NADM) like anal cancers, oropharyngeal cancers, Hodgkin lymphomas, hepatocarcinomas, and even lung cancers are occurring more often in PLWH. Therefore, there is still an urgent need to have a better understanding of the epidemiology of these cancers, the risk factors involved, the clinical presentations, the treatment, and their associate viral etiological agents, including the viral gene functions, and their effects on the host in leading to cellular transformation and oncogenesis.

This book *HIV/AIDS-Associated Viral Oncogenesis* edited by Meyers represents a must read material for clinicians, researchers, and students who are interested in this area. It consists of review chapters authored by leading experts in the field, covering all the known human oncogenic viruses and malignancies that are associated with AIDS and NADM. There are a total of nine comprehensive chapters; one chapter is on HIV/AIDS malignancies; two chapters on KSHV and KS; one chapter is on EBV and associated lymphomas. There are three chapters on HPV and its associating cancers, head and neck squamous cell carcinomas and oral cancers, the anal cancer, and cervical cancers. There is one chapter on MCPyV and Merkel cell carcinomas, and one chapter on HBV/HCV and hepatocarcinomas.

The chapter by Pyring provided a comprehensive review on the current status of AIDS/HIV associated malignancies and their associating viral etiological agents. There are two chapters on KSHV and KS. The first is by He et al. The authors provided a comprehensive review on the molecular biology of KSHV, the regulation of the viral gene expression, the host immune response against the virus infection, and the mechanisms of cellular transformation and tumorigenesis. The second KSHV chapter by Dittmer and Damania described the KSHV and its associate diseases. It also described the prevalence of infection, the molecular biology of the virus and the disease, and its treatment. The chapter on EBV and lymphomas by Lang et al provided a comprehensive review of EBV, its molecular biology and the regulation of viral and cellular gene expression in EBV-associated lymphomas. It also described the various types of lymphomas associated with EBV and its association with HIV infection. There are three chapters on the three cancers associate with HPV. The first chapter is on HPV associate cervical cancer by Du; it described the biology of HPV and the global burden of cervical cancer, and co-infection by HIV in women. It also reviewed the risk factors involved, screening for cervical cancers, and prevention of HPV infection. The second chapter on HPV is by Hagansee on oral cancer. It described the risks, the prevalence and prevention of the cancer. It also described the molecular mechanisms that underlie HPV-mediated oncogenesis to lead to cancer. The third chapter on HPV and anal cancer is by Wang and Polefsky who reviewed the current literatures on anal cancers, the virus, the epidemiology, the clinical characteristics, the prevention, as well as the treatment and outcome of the cancer. The chapter by Caprio on Merkel cell carcinoma reviewed the clinical disease, its etiological agent and the gene regulation of the virus and changes in the tumor at the molecular levels. Finally, the chapter by Hu et al on HBV/HCV liver cancers reviewed the epidemiology of HBV/HCV and HIV co-infections; also on the possible mechanisms of hepatocarcinogenesis as well as the management of the cancer. It also discussed the other hepatitis virus, the HGV.

AIDS-Associated Malignancies

1

Ramya Vangipuram and Stephen K. Tyring

Contents

Abstract

Malignancies were one of the earliest recognized manifestations that led to the description of the acquired immune deficiency syndrome (AIDS). The majority of cancers in AIDS patients are associated with coinfection with oncogenic viruses, such as Epstein–Barr virus, human herpesvirus 8, and human papillomavirus, with resulting malignancies occurring secondary to diminished immune surveillance against viruses and virus-infected tumor cells. Over 50% of AIDS lymphomas are associated with Epstein–Barr virus (EBV) and/or HHV8 infection. HHV8-associated diseases include Kaposi sarcoma (KS), primary effusion lymphoma (PEL), and multicentric Castleman disease (MCD). EBV is associated with several malignancies, including Hodgkin lymphoma (HL) and non-Hodgkin lymphoma (NHL). Coinfection with HIV and HPV is associated with an increased risk of various squamous cell carcinomas of epithelial tissues. HAART has significantly impacted the incidence, management, and prognosis of AIDS-related malignancies. In addition to changing the natural history of HIV infection in regard to incidence and survival, HAART has dramatically

R. Vangipuram · S. K. Tyring (✉)
Department of Dermatology, University of Texas Health Sciences Center at Houston, Houston, TX 77030, USA
e-mail: styring@ccstexas.com

C. Meyers (ed.), *HIV/AIDS-Associated Viral Oncogenesis*, Cancer Treatment and Research 177, https://doi.org/10.1007/978-3-030-03502-0_1

decreased the incidence of certain virally mediated HIV-associated malignancies such as KS and primary CNS lymphoma. The beneficial effects of HAART on these tumors are attributed to drug-mediated HIV suppression and immune reconstitution. However, HAART has had a less favorable impact on EBV- and HPV-related malignancies. This chapter presents an overview of HIV-associated malignancies mediated by HHV-8, EBV, and HPV, and reviews the effect of HAART on the epidemiology, presentation, treatment, and outcomes of these cancers.

Keywords
Human herpesvirus 8 · Epstein–Barr virus · Human papillomavirus
Human immunodeficiency virus · Kaposi sarcoma · AIDS-associated lymphoma · Anogenital cancer

1.1 Introduction

Malignancies were one of the earliest recognized manifestations that led to the description of the acquired immune deficiency syndrome (AIDS). The rising incidence of Kaposi sarcoma in young homosexual men, a rare skin cancer typically seen in elderly men of Eastern European and Mediterranean descent, was a harbinger of the AIDS epidemic in the early 1980s. This was followed by sporadic reports of high-grade B-cell non-Hodgkin's lymphoma (NHL), primary cerebral lymphoma, and systemic NHL. By 1985, both Kaposi's sarcoma and high-grade B-cell NHL were classified as "AIDS-defining" illnesses by the Centers for Disease Control (CDC). In subsequent years, the CDC listed invasive cervical cancer as an AIDS-defining illness, given its poorer prognosis in HIV-positive women. Research later showed that the majority of cancers in AIDS patients were associated with coinfection with oncogenic viruses, such as Epstein–Barr virus, human herpesvirus 8, and human papillomavirus, with resulting malignancies occurring secondary to diminished immune surveillance against viruses and virus-infected tumor cells.

Over 50% of AIDS lymphomas are associated with Epstein–Barr virus (EBV) and/or HHV8 infection. HHV8-associated diseases include Kaposi sarcoma (KS), primary effusion lymphoma (PEL), and multicentric Castleman disease (MCD). EBV is associated with several malignancies, including Hodgkin lymphoma (HL) and non-Hodgkin lymphoma (NHL). EBV is also implicated in cases of leiomyosarcoma, cervical, and anal cancer in patients with AIDS.

HAART has significantly impacted the incidence, management, and prognosis of AIDS-related malignancies. In addition to changing the natural history of HIV infection in regard to incidence and survival, HAART has dramatically decreased the incidence of certain virally mediated HIV-associated malignancies such as KS and primary CNS lymphoma. The beneficial effects of HAART on these tumors are attributed to drug-mediated HIV suppression and immune reconstitution. However, HAART has had a less favorable impact on EBV-related malignancies; NHLs remain the most common tumors in the HAART era. This chapter presents an overview of HIV-associated malignancies mediated by HHV-8, EBV, and HPV, and reviews the effect of HAART on the epidemiology, presentation, treatment, and outcomes of these cancers.

1.2 Gammaherpesvirus-Associated Malignancies

The human gammaherpesvirus family includes Epstein–Barr virus (EBV) and human herpesvirus (HHV8), previously known as Kaposi sarcoma-associated herpesvirus (KSHV). Gammaherpesviruses establish a persistent infection, especially in lymphoid cells. In the immunocompetent host, the clinical course is usually asymptomatic. In immunocompromised hosts, such as post-transplant patients on immunosuppression and HIV-infected patients, both EBV and HHV8 are implicated in the development of a wide range of lymphoproliferative disorders.

A. Human Herpes Virus-8-Related Tumors

HHV8 was first isolated from Kaposi sarcoma lesions in patients with AIDS by Chang and Moore in 1994 [1], and subsequent studies demonstrated its association with other lymphoproliferative disorders in this population [2, 3]. Human herpesvirus-8 virus is the etiologic agent of three AIDS-associated malignancies: Kaposi sarcoma, a plasmablastic variant of multicentric Castleman disease (HHV8-MCD), and primary effusion lymphoma. Unlike other herpesviruses, HHV8 is not ubiquitous: while HHV8 is highly prevalent in sub-Saharan Africa (>50%), it is quite rare in most European countries, Asia, and the United States (seroprevalence rate <10%) [4, 5]. The prevalence is elevated in men who have sex with men (MSM) [6–8]. HHV8 is mainly transmitted via saliva [9], and sexual risk factors are probably a surrogate marker for close physical contact [10, 11].

Five clinical variants have been described: classic, endemic, iatrogenic, AIDS-associated (epidemic), and non-epidemic KS [12]. Classic KS describes an indolent cutaneous disease among elderly men of Mediterranean, Eastern European (Ashkenazi) Jewish, and South American origin. Endemic KS is an aggressive HIV-unrelated form that is commonly seen in sub-Saharan Africa, and often presents with visceral involvement. Iatrogenic KS is seen in patients receiving immunosuppressive drugs, particularly those with solid-organ transplants [13]. KS was one of the first manifestations of the HIV/AIDS epidemic in the 1980s. Recently, non-epidemic KS was proposed as a fifth subtype in patients who are at high risk for HIV, but are HIV seronegative [12].

1. AIDS-Associated KS

The introduction of HAART in the mid-1990s has led to a significant reduction in the incidence of KS in developed countries. However, KS is still the most common AIDS-defining malignancy in parts of sub-Saharan Africa where the seroprevalences of both HIV and HHV8 are high. AIDS-related KS exhibits a wide spectrum of clinical presentations. KS is staged using the AIDS clinical trials group modified staging classification (Table 1.1). The prognosis depends on the stage of KS, the level of immunosuppression, and the response to anti-HIV therapy. HAART results in a decrease in the incidence of KS.

KS is a multicentric angioproliferative spindle cell tumor arising from HHV8-infected lymphatic endothelial cells. While HHV8 is the etiologic agent of KS, HIV-induced immunosuppression is also an important cofactor in the induction of this malignancy. Both absolute decreases in CD4+ counts and lack of HHV8-specific T-cell immunity are associated with KS [14, 15]. In addition, KS is independently associated with the degree of HIV viremia [16]. Before the widespread use of HAART, patients coinfected with HIV and HHV8 were estimated to be 400–2000 times more likely to develop KS than those with just HHV8 infection [17]. The implementation of HAART in the United States and Western Europe resulted in an initial 80% decrease in the incidence of KS [18]. However, further decreases after 2000 have been more modest, and KS remains the second most common tumor arising in HIV-infected persons in the United States, after non-Hodgkin lymphoma, with a cumulative incidence of approximately 2% in the HAART era [19].

KS lesions may involve the skin, oral mucosa, lymph nodes, and visceral organs, especially the pulmonary and gastrointestinal tract. Most patients present with painless cutaneous lesions, which may have a macular, papular, nodular, or plaque-like appearance. Lesions can range in color from pink to red or purple and in size from several millimeters to large confluent areas. Lesions are typically localized in the oral cavity, on the face, and lower extremities, but can involve almost any site. Visceral disease sometimes occurs in the absence of skin lesions.

Table 1.1 AIDS clinical trials group modified staging classification

TIS staging of KS	Good risk (all of the following)	Poor risk (any of the following)
Tumor (T)	Confined to skin, lymph nodes, or minimal oral disease	Tumor-associated edema or ulceration, extensive oral KS, gastrointestinal KS, KS in other non-nodal viscera
Immune status (I)	CD4 cell count >150 cells/uL	CD4 cell count <150 cells/uL
Systemic illness (S)	Karnovsky performance status >70	Karnovsky performance status <70 or other HIV-related illness

Oral lesions may lead to ulceration, dysphagia, and secondary infection. Gastrointestinal KS has been described in almost half of patients at the time of initial diagnosis [20]. Gastrointestinal involvement is often asymptomatic; however, bleeding, perforation, and obstruction may occur [21]. Pulmonary KS is more frequent among patients with extensive cutaneous disease and more advanced immunosuppression, though 15% of patients with pulmonary KS have no mucocutaneous lesions at diagnosis [22]. In contrast to KS at other visceral sites, pulmonary KS is frequently symptomatic, and patients may present with bronchospasm and/or dyspnea, which may be life-threatening [22].

AIDS-associated KS is staged by the classification developed by the AIDS Clinical Trials Group (ACTG) Oncology Committee [23]. This classification utilizes three variables: tumor extent (T), immune status (I), and systemic symptoms (S), which are classified as good risk (0) or poor risk (1). For tumor burden (T), poor risk (T_1) is defined by the presence of extensive cutaneous or oral disease, tumor-associated edema, ulceration or visceral disease; for immune status, poor risk (I_1) is defined by CD4+ <150 cells/μL; and for systemic illness, poor risk (S_1) is defined by the presence of other opportunistic infections, constitutional symptoms, or poor performance status. The ACTG staging system was developed and initially validated in the pre-HAART era. A survival analysis conducted after the introduction of HAART suggested that tumor extent and systemic illness, rather than CD4+ T-cell count were the most important predictors of survival [24]. It has been proposed that patients can be classified into two main risk categories: good risk (T_0S_0, T_1S_0, or T_0S_1) and poor risk (T_1S_1) [24]. The 3-year survival rate for patients at stage T_1S_1 is 53%, compared to the 3-year survival rates with T_0S_0, T_1S_0, and T_0S_1, which were 88, 80, and 81%, respectively [24].

The introduction of HAART has dramatically improved the overall survival of patients with KS. The incidence rate of KS declined from 15.2 per 1000 patient-years to 4.9 per 1000 patient-years after the introduction of HAART, with a relative risk (RR) for KS of 0.32 (99% confidence interval [CI] 0.26–0.4) in the HAART era compared with the pre-HAART era [25]. Effective control of HIV viremia with HAART is imperative in patients with AIDS-KS and in patients with limited KS, is often sufficient [26]. For HAART-naïve patients with early KS (T0), the administration of HAART alone was associated with disease regression in several studies [27, 28]. While there is some evidence that HIV protease inhibitors have specific anti-KS activity [29], most studies indicate that prevention or control of KS is related to the degree of control of HIV, rather than the specific HAART regimen utilized [30]. In addition to HAART, a wide variety of treatments appear able to inhibit KS growth, including antiretrovirals, cytotoxic chemotherapeutic agents, retinoids, thalidomide, and matrix metalloproteinase inhibitors [28–34].

2. HHV8-Associated Multicentric Castleman Disease

Castleman disease was originally described in 1956 as localized lymph node hyperplasia resembling a thymoma [35]. It is now understood to be not just a single disease but rather an uncommon, heterogeneous group of nonclonal lymphoproliferative disorders, which have a broad spectrum of clinical expression. There are generally two clinical variants: either localized to a single lymph node (unicentric) or with systemic involvement (multicentric). Multicentric Castleman disease (MCD) presents with generalized lymphadenopathy, multi-organ involvement, systemic symptoms of fever, fatigue, weight loss, and carries the potential for malignant transformation [36]. HHV8 is the etiologic agent of a plasmablastic form of MCD that is observed in HIV patients. MCD was first diagnosed in two homosexual men with AIDS in 1985 [37]. In individuals with AIDS, MCD is linked with malignant transformation to non-Hodgkin's lymphoma at 15-fold higher rate than those without MCD [38]. Unlike KS, HHV8-MCD appears to be becoming more frequent with the widespread use of HAART [39].

The clinical presentation of HHV8-MCD includes intermittent fevers, night sweats, fatigue, cachexia, edema, along with lymphadenopathy and/or hepatosplenomegaly [40]. Nonspecific respiratory and GI symptoms are common as well. Common laboratory abnormalities include anemia, cytopenias, hypoalbuminemia, hyponatremia, hypergammaglobulinemia, and elevated inflammatory markers such as C-reactive protein (CRP) [41]. HHV8-MCD symptoms are mediated by certain cytokines, especially human IL-6, HHV8 vIL-6, and human IL-10 [41]. vIL-6 is believed to play an important role in pathogenesis of HHV8-MCD, which may be independent or complementary to that of human IL-6, through autocrine and paracrine mechanisms of action [42]. HHV8-MCD is diagnosed via biopsy, whereby affected lymph nodes demonstrate involuted germinal centers with hyperplasia of vasculature and expansion of HHV8-infected plasmablasts in the mantle zone of the follicles [43].

Patients may have a waxing and waning course with exacerbations and subsequent remissions. At times, symptom flares can be severe and fatal. Flares are typically associated with high HHV8 viral loads [44]. There is no single consensus definition of HHV8-MCD flare or symptomatic activity; however, the French ANRS (Agence Nationale de Recherche sur le SIDA Castleman B trial group) have described criteria to define an attack of HIV MCD, based on fever, a C-reactive protein greater than 20 mg/L in the absence of any other cause, and 3 of 12 additional clinical findings (Table 1.2) [45]. HHV8-viral load has at times been used to assess symptomatic patients with HHV8-MCD, although assays vary between groups, and elevated HHV8 viral load is not specific for HHV8-MCD [46]. CT imaging in patients with HHV8-MCD generally shows diffuse, symmetric adenopathy, and hepatosplenomegaly [43]. Hemophagocytic syndromes also have been described [43]. Concomitant KS is present in up to 70% of individuals [43].

There is no standard therapy for HHV8-MCD. HIV-positive patients with HHV8-MCD generally are treated with concurrent HAART in addition to various therapies such as immune modulators, chemotherapy, and antiviral agents [47]. Rituximab, an anti-CD20 monoclonal antibody, given alone or in conjunction with chemotherapy, is thought to confer a beneficial effect by eliminating reactive B-cells, thus depriving the HHV8-infected plasmablasts of proliferation and survival signals by breaking virus and cytokine-driven feedback loops with the

Table 1.2 French ANRS criteria for HIV MCD flare

1. Fever
2. Serum C-reactive protein level >20 mg/L in the absence of any other etiology
3. At least, three of the following symptoms:
– Peripheral lymphadenopathy
– Splenomegaly
– Edema
– Pleural effusion
– Ascites
– Cough
– Nasal obstruction
– Xerostomia
– Rash
– Central nervous system symptoms
– Jaundice
– Autoimmune hemolytic anemia

Table 1.3 Expression of EBV latent genes and association with lymphomas

Latency pattern	EBNA-1	EBNA-2	EBNA-3	LMP-1	LMP-2	EBER	Disease
Type I	+	–	–	–	–	+	Burkitt lymphoma, primary effusion lymphoma
Type II	+	–	–	+	+	+	Hodgkin lymphoma
Type III	+	+	+	+	+	+	Primary CNS lymphoma

reactive B-cells [48, 49]. However, rituximab is associated with exacerbations of cutaneous KS [50].

Human IL-6 is important to the pathogenesis of MCD, and the use of monoclonal antibodies directed against IL-6 (siltuximab) or its receptor (tocilizumab) has shown clinical efficacy in HIV-negative HHV8 negative MCD [51–53]. However, because vIL-6 is antigenically different from human IL-6, a potential role for siltuximab in the treatment of HHV8-MCD remains to be explored. While human IL-6 is elevated in HHV8-MCD and contributes to symptoms and disease pathogenesis, given the additional role of vIL-6 and other HHV8 genes, it is unknown whether antihuman IL-6 therapy alone will be sufficient.

Even though life expectancy in multicentric Castleman disease has improved in the HAART era, it continues to have a poor prognosis and an increased incidence of non-Hodgkin lymphoma in the HIV context. Infection, multi-organ failure, Kaposi sarcoma, non-Hodgkin lymphoma, and progressive multicentric Castleman disease were the most often reported causes of death [43].

3. Primary Effusion Lymphoma

PEL is a rare lymphoproliferative disorder, accounting for 1–4% of all AIDS-related lymphomas, and even fewer cases in the HIV-negative individual. It was first reported in 1989 by Knowles et al. as a lymphoma syndrome characterized by malignant effusions in HIV-positive individuals. PEL is divided into classic and solid variants. Lymphomatous effusions are characteristic of classic PEL. Pleural involvement is seen in 60–90% of patients, followed by peritoneal (30–60%), pericardial (up to 30%), joint spaces, and rarely, meninges [54–56]. The solid variant of PEL presents with tissue-based tumors involving the GI tract, lung, central nervous system (CNS), skin, and lymph nodes [57]. According to the World Health Organization (WHO) classification of tumors of hematopoietic and lymphoid tissues, the presence of HHV8 is considered an essential criterion for the diagnosis of PEL, though few cases of HHV8-negative PEL have been described. PEL is often associated with other HHV8-associated malignancies, such as KS and HHV8-MCD. Along with HHV8 infection, 70–80% of cases have coexisting EBV infection and latency I gene expression (90%) [58].

PEL usually presents in HIV-infected young to middle-aged men, with a median age at diagnosis of 41 years. The male to female ratio is 8:1. Patients are usually severely immunosuppressed and present with advanced disease (stages III and IV) at diagnosis. The diagnosis of PEL requires the demonstration of HHV8 in the neoplastic cells. EBV coinfection can be demonstrated through in situ staining for EBV-encoded small RNAs. Neoplastic cells have a unique phenotype characterized by CD45, CD30, CD38, CD138, and MUM1 coexpression [55]. Classic B-cell and T-cell markers are typically not seen.

There is no standard therapy for PEL and prognosis is poor. Historically, median survival ranges between 3 and 9 months; 2-year overall survival rates of 33–39% are reported in studies using CHOP (cyclophosphamide, doxorubicin, vincristine, and prednisone) or CHOP-like regimens [56]. Bortezomib, a proteasome inhibitor that induces HHV8-lytic activation, along with lenalidomide, an immunomodulatory and antineoplastic agent, is commonly used in combination with chemotherapy [59]. Other targeted agents that have demonstrated activity in mouse models include the mTOR inhibitor sirolimus and the anti-CD30 immunotoxin, brentuximab vedotin [60, 61]. Highly active antiretroviral therapy should be continued or initiated in patients with HIV infection. Although PEL is a CD20-negative tumor, advances in understanding the mechanism of HHV8-infection of B-cells, and clinical overlap with HHV8-MCD support the use of rituximab in the treatment of PEL, especially in patients with concurrent HHV8-MCD.

B. Epstein–Barr Virus

EBV is the most common persistent virus infection in humans, and approximately 95% of the world′s population has an asymptomatic lifelong carrier status. Acute infection occurs in the epithelium of the oropharynx and may be asymptomatic or cause infectious mononucleosis. In immunocompetent persons, the virus then generally forms an asymptomatic latent chronic infection primarily in B-cells [62].

EBV encodes several latency-associated genes that are variably expressed during primary and chronic infection, and which may contribute to oncogenesis. The level of immunosuppression defines the type of lymphoma that will develop in the HIV$^+$ setting [63]. EBV-associated lymphomas can be classified into 3 different categories based on latency patterns, which provide insight into disease pathogenesis. Three EBV latency patterns are recognized (Table 1.3).

Latency 1 tumors generally occur at relatively preserved CD4+ T-cell counts. The tumor cells express EBV nuclear antigens (EBNA) 1, EBV-encoded RNA (EBER), and several microRNAs, and are associated with monomorphic DLBCL, Burkitt lymphoma, and plasmablastic lymphoma [64]. EBNA1 expressed in all latently infected cells is responsible for the maintenance and replication of the episomal EBV genome. It can also induce oxidative stress as well as promote telomere dysfunction [65]. EBNA1 may inhibit c-Myc induced apoptosis, and thereby contribute to oncogenesis [65].

Latency II is an intermediate pattern with expression of many proteins except for EBNA2. The expression of latency membrane protein 1 (LMP1) in the absence of EBNA2 is used to demonstrate latency type II. LMP1 functions as a classic oncogene and is essential for EBV-induced B-cell transformation in vitro [66]. Classical Hodgkin lymphoma has latency 2 pattern (Table 1.3).

EBV type 3 latency pattern is the most immunogenic and is characterized by the expression of all six EBV nuclear antigens (EBNA) and all three LMPs. This latency pattern can be found in individuals with severe immunosuppression and is typical of EBV-infected immortalized B-cells. The viral proteins are highly immunogenic and trigger a strong cytotoxic T-cell reaction. The expression of EBNA2 is important to demonstrate latency type III. Latency 3 tumors occur at the lowest CD4+ T-cell counts and include primary CNS lymphoma. CD4+ T-cell immune reconstitution with HAART is most important in the prevention and treatment of this category of EBV-associated lymphomas.

1. AIDS-Related Lymphomas

AIDS-related lymphoma remains a significant cause of morbidity and mortality in HIV-infected individuals. The increased risk for lymphoma among HIV-infected individuals is related to duration and degree of immunosuppression, induction of cytokines leading to B-cell proliferation, and opportunistic infections with oncogenic herpesviruses such as EBV and HHV8 [67]. The relative risk of AIDS-associated malignancies increases as CD4+ T-cell counts decline [68]. Over 80% of all cases are associated with EBV. The HIV-associated lymphoma subtypes, which are related to EBV infection, include both classic Hodgkin's lymphoma and non-Hodgkin's lymphomas, such as diffuse large B-cell lymphoma, primary central nervous system lymphoma, Burkitt lymphoma, and plasmablastic lymphoma. EBV is also associated with rare cases of leiomyosarcoma in children with HIV [69].

Compared with the types of lymphomas that occur in HIV-negative individuals, AIDS-associated lymphomas usually comprise the more aggressive histological subtypes and have a higher incidence of extranodal involvement and an aggressive clinical course [70]. The incidence of high-grade B-cell non-Hodgkin lymphoma (NHL) in the pre-HAART was 60–200 times higher in HIV-infected individuals than in HIV-uninfected persons [70]. HAART is associated with a decrease in incidence of opportunistic infections and AIDS-associated malignancies, including NHL; nevertheless, the incidence ratio of NHL still remains relatively high in HIV-infected patients [70–72]. However, the incidence of PCNSL has dramatically decreased since the introduction of HAART [73]. Although the incidencc of HIV-associated NHLs has significantly decreased after the introduction of HAART, with the most dramatic decline observed in PCNSL, this decline is less marked than other HIV-associated morbidities [74]. The overall prevalence of HIV-associated lymphoma is significantly higher compared to that of the general population and it continues to be relevant even after the wide availability of HAART.

a. Non-hodgkin Lymphomas

NHL has been considered as an AIDS-defining cancer since 1985 and still remains one of the major causes of death in HIV-infected patients [75]. Diffuse large B-cell lymphomas (DLBCL), which include including primary central nervous system lymphoma (PCNSL) and Burkitt lymphoma (BL), constitute 90% of HIV-related non-Hodgkin lymphomas (NHL) with relative frequencies of 50 and 40%, respectively [76]. After the implementation of HAART, the risk of developing aggressive B-cell NHL has decreased, while the risk of developing plasmablastic lymphoma, primary effusion lymphoma (PEL), and classical Hodgkin lymphoma has increased [18].

i. Diffuse Large B-Cell Lymphoma

HIV-associated diffuse large B-cell lymphomas can involve lymph nodes, or present in virtually any extranodal site. The brain is the most common extranodal site, with primary CNS lymphomas accounting for 15–30% of HIV-associated NHL lymphomas [76]. Other frequently involved extranodal sites in HIV-infected patients include the gastrointestinal tract, liver, and bone marrow.

ii. Primary CNS Lymphoma

PCNSL is defined as the involvement of the brain, leptomeninges, eyes or spinal cord by a lymphoma. It occurs late in the course of HIV disease and is associated with extremely low CD4 cell counts (<50 cells/μl). HIV^+ patients present at a younger age, worse performance status, higher lactic dehydrogenase (LDH) at presentation and shorter overall survival compared to HIV^- patients [77].

Clinical presentation includes headaches, focal neurological signs, changes in mental status, confusion, memory loss, and seizures [78]. Radiography shows solitary or multiple contrast-enhanced lesions, often with periventricular and central necrosis, along with prominent mass effect and edema. It is often difficult to differentiate cerebral mass lesion from toxoplasmosis as both are seen in advanced immunodeficiency (CD4 cell counts <50 cells/mm [3]) and present with headaches, focal neurologic deficits, and similar radiological findings.

All cases of HIV-associated PCL are associated with EBV, which may be detected by immunohistochemical staining of biopsy tissue or by PCR amplification of cerebrospinal fluid. Histopathology reveals high-grade, diffuse large B-cell or immunoblastic non-Hodgkin lymphoma cells. Despite a high response rate of around 50%, radiotherapy, when used alone, does not provide a substantial survival benefit in patients with primary CNS lymphoma, with a median overall survival of 10–18 months and 5-year overall survival of 5% [79]. High-dose methotrexate-based chemotherapy is recommended for first-line treatment of primary CNS lymphoma [79]. Chemotherapeutic treatments to be combined with high-dose methotrexate should be selected from active drugs known to cross the blood–brain barrier.

iii. Burkitt Lymphoma

BL is a highly aggressive B-cell NHL. BL constitutes 40% of the HIV-related lymphomas. In contrast to DLBCL, BL tends to occur in patients with relatively preserved immune function; CD4+ T-cell counts are relatively normal (usually >200/μl) [80].

Patients typically present with advanced stage disease, B symptoms such as fever, night sweats, weight loss, and poor performance status. Nodal and extranodal involvement is common, with lesions frequently seen in the gastrointestinal tract, bone marrow (BM), or central nervous system (CNS) [81]. Symptoms are related to either abdominal mass or extensive bone marrow infiltration. Morphologically, BL in the HIV^+ setting demonstrates more variation in cell size and shape with more plasmacytoid morphology with eccentrically located nuclei and amphophilic cytoplasm as compared to sporadic cases [81]. The clinical course is rapidly progressive, with a propensity to involve the CNS. Prognostic factors include CD4+ T-cell count, PS, and bone marrow involvement [82]. HIV-positive patients are currently treated with the same intensive chemotherapy regimens used for immunocompetent patients. Survival of BL patients has remained poor in the HAART era [83].

iv. Plasmablastic Lymphoma

Plasmablastic lymphoma (PBL), a distinct subtype of DLBCL, constitutes approximately 3% of HIV-related lymphomas [84]. It is considered an AIDS-defining illness and was first described in 1997 as an HIV-related lymphoma involving the jaw and oral cavity [85]. Subsequent studies have shown its association with other types of immunodeficiency including post-transplant and iatrogenic, and age-related immunosenescence [86]. HIV-associated PBL has a predilection for the oral cavity (50%); however, in 45% of the cases there is extraoral involvement; most commonly the gastrointestinal tract, followed by the sinonasal cavity, skin, soft tissue, lung, bones, and less frequently lymph nodes [84]. PBL occurs at all ages, but patients with HIV-related disease are significantly younger than those with other types of immunosuppression, tend to be male, and have advanced stage (III/IV) at presentation. The prognosis is generally poor with a high mortality. Studies have shown poor prognosis regardless of the therapy received; in many cases, even a complete response to chemotherapy did not significantly improve survival, with median time of 6–7 months [87].

b. Hodgkin Lymphoma

HL is the most common non-AIDS-defining cancer [88]. HIV-infected patients have a fivefold to 25-fold higher chance of developing HL in comparison to the general population, with HAART therapy having no impact in its incidence [89]. HIV-associated HL has an aggressive clinical presentation with systemic B symptoms, disseminated extranodal disease, and bone marrow involvement in roughly 50% of cases [90]. The predominant subtypes of HL in HIV-positive are the mixed cellularity (MC) and lymphocyte-depleted (LD) individuals, while in the general population the most common is the nodular sclerosis subtype [88].

The combined use of HAART and chemotherapy has significantly improved the prognosis of patients with HIV-HL. The post-HAART era is associated with an improvement in survival which was attributed to virological response to antiretroviral therapy and a reduction in HIV-associated mortality [91, 92]. Current recommendations are to treat HIV-infected individuals as aggressively as HIV-negative patients, along with providing supportive therapy such as hematopoietic growth factors, and prophylaxis against opportunistic infections.

1.3 Human Papillomavirus-Associated Malignancies

Human papillomavirus (HPV) is the most common sexually transmitted infection worldwide, with 14 million persons infected annually and a prevalence of 79 million persons worldwide [93]. It is the etiologic agent of cervical cancer and is related to a subset of cancers of the anus, penis, vagina, and vulva, as well as a proportion of head and neck squamous cell cancers in the immunocompetent patient and in individuals with AIDS [94]. Persons with AIDS have an increased risk of developing these cancers, which is related to the high incidence and persistence of HPV infections in this population [95]. HPV prevalence in HIV-positive men has been reported to be as high as 93% [93].

Human papillomaviruses are small DNA viruses that infect squamous epithelial tissues. The highest prevalence of HPV is found in the skin (61%), followed by the mucosal surfaces, including vagina (41.5%), and mouth (30%) [96]. Over 150 types of HPV have been identified; approximately one-third of these infect the squamous epithelia of the genital tract and are sexually transmitted. HPV subtypes are divided into two groups: low-risk, non-oncogenic HPV types which are associated with anogenital warts, and high-risk oncogenic types which are associated with genital and oropharyngeal cancers. Approximately, 15 genital HPVs are categorized as high-risk genotypes and cause most cervical cancers, with over 99% of cervical lesions containing HPV viral sequences [97]. High-risk HPVs are also associated with many penile, vulvar, and anal carcinomas, and contribute to over 40% of oral cancers [98]. High-risk genotypes include 16, 18, 31, 35, 45, 51, 52 and 58 (Table 1.4) [99, 100]. High-risk HPV are more prevalent among HIV-infected individuals [101].

HPV oncoproteins disrupt multiple cellular signaling pathways to maintain infected cells in a proliferative state to facilitate viral replication and persistence. Consequently, mutations in cellular genes accumulate, leading to increased genomic instability, which results in full transformation. The primary viral factors responsible for altering these pathways and mediating progression to malignancy are the E5, E6, and E7 proteins [102]. The efficient disruption of p53 and Rb function by E6 and E7 is crucial for this process [102].

HPV-related tumors in HIV-positive patients tend to occur at a younger age and at a more advanced stage than in HIV-negative patients [103–105]. In a multivariate analysis, low CD4+ counts (≤ 200 cells/μL) were shown to be the strongest independent predictor of infection with high-risk HPV genotypes and genital warts [105]. Furthermore, HIV-positive patients with genital warts have greater resistance to standard treatment and HIV-positive women being treated for CIN are more likely to relapse, as compared to the general population [106, 107]. Similar to other sexually transmitted infections, HPV is thought to confer greater susceptibility to the acquisition of HIV. One randomized control trial of 2168 young men in Kenya demonstrated that HPV infection was independently associated with HIV acquisition [108]. A similar study in the US showed comparable results with a 3.5-fold increased risk of HIV seroconversion in HPV-positive MSM [109]. The impact of HAART on HPV infection and HPV-associated diseases is not well understood. HPV-associated malignancies, such as anal carcinoma and invasive cervical cancer, have remained stable or have even increased [104].

Table 1.4 Classification of oncogenic risk by HPV genotype

Low risk	6, 11, 40, 42, 43, 44, 54, 61, 70, 72, 81, 89
High risk	16, 18, 31, 33, 35, 39, 45, 51, 52, 56, 58, 59, 66, 68, 69, 73, 82

A. Cervical Cancer

In 1993, the CDC designated squamous cell carcinoma (SCC) of the cervix as an AIDS-defining cancer, as its incidence is estimated to be five times greater among female AIDS patients [104]. This is attributed to high-risk sexual behavior which underlies both HIV and HPV transmission, as well as the fact that HPV increases the efficiency of HIV sexual acquisition and the impact of immunosuppression on HPV persistence [110]. HIV is associated with a high frequency of multiple HPV genotypes, a higher prevalence and persistence of HPV in the cervix, as well as the higher prevalence of cervical intraepithelial neoplasia (CIN)/squamous intraepithelial lesion (SIL), a higher progression from low-grade SIL (LGSIL) to high-grade SIL (HGSIL), and a greater likelihood of relapse of CIN II/III after therapy [111]. The risk for SIL is greatest among women with CD4 counts <200 cells/mm^3 [111]. HIV-infected women are more likely than uninfected women to develop cervical HPV infections, have multiple infecting HPV types, be less likely to clear HPV, and be more likely to progress through preneoplastic stages to develop cervical cancer. Studies have shown that HIV is associated with a 27% increase in cancer-specific mortality among women with cervical cancer [104]. In most centers, HIV-positive women with invasive cervical cancer are treated using the same protocols as are used in immunocompetent women, which includes a combination of surgery, chemotherapy, and radiotherapy [111].

B. HPV-Associated Squamous Cell Cancer of the Anal Cancer

HPV infection is associated with SCC of the anal canal (AC), which is relatively low in the general population but is substantially elevated in HIV-positive patients, especially men who have sex with men [110]. HPV16 has a well-documented association with AC and is found in approximately 70% of AC lesions [103]. Anal carcinoma has been included in the non-AIDS-defining cancers, which cumulatively still represent a leading cause of death among virologically suppressed individuals with high CD4+ cell counts [103]. The relative risk for developing AC is 37 times greater among HIV-positive MSM and 10 times greater among renal transplant recipients than that of the general population [109]. Its incidence continues to increase despite the introduction of HAART [103]. In addition, HIV-positivity is associated with higher recurrence rates after treatment and worse recurrence-free survival [111].

HIV-positive patients should be treated similarly to non-HIV-positive individuals. HPV-associated SCC is treated based on stage. In local and locally advanced AC, concomitant chemoradiation therapy based on mitomycin C and 5-fluorouracil (5-FU) is the current best treatment, with metastatic AC, chemotherapy with 5-FU and cisplatin remains the gold standard [112, 113]. There are no indications for induction or maintenance therapies in locally advanced tumors. Many novel strategies such as targeted therapies, vaccination, immunotherapy, and photodynamic therapy are in clinical trials for the treatment of AC.

C. Intraepithelial Neoplasia

Persistent infection with oncogenic HPV types has also been associated with vaginal intraepithelial neoplasia (VIN), penile intraepithelial neoplasia (PIN), and their progression to invasive squamous cell carcinoma. The incidence of precursor lesions and their subsequent progression to cancer is markedly higher in HIV-positive men and women compared with HIV-negative counterparts [113]. HIV-positive women have increased incidence and prevalence of both VIN and vulvovaginal carcinoma, with VIN occurring 29 times more frequently in HIV-infected compared to non-HIV-infected women [114]. VIN in immunosuppressed patients often presents as multifocal, extensive disease with tendency to recur after treatment [115]. HIV-positive men have a twofold to threefold increased risk for penile cancer compared to their HIV-negative counterparts and higher rates of PIN [116].

D. Squamous Cell Cancers of the Head and Neck

Human papillomavirus has been identified as a causal factor in a subset of head and neck squamous cell cancers, primarily involving the oropharynx. The incidence of oropharyngeal cancers has increased over the past several decades, while other head and neck cancers have decreased. The estimated oral HPV prevalence is 40% in HIV-positive individuals [117]. Most HPV-associated HNSCC involves the oropharynx, and a recent large case series of head and neck cancers in HIV-infected individuals found that 64% of HIV oropharyngeal cases were HPV positive [118].

1.4 Conclusion

Malignancies continue to be a significant cause of morbidity and mortality in patients with AIDS. The majority of cancers in these patients are associated with coinfection with oncogenic viruses such as HHV8, EBV, and HPV. For certain AIDS-defining malignancies such as KS and PCNSL, the impact of HAART has been dramatic, with significant decreases in incidence and improvement in treatment outcomes. For other malignancies such as AIDS-associated lymphoma and HPV-associated SCC, HAART appears to have no effect on the natural history of the malignancy. Because the life expectancy of patients with HIV has increased, the vulnerability of this population to comorbidities such as cancer has also risen. While there have been dramatic advances in increasing the longevity and survival of patients with HIV/AIDS, further research is necessary to continue preventing and treating malignancies in HIV/AIDS.

References

1. Moore PS, Chang Y (1995) Detection of herpesvirus-like DNA sequences in Kaposi's sarcoma in patients with and without HIV infection. N Engl J Med 332:1181–1185
2. Cesarman E, Chang Y, Moore PS, Said JW, Knowles DM (1995) Kaposi's sarcoma-associated herpesvirus-like DNA sequences in AIDS-related body-cavity-based lymphomas. N Engl J Med 332:1186–1191
3. Soulier J, Grollet L, Oksenhendler E et al (1995) Kaposi's sarcoma-associated herpesvirus-likeDNAsequences in multicentric Castleman's disease. Blood 86:1276–80
4. Grossman Z, Isovich J, Schwartz F et al (2002) Absence of Kaposi sarcoma among Ethiopian immigrants to Israel despite high seroprevalence of human herpesvirus 8. Mayo Clin Proc 77:905–909
5. Alblashi D, Chatlynne L, Cooper H et al (1999) Seroprevalence of human herpesvirus-8 (HHV-8) in countries of southeast Asia compared to the USA, the Caribbean and Africa. Br J Cancer 81:893–897
6. Martin N, Ganem D, Osmond D, Page-Shager D et al (1998) Sexual transmission and the natural history of human herpesvirus 8 infection. N Engl J Med 338:948–954
7. Smith N, Sabin C, Gopal R et al (1999) Serologic evidence of human herpesvirus 8 transmission by homosexual but not heterosexual sex. J Infect Dis 180:600–606
8. Martro E, Esteve A, Schulz T et al (2007) Risk factors for human herpesvirus 8 infection and AIDS-associated Kaposi's sarcoma among men who have sex with men in a European multicentre study. Int J Cancer 120:1129–1135
9. Casper C, Krantz E, Selke S et al (2007) Frequent and asymptomatic oropharyngeal shedding of human herpesvirus 8 among immunocompetent men. J Infect Dis 195:30–36
10. Martin J, Osmond D (2000) Invited commentary: determining specific sexual practices associated with human herpesvirus 8 transmission. Am J Epidemiol 151:225–229
11. Campbell T, Borok M, Ndemara B et al (2009) Lack of evidence for frequent heterosexual transmission of human herpesvirus 8 in Zimbabwe. Clin Infect Dis 48:1601–1608
12. Hinojosa T, Lewis DJ, Liu M, Garza G, Vangipuram R, Ramos E, Salas-Alanis JC, Nawas ZY, Tyring SK (2017) Nonepidemic Kaposi sarcoma: a recently proposed category. JAAD Case Rep 3(5):441–443
13. Siegel JH, Janis R, Alper JC et al (1969) Disseminaed visceral Kaposi's sarcoma. Apperance after human renal homograft operation. JAMA 207(8):1493–1496
14. Biggar RJ, Chatruvedi AK, Goederr JJ, Engels EA (2007) AIDS-related cancer and severity of immunosuppression in persons with AIDS. J Natl Cancer Inst 99:962–972
15. Guihot A, Dupin N, Marcelin AG et al (2006) Low T cell responses to human herpesvirus 8 in patients with AIDS-related and classic Kaposi sarcoma. J Infect Dis 194:1078–1088
16. Silverberg MJ, Chao C, Leyden WA et al (2011) HIV infection, immunodeficiency, viral replication, and the risk of cancer. Cancer Epidemiol Biomarkers Prev 20:2551–2559
17. Mbulaiteye SM, Biggar RJ, Goedert JJ, Engels EA (2003) Immune deficiency and risk for malignancy among persons with AIDS. J Acquir Immune Defic Syndr 32:527–533
18. Engels EA, Pfeiffer RM, Goedert JJ et al (2006) Trends in cancer risk among people with AIDS in the United States 1980–2002. AIDS 20:1645–1654
19. Simard EP, Pfeiffer RM, Engels EA (2011) Cumulative incidence of cancer among individuals with acquired immunodeficiency syndrome in the United States. Cancer 117:1089–1096
20. Dezube BJ, Pantanowitz L, Aboulafia DM (2004) Management of AIDS-related Kaposi sarcoma advences in target discover and treatment. AIDS Read 14(5):236–351
21. Danzig JB, Brandt LJ, Reinus JF, Klein RS (1991) Gastrointestinal malignancy in patients with AIDS. Am J Gastrotenterol 86(6):715–718
22. Sigel K, Pitts R, Crothers K (2016) Lung Malignancies in HIV Infection. Semin Respir Crit Care Med 37(2):267–276

23. Krown SE, Testa MA, Huang J (1997) AIDS-related Kaposi's sarcoma: prospective validation of the AIDS Clinical Trials Group staging classification. AIDS clinical trials group oncology committee. J Clin Oncol 15:3085–3092
24. Nasti G, Talamini R, Antinori A et al (2003) AIDS-related Kaposi's Sarcoma: evaluation of potential new prognostic factors and assessment of the AIDS clinical trial group staging system in the Haart Era—the Italian cooperative group on AIDS and tumors and the Italian Cohort of Patients Naive from Antiretrovirals. J Clin Oncol 21:2876–2882
25. International Collaboration on HIV and Cancer (2000) Highly active antiretroviral therapy and incidence of cancer in human immunodeficiency virus-infected adults. J Natl Cancer Inst 92:1823–1830
26. Bower M, Fox P, Fife J et al (1999) Highly active anti-retroviral therapy (HAART) prolongs time to treatment failure in Kaposi's sarcoma. AIDS 13:2105–2111
27. Krown SE (2004) Highly active antiretroviral therapy in AIDS-associated Kaposi's sarcoma: implications for the design of therapeutic trials in patients with advanced, symptomatic Kaposi's sarcoma. J Clin Oncol 22:399–402
28. Bower M, Weir J, Francis N et al (2009) The effect of HAART in 254 consecutive patients with AIDS-related Kaposi's sarcoma. AIDS 23:1701–1706
29. Sgadari C, Monini P, Brillari F, Ensoli B (2003) Use of HIV protease inhibitors to block Kaposi's sarcoma and tumour growth. Lancet Oncol 4(9):537–547
30. Martinez V, Caumes E, Gabotti L et al (2006) Remission from Kaposi's sarcoma on HAART is associated with suppression of HIV replication and is independent of protease inhibitor therapy. Br J Cancer 94:1000–1006
31. Nasti G, Errante D, Talamini R et al (2000) Vinorelbine is an effective and safe drug for AIDS-related Kaposi's sarcoma: results of a phase II study. J Clin Oncol 18:1550–1557
32. Saville MW, Lietzau J, Pluda JM, Feuerstein I, Odom J, Wilson WH, Humphrey RW, Feigal E, Steinberg SM, Broder S et al (1995) Treatment of HIV-associated Kaposi's sarcoma with paclitaxel. Lancet 346(8966):26–28
33. Little RF, Wyvill KM, Pluda JM et al (2000) Activity of thalidomide in AIDS-related Kaposi's sarcoma. J Clin Oncol 18:2593–2602
34. Dezube BJ, Krown SE, Lee JY, Bauer KS, Aboulafia DM (2006) Randomized phase II trial of matrix metalloproteinase inhibitor COL-3 in AIDS-related Kaposi's sarcoma: an AIDS Malignancy Consortium Study. J Clin Oncol 24:1389–1394
35. Castleman B, Iverson L, Menendez VP (1956) Localized mediastinal lymph node hyperplasia resembling thymoma. Cancer 9(4):822–830
36. Peterson BA, Frizzera G (1993) Multicentric Castelman's disease. Semin Oncol 20(6): 636–647
37. Lachant NA, Sun NC, Leong LA et al (1985) Multicentric angiofollicular lymph node hyperplasia (Casttleman's disease) followed by Kaposi's sarcoma in two homosexual males with the acquired immunodeficiency syndrome (AIDS). Am J Clin Pathol 83(1):27–33
38. Oksenhendler E, Boulanger E, Galicier E et al (2002) High incidence of Kaposi sarcoma-associated herpesvirus-related non-Hodgkin lymphoma in patients with HIV infection and multicentric Castleman disease. Blood 99(7):2331–2336
39. Powles T, Stebbing J, Bazeos A et al (2009) The role of immune suppression and HHV-8 in the increasing incidence of HIV-associated multicentric Castleman's disease. Ann Oncol 20:775–779
40. Oksenhendler E, Duarte M, Soulier J et al (1996) Multicentric Castleman's disease in HIV infection: a clinical and pathological study of 20 patients. AIDS 10:61–67
41. Oksenhendler E, Carcelain G, Aoki Y et al (2000) High levels of human herpesvirus 8 viral load, human interleukin-6, interleukin-10, and C reactive protein correlate with exacerbation of multicentric castleman disease in HIV-infected patients. Blood 96:2069–2073
42. Polizzotto MN, Uldrick TS, Wang V et al (2013) Human and viral interleukin-6 and other cytokines in Kaposi sarcoma herpesvirus-associated multicentric Castleman disease. Blood 122:4189–4198

43. Bhutani M, Polizzotto MN, Uldrick TS, Yarchoan R (2015) Kaposi Sarcoma–associated Herpesvirus-associated Malignancies: epidemiology, pathogenesis, and advances in treatment. Semin Oncol 42(2):223–246
44. Polizzotto MN, Uldrick TS, Hu D, Yarchoan R (2012) Clinical manifestations of Kaposi sarcoma herpesvirus lytic activation: multicentric Castleman disease (KSHV-MCD) and the KSHV inflammatory cytokine syndrome. Front Microbiol 3:73
45. Gerard L, Berezne A, Galicier L et al (2007) Prospective study of rituximab in chemotherapy-dependent human immunodeficiency virus associated multicentric Castleman's disease: ANRS 117 CastlemaB Trial. J Clin Oncol 25(22):3350–3356
46. Bower M, Pria AD, Coyle M et al (2014) Diagnostic criteria schemes for multicentric Castleman disease in 75 cases. J Acquir Immune Defic Syndr 65:e80–e82
47. Mylona EE, Baranoutis IG, Lekakis LJ et al (2008) Castleman's disease in HIV infection: a systematic review of the literature. AIDS Rev 10:25–35
48. Bower M, Newsom-Davis T, Naresh K et al (2011) Clinical features and outcome in HIV-associated multicentric Castleman's disease. J Clin Oncol 29:2481–2486
49. Ide M, Kawahi Y, Izumi Y et al (2006) Long-term remission in HIV-negative patients with multicentric Castleman's disease using rituximab. Eur J Haematol 76:119–123
50. Pantanowitz L, Fruh K, Marconi S et al (2008) Pathology of rituximab-induced Kaposi sarcoma flare. BMC Clin Pathol 8:7
51. Van Rhee F, Wong RS, Munshi N et al (2014) Siltuximab for multicentric Castleman's disease: a randomised, double-blind, placebo-controlled trial. Lancet Oncol 15:966–974
52. Kawabata H, Kadowaki N, Nishikori M et al (2013) Clinical features and treatment of multicentric castleman's disease: a retrospective study of 21 Japanese patients at a single institute. J Clin Exp Hematop 53:69–77
53. Kurzrock R, Voorhees PM, Casper C et al (2013) A phase I, open-label study of siltuximab, an anti-IL-6 monoclonal antibody, in patients with B-cell non-hodgkin lymphoma, multiple myeloma, or castleman disease. Clin Cancer Res 19:3659–3670
54. Knowles DM, Inghirami G, Ubriaco A et al (1989) Molecular genetic analysis of three AIDS-associated neoplasms of uncertain lineage demonstrates their B-cell derivation and the possible pathogenetic role of the Epstein-Barr virus. Blood (73)3:792–799
55. Nador RG, Cesarman E, Chadburn A et al (1996) Primary effusion lymphoma: a distinct clinicopathologic entity associated with the Kaposi's sarcoma-associated herpes virus. Blood 88:645–656
56. Boulanger E, Gerard L, Gabarre J et al (2005) Prognostic factors and outcome of human herpesvirus 8-associated primary effusion lymphoma in patients with AIDS. J Clin Oncol 23:4372–4380
57. Chadburn A, Hyjek E, Mathew E et al (2004) KSHV-positive solid lymphomas represent an extra-cavitary variant of primary effusion lymphoma. Am J Surg Pathol 28:1401–1416
58. Cesarman E, Chang Y, Moore PS et al (1995) Kaposi's sarcoma-associated herpesvirus-like DNA sequences in AIDS-related body-cavity-based lymphomas. N Engl J Med 332:1186–1191
59. Arora N, Gupta A, Sadeghi N (2017) Primary effusion lymphoma: current concepts and management. Curr Opin Pulm Med 23(4):365–370
60. Gasperini P, Tosato G (2009) Targeting the mammalian target of Rapamycin to inhibit VEGF and cytokines for the treatment of primary effusion lymphoma. Leukemia 23:1867–1874
61. Bhatt S, Ashlock BM, Natkunam Y et al (2013) CD30 targeting with brentuximab vedotin: a novel therapeutic approach to primary effusion lymphoma. Blood 122:1233–1242
62. Young LS, Yap LF, Murray PG (2016) Epstein-Barr virus: more than 50 years old and still providing surprises. Nat Rev Cancer 16:789–802
63. Bellan C, Lazzi S, DeFalco G et al (2003) Burkitt's lymphoma: new insights into molecular pathogenesis. J Clin Pathol 56:188–193

64. Ramalingam D, Kieffer-Kwon P, Ziegelbauer JM (2012) Emerging themes from EBV and KSHV microRNA targets. Viruses 4(9):1687–1710
65. Nagy N, Klein G, Klein E (2009) To the genesis of Burkitt lymphoma: regulation of apoptosis by EBNA-1 and SAP may determine the fate of Ig-myc translocation carrying B lymphocytes. Semin Cancer Biol 19(6):407–410
66. Wang D, Liebowitz D, Kieff E (1985) An EBV membrane protein expressed in immortalized lymphocytes transforms established rodent cells. Cell 43:831–840
67. Knowles D (2003) Etiology and pathogenesis of AIDS-related non-Hodgkin's lymphoma. Hematol Oncol Clin N Am 17:785–820
68. Mbulaiteye SM, Biggar RJ, Goedert JJ, Engels EA (2003) Immune deficiency and risk for malignancy among persons with AIDS. J Acquir Immune Defic Syndr 32:527–533
69. McClain KL, Joshi VV, Murphy SB (1996) Cancers in children with HIV infection. Hematol Oncol Clin North Am 10:1189–1201
70. Ledergerber B, Telenti A, Egger M (1999) Risk of HIV related Kaposi's sarcoma and Non-Hodgkin's lymphoma with potent antiretroviral therapy: prospective cohort study. Br Med J 319:23–24
71. Stebbing J, Gazzard B, Mandalia S et al (2004) Antiretroviral treatment regimens and immune parameters in the prevention of systemic AIDS-related non-Hodgkin's lymphoma. J Clin Oncol 22:2177–2183
72. Besson C, Goubar A, Gabarre J et al (2001) Changes in AIDS-related lymphoma since the era of highly active antiretroviral therapy. Blood 98:2339–2344 [PubMed]
73. Engels EA, Pfeiffer RM, Goedert JJ et al (2006) Trends in cancer risk among people with AIDS in the United States 1980–2002. AIDS 20:1645–1654
74. Simard EP, Pfeiffer RM, Engels EA (2011) Cumulative incidence of cancer among individuals with acquired immunodeficiency syndrome in the United States. Cancer 117:1089–1096
75. Simard EP, Engels EA (2010) Cancer as a cause of death among people with AIDS in the United States. Clin Infect Dis 51(8):957–962
76. Grogg KL, Miller RF, Dogan A (2007) HIV infection and lymphoma. J Clin Pathol 60:1365–1372
77. Bayraktar S, Bayraktar UD, Ramos JC et al (2011) Primary CNS lymphoma in HIV positive and negative patients: comparison of clinical characteristics, outcome and prognostic factors. J Neuro-Oncol 101:257–265
78. Raez LZ, Patel P, Feun L et al (1998) Natural history and prognostic factors for survival in patients with acquired immune deficiency syndrome (AIDS)-related primary central nervous system lymphoma (PCNSL). Crit Rev Oncog 9:199–208
79. Hoang-Xuan K, Bessell E, Bromberg J (2015) Diagnosis and treatment of primary CNS lymphoma in immunocompetent patients: guidelines from the European Association for Neuro-Oncology. Lancet Oncol 16(7):e322–e332. https://doi.org/10.1016/S1470-2045(15)00076-5
80. Guech-Ongey M, Simard EP, Anderson WF et al (2010) AIDS-related Burkitt lymphoma in the United States: what do age and CD4 lymphocyte patterns tell us about etiology and/or biology? Blood 116:5600–5604
81. Davi F, Delecluse HJ, Guiet P et al (1998) Burkitt-like lymphomas in AIDS patients: characterization within a series of 103 human immunodeficiency virus-associated non-Hodgkin's lymphomas. J Clin Oncol 16:3788–3795
82. Xicoy B, Ribera JM, Müller M et al (2014) Dose-intensive chemotherapy including rituximab is highly effective but toxic in human immunodeficiency virus-infected patients with Burkitt lymphoma/leukemia: parallel study of 81 patients. Leuk Lymphoma 55:2341–2348
83. Lim ST, Karim R, Nathwani BN (2005) AIDS-related Burkitt's lymphoma versus diffuse large-cell lymphoma in the pre-highly active antiretroviral therapy (HAART) and HAART eras: significant differences in survival with standard chemotherapy. J Clin Oncol 23(19): 4430–4438 Epub 2005 May 9

84. Bibas M, Castillo JJ (2014) Current knowledge on HIV-associated Plasmablastic Lymphoma. Mediterr J Hematol Infect Dis 6(1):e2014064. https://doi.org/10.4084/MJHID.2014.064
85. Delecluse HJ, Anagnostopoulos I, Dallenbach F et al (1997) Plasmablastic lymphomas of the oral cavity: a new entity associated with the human immunodeficiency virus infection. Blood 89:1413–1420
86. Teruya-Feldstein J, Chiao E, Filippa DA et al (2004) CD20-negative large-cell lymphoma with plasmablastic features: a clinically heterogenous spectrum in both HIV-positive and -negative patients. Ann Oncol 15:1673–1679
87. Castillo JJ, Furman M, Beltrán BE et al (2012) Human immunodeficiency virus-associated plasmablastic lymphoma: poor prognosis in the era of highly active antiretroviral therapy. Cancer 118:5270–5277
88. Pinzone MR, Fiorica F, Di Rosa M et al (2012) Non-AIDS-defining cancers among HIV-infected people. Eur Rev Med Pharmacol Sci 16:1377–1388
89. Bohlius J, Schmidlin K, Boué F et al (2011) HIV-1-related Hodgkin lymphoma in the era of combination antiretroviral therapy: incidence and evolution of CD4+ T-cell lymphocytes. Blood 117:6100–6108
90. Serrano M, Bellas C, Campo E et al (1990) Hodgkin's disease in patients with antibodies to human immunodeficiency virus: a study of 22 patients. Cancer 65:2248–2254
91. Palella FJ Jr, Delaney KM, Moorman AC et al (1998) Declining morbidity and mortality among patients with advanced human immunodeficiency virus infection. HIV outpatient study investigators. N Engl J Med 338:853–860
92. Xicoy B, Ribera JM, Miralles P et al (2007) Results of treatment with doxorubicin, bleomycin, vinblastine and dacarbazine and highly active antiretroviral therapy in advanced stage, human immunodeficiency virus-related Hodgkin's lymphoma. Haematologica 92:191–198
93. Park IU, Introcaso C, Dunne EF (2015) Human papillomavirus and genital warts: a review of the evidence for the 2015 centers for disease control and prevention sexually transmitted diseases treatment guidelines. Clin Infect Dis 61(suppl 8):S849–S855
94. Grulich AE, van Leeuwen MT, Falster MO, Vajdic CM (2007) Incidence of cancers in people with HIV/AIDS compared with immunosuppressed transplant recipients: a meta-analysis. Lancet 370(9581):59–67
95. Cameron JE, Hagensee ME (2007) Human papillomavirus infection and disease in the HIV+ individual. Cancer Treat Res 133:185–213
96. Ma Y, Madupu R, Karaoz U et al (2014) Human papillomavirus community in healthy persons, defined by metagenomics analysis of human microbiome project shotgun sequencing data sets. J Virol 88:4786–4797
97. Walboomers JM et al (1999) Human papillomavirus is a necessary cause of invasive cervical cancer worldwide. J Pathol 189:12–19
98. Parkin DM, Bray F (2006) Chapter 2: the burden of HPV-related cancers. Vaccine 24(Suppl. 3): 11–25
99. Hariri S, Unger ER, Powell SE et al (2012) Human papillomavirus genotypes in high-grade cervical lesions in the United States. J Infect Dis 206(12):1878–1886 [PubMed]
100. Robbins HA, Pfeiffer RM, Shiels MS, Li J, Hall HI, Engels EA (2015) Excess cancers among HIV-infected people in the United States. J Natl Cancer Inst 107(4)
101. Ellerbrock TV, Chiasson MA, Bush TJ et al (2000) Incidence of cervical squamous intraepithelial lesions in HIV-infected women. JAMA 283(8):1031–1037
102. Moody CA, Laimins LA (2010) Human papillomavirus oncoproteins: pathways to transformation. Nat Rev Cancer 10:550–560. https://doi.org/10.1038/nrc2886
103. Piketty C, Selinger-Leneman H, Bouvier AM (2012) Incidence of HIV-related anal cancer remains increased despite long-term combined antiretroviral treatment: results from the French hospital database on HIV. J Clin Oncol 30:4360–4366

104. Frisch M, Biggar RJ, Goedert JJ (2000) Human papillomavirus-associated cancers in patients with human immunodeficiency virus infection and acquired immunodeficiency syndrome. J Natl Cancer Inst 92:1500–1510. https://doi.org/10.1093/jnci/92.18.1500
105. Lacey CJ (2005) Therapy for genital human papillomavirus-related disease. J Clin Virol 32 (Suppl. S1):82–90. https://doi.org/10.1016/j.jcv.2004.10.020
106. Heard I (2009) Prevention of cervical cancer in women with HIV. Curr Opin HIV AIDS 4:63–73. https://doi.org/10.1097/coh.0b013e328319bcbe [PubMed] [Cross Ref]
107. Smith JS, Moses S, Hudgens M, Parker CB, Agot K, Maclean I et al (2010) Increased risk of HIV acquisition among Kenyan men with human papillomavirus infection. J Infect Dis 201:1677–1685
108. Chin-Hong PV, Husnik M, Cranston RD et al (2009) Anal human papillomavirus infection is associated with HIV acquisition in men who have sex with men. AIDS 23:1135–1142
109. Chaturvedi AK, Madeleine MM, Biggar RJ, Engels EA (2009) Risk of human papillomavirus-associated cancers among persons with AIDS. J Natl Cancer Inst 101:1120–1130
110. Reusser NM, Downing C, Guidry J, Tyring SK (2015) HPV carcinomas in immunocompromised patients. J Clin Med 4(2):260–281. https://doi.org/10.3390/jcm4020260
111. Meyer JE, Panico VJA, Marconato HMF et al (2013) HIV positivity but not HPV/p16 status is associated with higher recurrence rate in anal cancer. J Gastrointest Cancer 44:450–455
112. Ghosn M, Kourie HR, Abdayem P, Antoun J, Nasr D (2015) Anal cancer treatment: current status and future perspectives. World J Gastroenterol 21(8):2294–2302
113. van der Burg SH, Palefsky JM (2009) Human immunodeficiency virus and human papilloma virus—why HPV-induced lesions do not spontaneously resolve and why therapeutic vaccination can be successful. J Transl Med 7:108
114. Abercrombie PD, Korn AP (1998) Vulvar intraepithelial neoplasia in women with HIV. AIDS Patient Care STDS 12:251–254
115. Jamieson DJ, Paramsothy P, Cu-Uvin S et al (2006) HIV epidemiology research study group. Vulvar, vaginal, and perianal intraepithelial neoplasia in women with or at risk for human immunodeficiency virus. Obstet Gynecol 107:1023–1028
116. Sirera G, Videla S, Pinol M et al (2006) High prevalence of human papillomavirus infection in the anus, penis and mouth in HIVpositive men. AIDS 20:1201–1204
117. Beachler DC, Weber KM, Margolick JB et al (2012) Risk factors for oral HPV infection among a high prevalence population of HIV positive and at-risk HIV-negative adults. Cancer Epidemiol Biomarkers Prev 21:122–133
118. Lacey CJ (2005) Therapy for genital human papillomavirus-related disease. J Clin Virol 32(Suppl. S1):82–90

Molecular Biology of KSHV in Relation to HIV/AIDS-Associated Oncogenesis

2

Meilan He, Fan Cheng, Suzane Ramos da Silva, Brandon Tan, Océane Sorel, Marion Gruffaz, Tingting Li and Shou-Jiang Gao

Contents

M. He · F. Cheng · S. R. da Silva · B. Tan · O. Sorel · M. Gruffaz · T. Li · S.-J. Gao (✉)
Department of Molecular Microbiology and Immunology, Keck School of Medicine, University of Southern California, Los Angeles, USA
e-mail: shoujiag@usc.edu

C. Meyers (ed.), *HIV/AIDS-Associated Viral Oncogenesis*, Cancer Treatment and Research 177, https://doi.org/10.1007/978-3-030-03502-0_2

Abstract

Discovered in 1994, Kaposi's sarcoma-associated herpesvirus (KSHV) has been associated with four human malignancies including Kaposi's sarcoma, primary effusion lymphoma, a subset of multicentric Castleman's disease, and KSHV inflammatory cytokine syndrome. These malignancies mostly occur in immunocompromised patients including patients with acquired immunodeficiency syndrome and often cause significant mortality because of the lack of effective therapies. Significant progresses have been made to understand the molecular basis of KSHV infection and KSHV-induced oncogenesis in the last two decades. This chapter provides an update on the recent advancements focusing on the molecular events of KSHV primary infection, the mechanisms regulating KSHV life cycle, innate and adaptive immunity, mechanism of KSHV-induced tumorigenesis and inflammation, and metabolic reprogramming in KSHV infection and KSHV-transformed cells.

2.1 Introduction

Discovered in 1994, KSHV is a human oncogenic gammaherpesvirus [1]. KSHV is causatively associated with several malignancies, including Kaposi's sarcoma (KS), primary effusion lymphoma (PEL), a subset of multicentric Castleman's disease (MCD), and KSHV inflammatory cytokine syndrome (KICS), most of which are commonly found in HIV-1-infected individuals [1–4].

KS is a multifocal mesenchymal neoplasm characterized by neo-angiogenesis, inflammatory infiltration, and spindle-shaped tumor cells that express mixed cellular markers, including vascular and lymphatic endothelial, mesenchymal, and hematopoietic precursor cells [5]. Early stage of KS primarily affects mucocutaneous tissues but advanced stage of KS is often involved with visceral organs [5]. KS is one of the most common malignancies in AIDS patients. While the advent of antiretroviral therapy has substantially reduced the incidence of KS in Western countries, it has stabilized or even rebound in recent years in some populations, and continues to be the most common cancer in some African regions [6]. Hence, KS remains to be one of the most important malignancies in AIDS patients causing significant morbidity and mortality.

PEL is a rare and aggressive non-Hodgkin's B cell lymphoma clinically characterized by lymphomatous effusions in body cavities usually without tumor masses [7]. PEL often occurs in advanced AIDS patients with a decreased CD4 T cell count at diagnosis. Approximately, half of PEL patients have KS or are at risk for developing KS. PEL is resistant to conventional chemotherapy with a short median survival of less than 6 months [7].

MCD is a polyclonal B cell lymphoproliferative disorder characterized by inflammatory symptoms, including fever, cachexia, lymphadenopathy, splenomegaly, cytopenia, and hypoalbuminemia [8]. MCD in the setting of HIV is typically associated with KSHV infection and is usually fatal without treatment. Furthermore, there is no established standard of treatment for KSHV-associated MCD [8].

KICS is a newly described severe systemic inflammatory symptom associated with elevated viral loads and cytokine production [4]. The symptoms of KICS are similar to MCD but without any pathological evidence of MCD. KICS patients have poor prognosis, stressing the need for better understanding of its biology [9].

To dissect the biology of KSHV-associated malignancies and discover new approaches for potential therapy, extensive studies of KSHV from the aspects of virology to its associated pathogenesis have been done in the last three decades. Here, we present an update of literature review of KSHV in the following topics: (1) primary infection, (2) life cycle, (3) immunity, (4) tumorigenesis, (5) inflammation, and (6) metabolism. Because of space constraint, we can't describe all studies in detail and cite every reference. However, several excellent reviews have been published in the last few years and readers are advised to refer to those articles and the previous edition of this book chapter for additional information [5].

2.2 KSHV Primary Infection

KSHV has a broad cellular tropism and infects numerous cell types in vivo and in vitro, including endothelial cells, B cells, monocytes, macrophages, epithelial cells, keratinocytes, mesenchymal stem cells, and neurons [10–14]. Following primary infection, KSHV eventually establishes latency in all the cell types examined so far. While KSHV establishes latent infection without any active lytic replication in some cell types, it has an early full productive replication phase shortly after primary infection in others [15, 16]. To better understand the mechanism that controls KSHV latency and lytic replication following primary infection, it is necessary to identify cell types and conditions that support early lytic replication and the associated cellular pathways.

2.2.1 Attachment, Entry, and Cellular Receptors

KSHV enters the host cell and delivers its genome into the nucleus through a series of events tightly regulated by diverse viral and host factors [17, 18]. These events include attachment to the host cell surface, binding to specific entry receptors, and internalization through fusion of viral envelope with the membrane of intracellular vesicles following receptor-mediated endocytosis [17, 18].

The attachment of KSHV to the host cell is through interactions between viral glycoproteins (gB, gH, and gpK8.1) and cell surface molecule heparan sulfate, a linear polysaccharide ubiquitously expressed at the extracellular matrix [19–22]. Following attachment, KSHV binds to the specific entry receptors, including integrins, DC-SIGN, xCT, and ephrin type-A receptor 2 (EphA2), and activates a cascade of signaling pathways to promote receptor-mediated endocytosis [17, 18].

Integrins are a large family of cell adhesion receptors, widely expressed in various cell types, including endothelial cells and B cells. KSHV was the first herpesvirus demonstrated to utilize integrins as entry receptors [13]. An integrin binding RGD motif (arginine–glycine–aspartic acid) of glycoprotein gB mediates its interactions with integrins $\alpha3\beta1$, $\alpha V\beta3$, and $\alpha V\beta5$ expressed on the surface of human foreskin fibroblasts (HFF), human dermal microvascular endothelial cells (DMVEC), human monocytic THP-1 cells, human fibrosarcoma HT1080, Vero cells, and HEK-293T cells [13, 23–25].

Dendritic cell-specific intercellular adhesion molecule-3-grabbing non-integrin (DC-SIGN) is a C-type lectin expressed by dendritic cells (DCs), macrophage subpopulations, and activated B lymphocytes. KSHV uses DC-SIGN as a binding and entry receptor to infect human myeloid DCs, macrophages, and activated B cells [26, 27]. While blocking binding of KSHV to DC-SIGN does not affect virus attachment to the cells, it inhibits KSHV infection in human monocytic THP-1 cells [23].

Human cysteine/glutamate exchange transporter system x_c^- (xCT) is an amino acid transporter that imports L-cystine and exports L-glutamate across plasma membrane [28]. xCT mediates KSHV cell fusion and virion entry [29]. xCT interacts with α3β1 integrin to form a complex, which triggers downstream signaling cascades essential for viral gene expression during primary infection of DMVEC [24].

Eph2A, a receptor protein tyrosine kinase (RTK), serves as an entry receptor through direct interaction with gH/gL glycoprotein complex [12, 30]. Eph2A plays an important role in regulating macropinocytosis and trafficking of KSHV through its association with signaling molecules (e.g., FAK, Src, and c-Cbl) in the lipid raft (LR) regions during primary infection of DMVEC [31]. In contrast, KSHV infection of HFF induces association of integrins with Eph2A in non-LR regions, suggesting a crucial role of Eph2A in KSHV entry through clathrin-mediated endocytosis [32].

2.2.2 Internalization and Intracellular Trafficking

KSHV infects most types of cells through clathrin-mediated endocytosis and macropinocytosis. Clathrin-mediated endocytosis is an endocytic portal into cells through which cargos are taken up using clathrin-coated vesicles. KSHV enters human umbilical vein endothelial cells (HUVEC), HFF, HEK293 cells, and BJAB cells via clathrin-mediated endocytosis [33–35]. During infection of HUVEC, KSHV particles are co-localized with early endosome antigen (EEA1) and late endosome/lysosome marker (LAMP1) [34]. By electron microscopy, KSHV virions are present in the endocytic vesicles in HFF cells [33], and KSHV entry is sensitive to inhibitors of clathrin-mediated endocytosis [33, 34].

KSHV utilizes macropinocytosis as the major route to enter DMVEC [36]. Inhibition of membrane blebbing, an important event in macropinocytosis, significantly blocks KSHV entry [37]. It is identified that ESCRT-0 component Hrs regulates KSHV entry and ESCRT-I protein Tsg101 plays a role in the trafficking of virus particles in DMVEC [38, 39].

Studies on other cell types further suggest that KSHV entry is cell type-dependent. KSHV enters THP-1 via clathrin- and caveolae-mediated endocytosis but not macropinocytosis [23] while KSHV enters primary B lymphocytes by DC-SIGN-mediated endocytosis [26].

Upon internalization, the intracellular transport of KSHV particles relies on the cytoskeletons. In HUVEC, KSHV is co-localized with actin filaments during early infection and induces dynamic actin cytoskeleton rearrangements. Disruption of actin dynamics significantly inhibits KSHV trafficking [34]. In addition, KSHV infection modulates microtubule polymerization to promote the trafficking of viral capsids in HFF [40]. Disruption of microtubule formation or impairing dynein-directed retrograde microtubule transport strongly reduces KSHV trafficking [40].

2.2.3 Regulation of Cellular Signaling Pathways During Primary Infection

KSHV dysregulates multiple signaling pathways to promote primary infection [18]. Interactions between KSHV and cell surface receptors activate focal adhesion kinase (FAK) signaling in several cell types [41]. Activated FAK is vital for many processes including cytoskeleton rearrangement and endocytosis, which facilitate virus entry [42]. Calcium and integrin binding protein 1 (CIB1), an enhancer of FAK, ERK1/2, and PAK kinases [43, 44], facilitates Eph2A-related signaling and regulates KSHV entry and macropinocytosis [45]. c-Cbl, a multifunctional E3 ubiquitin ligase, is induced by KSHV to promote virus entry in endothelial cells [37, 46, 47]. In addition, KSHV infection induces reactive oxygen species (ROS) to promote virus entry and subsequent viral gene expression [48].

Primary KSHV infection activates ERK, JNK, p38 MAPK pathways to promote virus entry, viral gene expression, and productive viral replication [49–51]. KSHV infection suppresses dual-specificity phosphatase-1 (DUSP1) to activate MAPK signaling, facilitating viral gene expression, pro-inflammatory factor secretion, and cell invasiveness [52]. In HUVEC, KSHV activates MSK/CREB1 signaling pathway in an ERK- and p38-dependent manner to regulate viral lytic replication at the postentry stage [53]. Endogenous activity of AMPK, which maintains cellular homeostasis, inhibits KSHV lytic replication [54]. Activation of AMPK activity decreases while inhibition of AMPK increases KSHV lytic replication during primary infection of HUVEC [54]. In addition, KSHV infection leads to sustained NF-κB induction, which regulates viral and host cell gene expression and possibly affects the establishment of latent infection [55]. Nuclear factor erythroid 2-related factor 2 (Nrf2) is activated by KSHV infection through an ROS-dependent pathway [56]. Knockdown of Nrf2 decreases early lytic gene expression but increases latency-associated nuclear antigen (LANA) expression in the infected cells, indicating its crucial role in viral gene expression [56].

2.2.4 Viral Gene Expression During Primary Infection and the Establishment of Viral Latency

Viral gene expression profiles during KSHV primary infection are heavily dependent on the types of cells infected [57–59]. In cells that support productive KSHV infection (e.g., HUVEC), the expression of latent transcripts precedes the cascade of lytic genes [57]. Latent transcripts are sustained at high levels throughout infection. The lytic transcripts are expressed in the order of immediate early (IE), early (E), and late (L) transcripts, and reach peaks at around 54 h post infection (hpi). After 54 hpi, the levels of lytic transcripts decline while latent transcripts continue to increase, leading to the switch from lytic replication to viral latency [57]. In cells such as CD14+ monocytes, HFF, and DMVEC that support minimal lytic activities without producing infectious virions during KSHV primary infection, the expression of lytic transcripts is weak and transient (within 24 hpi) while latent transcripts are expressed throughout the infection process [58, 59].

The establishment of latency is an essential step for persistent infection and induction of KSHV-associated malignancies. One hallmark of KSHV latency is the global repression of viral lytic genes. During primary infection, the chromatin-free KSHV genome undergoes biphasic chromatinization with an initial transcriptionally active euchromatin phase characterized by high levels of the H3K4me3 and H3K27ac activating histone marks, followed by a heterochromatinization phase featured by decreased levels of activating histone marks and increased levels of repressive marks H3K27me3 and H2AK119ub [60–62]. The euchromatin-to-heterochromatin transition corresponds with the expression switch of viral lytic to latent genes and depends on the recruitment of polycomb repressive complexes 1 and 2 (PRC1 and PRC2) to lytic promoters by LANA [63].

KSHV triggers DNA damage response (DDR) signaling inducing phosphorylation of DDR-associated proteins, ataxia telangiectasia mutated (ATM) and H2AX, during primary infection of endothelial cells [64]. Inhibition of ATM or H2AX activation leads to over 80% reduction in the nuclear viral DNA copy number, indicating an essential role of the DDR proteins for the establishment of KSHV latency during primary infection [64].

2.3 KSHV Life Cycle

Following an acute phase of infection with or without active lytic replication, KSHV enters latency, which is essential for the development of KSHV-associated malignancies [65]. Upon stimulation by specific intracellular and extracellular signals, latent KSHV is reactivated into lytic replication which culminates in virion production and cell death [65].

2.3.1 The Latency Locus

The KSHV latency locus encodes LANA, viral homologues of the cellular FLICE-like inhibitory protein (vFLIP) and cyclin D (vCyclin), Kaposin A, B, and C, and 12 precursor microRNAs (pre-miRNAs). Transcription of the latent locus occurs from LANA promoters (LT1 and LT2) and Kaposin promoter. LT1 drives the expression of LANA, vCyclin, and vFLIP whereas LT2 drives the expression of vCyclin and vFLIP [66–68]. The LANA promoter is bidirectional and can drive the expression of upstream lytic genes such as surface glycoprotein vOX2 and viral G-protein coupling receptor (vGPCR) during reactivation [69]. Of the 12 pre-miRNAs, miR-K1 to -K9 and -K11 form a cluster region located between vFLIP and Kaposin while miR-K10 and miR-K12 are in the Kaposin coding region or 3’UTR, respectively [70]. In addition to the latency locus, KSHV encodes another latent protein viral interferon regulatory factor 3 (vIRF-3) located outside the latency locus, which is expressed in PEL but not in KS cells [71].

2.3.2 KSHV Latency and Latent Nuclear Antigen (LNA) or Latency-Associated Nuclear Antigen (LANA)

LANA (LNA) is a KSHV latent protein discovered as an immunodominant antigen and has been used for detecting KSHV infection [72–74]. LANA is approximately 1162aa in length with a proline-rich N terminal and repeats regions (CRs) composed of glutamine (Q), glutamic acid (E), and aspartic acid (D) [75]. The CRs can be further divided into three distinct regions: CR1 (aa 321–429), CR2 (aa 430–768), and CR3 (aa 769–937), with CR3 containing a leucine zipper domain. CR1 is involved in immune evasion by inhibiting major histocompatibility complex class I (MHC-I) peptide presentation in *cis* [76] while CR2 and CR3 decrease LANA synthesis and enhance its stability [77]. Although its predicted size is 135 kDa, LANA is resolved as double bands of 226–234 kDa in SDS-PAGE [72]. The second band is the result of a 76 aa truncation in the C-terminal region [78]. Besides the two bands, multiple bands between 150 and 180 kDa are present in KSHV-infected cells due to noncanonical translation initiation [79].

LANA is a multifunctional protein and its key function is to maintain the cellular persistence of KSHV episome [65]. During latency, the KSHV genome replicates once in each cell cycle, and the copy number is stable (40–150 copies/cell in PEL cells) [80–82]. Without LANA, KSHV is unable to establish and maintain its episome in mammalian cells [83]. LANA has an essential nuclear localization signal (NLS) at its N-terminus (aa 24–30) and a second one at the C-terminus, and is detected in dot-like pattern by immunohistochemistry or immunofluorescence [73, 74, 84]. The N-terminus also has a chromosome binding site (CBS) (aa 5–13), which interacts with histones H2A/H2B whereas the C-termini has a DNA binding and a dimerization domain (DBD), which allows LANA to bind to LANA-binding sites (LBS) located within KSHV terminal repeat (TR) region [85–88].

LANA interacts with chromatin-associated proteins such as heterochromatin protein 1α (HP1), KSHV LANA-interacting protein 1 (KLIP1), methyl CpG-binding protein (MeCP2), bromodomain protein 4 (Brd4), RING3/Brd2, kinetochores-associated proteins such as centromere protein F (CENP-F), budding uninhibited by benzimidazoles (Bub-1), and nuclear mitotic apparatus protein (NuMA) [89–94]. Furthermore, LANA is associated with nucleophosmin (NPM), and the origin recognition complexes (ORCs) [95]. Some of these interactions are essential for KSHV genome segregation to daughter cells and repression of KSHV lytic replication. LANA silences the replication and transcriptional activator (RTA) promoter and interacts with RTA to inhibit its transactivation function [96]. Deletion or disruption of LANA abolishes the establishment of KSHV latency and increases the expression of KSHV lytic genes and production of infectious virions [83, 97]. Hence, LANA is the predominant regulator in maintaining latency by mediating episome replication, proper segregation to daughter cells, and repressing KSHV lytic replication program [65]. LANA also contributes to KSHV latency by promoting host cell proliferation and survival, which will be detailed in a later section.

2.3.3 Epigenetic Silencing and Regulation of KSHV Latency

To silence the expression of viral lytic genes, the KSHV genome undergoes epigenetic remodeling during latency. The KSHV genome is heavily methylated and contains histone repressive marks and HDACs [61, 62, 98–102]. To mediate viral genome replication, LANA binds to the latent origin of replication in the TR, which also harbor ORC, MCM, CDC6, PARP1, and hyperacetylated histones [100, 103–105]. During latency, the spread of transcription beyond the latent locus is arrested by H19/Lgf2 insulators recruited to the CTCF-binding site, which also harbors CTCF, cohesins, RAD21, SMC1 and SMC 3 [106, 107], mediating viral chromosome conformation, expression of latent genes, and silencing of lytic genes [108–110]. In addition to LANA, vFLIP and miR-K1 promote KSHV latency by activating the NF-κB pathway [111, 112]. Several KSHV miRNAs inhibit RTA expression by direct targeting and silencing or indirect activation of cellular pathways including Rbl2, DNMTs, NFIB, and IKKε [113–117]. These cellular factors could cause chromatin remodeling of KSHV genome. KSHV miRNAs also target several other viral genes, which could regulate viral latency [118].

2.3.4 Reactivation of KSHV from Latency

The mechanism of KSHV reactivation is involved with complex interactions of viral genes, cellular factors, and extracellular signals. During reactivation, the quiescent state of the KSHV genome is disrupted and undergoes epigenetic remodeling, resulting in the expression of viral lytic genes and production of infectious viral particles [119].

2.3.5 Viral Genes Required for Reactivation

KSHV lytic genes can be broadly divided into three classes: IE, E, and L genes. IE gene expression is not dependent on de novo translation of any proteins, whereas E and L genes require de novo expression of proteins. Late genes are also dependent on viral DNA replication. Here, we will discuss several viral lytic genes that are important for viral lytic replication.

RTA is an IE gene. Expression of RTA is essential and sufficient for KSHV reactivation [120, 121]. RTA transactivates numerous viral genes, including itself, by binding to the palindromic RTA-responsive element (RRE) [122–126]. RTA cooperates with cellular factors such as Sp1, Oct-1, XBP-1, RBP-Jk, and C/EBPα to transactivate genes [127–131]. As an E3 ubiquitin ligase, RTA targets numerous repressors of viral lytic replication for degradation [132–135]. RTA binding to origin of lytic replication (oriLyts) is required for viral DNA replication [136, 137]. Several isoforms of RTA, which possess transactivation activities weaker than the canonical isoform, have been identified but the regulation of their expression as well as their specific target genes remain unclear [138]. Hence, RTA's complex functions are not fully understood despite it is known as the master regulator of KSHV lytic replication.

mRNA transcript accumulation (MTA) is an E gene required for KSHV reactivation [139, 140]. MTA interacts with RTA to enhance RTA expression [141]. Importantly, MTA mediates viral transcript processing by hijacking splicing and nuclear export factors such as TREX for efficient viral gene expression, particularly for intronless viral transcripts [142]. However, MTA's role in nuclear export is controversial [143]. MTA interacts with an RNA stem-loop structure termed the MTA-responsive element (MRE) [144, 145]. Of interest, MTA protects vIL6 from miRNA-mediated degradation though the exact mechanism remains unknown [146]. To promote translation, MTA interacts with PYM to shuttle transcripts to the 48S transcription pre-initiation complex [147]. Taken together, MTA enhances viral gene expression by hijacking cellular RNA processing events and translation.

K-bZip encoded by ORF-K8 is an E gene. K-bZip is a leucine zipper protein with multiple functions [148]. It interacts with RTA and inhibits RTA transactivation of several viral genes, notably ORF57, ORF-K15, itself, and RTA autoactivation [149]. K-bZIP interacts with HDAC1/2 to silence viral promoters [150] and this repressive function depends on its SUMO modification of KSHV genome and heterochromatin histone demethylase JMJD2 [151–154]. K-bZIP supports lytic DNA replication by overcoming LANA's repression of the OriLyts [136, 155]. Furthermore, viral protein kinase (vPK/ORF36) co-localizes with K-bZIP at oriLyts and phosphorylates K-bZIP to prevent its sumoylation, thus reducing its transcription repression activity [156]. Taken together, K-bZIP contributes to viral DNA replication and repression of lytic genes during lytic replication.

2.3.6 Factors Involved in KSHV Reactivation

KSHV reactivation from latency is accompanied by dynamic chromatin remodeling [101, 102]. Inhibition of class II HDACs, EZH2, or DNA methylation with small molecules is sufficient to induce KSHV reactivation [101, 102, 157]. During reactivation, the KSHV episome gains activating histone marks (H3K4me3, acH3) and loses repressive histone marks (H3K9me3, H3K27me3, H4K20me3) [157]. This is facilitated by RTA, which recruits CBP/p300 and SWI/SNF to promote H3K27Ac on lytic promoters [158]. Inhibition of SIRT1, a class III HDAC and NAD^+ sensor, results in expression of lytic genes thus linking epigenetics to the cellular metabolic state [98, 99]. In fact, high glucose suppresses SIRT1 leading to KSHV reactivation [159].

KSHV infection often occurs in the context of immunosuppression [160]. How KSHV interacts with other pathogens is poorly understood in vivo but several in vitro studies have attempted to delineate these events. HIV Tat alone can induce RTA expression and enhance KSHV entry into endothelial cells [161–163]. Coinfection of PEL cells with EBV favors viral persistence of both viruses [164, 165]. KSHV RTA, EBV ZTA, and EBV LMP1 prevent reactivation of both viruses. Additionally, HCMV, HSV-1, HSV-2, HHV-6 and HHV-7 can induce KSHV reactivation [166–171]. Bacterial metabolic products such as LPS, short-chain fatty acids, and lipoteichoic acid enhance infectivity and reactivation [172–174].

Hypoxia plays a critical role in reactivation [175]. Binding of HIF-1α to hypoxia-responsive elements (HRE) in the promoters of KSHV lytic genes enhances their expression and lytic replication [175–177]. Furthermore, LANA cooperates with HIF-1α at the HRE to enhance RTA expression [178, 179]. Another hypoxia-inducible gene, XBP-1, binds to the RTA promoter to enhance reactivation [180, 181]. Cross talk between hypoxia and epigenetics could occur through KAP1, which is recruited to the KSHV genome through LANA, and decreased levels of KAP1 during hypoxia enhance lytic gene expression [182, 183].

Moreover, oxidative stress also contributes to KSHV reactivation. In fact, hydrogen peroxide (H_2O_2) is necessary and sufficient for inducing KSHV reactivation [184]. Since KS is a highly inflammatory tumor, the abundant infiltrating immune cells and inflammatory cytokines in KS tumors could secrete or induce H_2O_2, respectively, leading to KSHV reactivation [184, 185]. H_2O_2 activates MAPK leading to phosphorylation of ERK1, JNK, p38, and c-Jun, which is sufficient for KSHV reactivation and this can be reversed by the antioxidant N-acetyl-cysteine (NAC) [184, 185]. Furthermore, ROS induced by anticancer drugs such as cisplatin and arsenic trioxide reactivate KSHV and cause cell death in PEL cells [185].

Since KSHV utilizes the host machinery for transcription, viral transcripts are similarly modified with epitranscriptomic marks such as methyl-6-adenosine (m^6A) [186, 187]. Viral transcripts containing m^6A enhances transcript degradation by the m^6A binding protein YTHDF2, which recruits the CCR4-NOT complex [188, 189]. It is possible that the host cell utilizes m^6A as an antiviral mechanism to limit viral reactivation [187].

2.4 KSHV and Immunity

While the immune system is dedicated to protecting the host from invading pathogens such as viruses, KSHV has evolved various strategies to counteract both the innate and adaptive immune responses, which are essential for viral replication and persistent infection.

2.4.1 KSHV and Innate Immunity

Several KSHV-encoded proteins interfere with both type I (IFN-α and IFN-β) and type II (IFN-γ) interferon responses. KSHV was the first virus found to encode viral homologs of cellular interferon regulatory factors (vIRFs) [190]. Each of the four vIRFs blocks the IRF-mediated transcription of type I IFN by a distinct mechanism [190–192]. Moreover, ORF45 and RTA inhibit IRF7-dependent type I IFN response [193, 194] while K8 inhibits IRF3-mediated IFN-β transcription [195]. ORF-K3 and ORF-K5, which are viral E3 ligases, repress the IFN-γ-mediated JAK/STAT signaling pathway by inducing the degradation of IFN-γ [196, 197].

The pattern recognition receptors (PRRs) sense various pathogen-associated molecular patterns (PAMPs) and trigger the type I IFN signaling and production of inflammatory cytokines during pathogen infection. KSHV stimulates TLR3 expression at the early stages of de novo infection; however, the expression of vIRFs inhibits TLR3-mediated immune responses at later time points [198–200]. RTA, ORF21, and ORF31 inhibit both TLR2 and TLR4 signaling [198, 201]. In addition to modulating the TLR signaling pathway, KSHV ORF63 blocks the cellular NOD-like receptor (NLR)-mediated pathway whereas ORF64 inhibits the activation of retinoic acid-inducible gene-I (RIG-I) [202]. KSHV DNA is sensed by IFI16 and cGAS-STING pathways leading to the activation of inflammasome [203–206]. To ensure efficient viral lytic replication, KSHV encodes numerous proteins including vIRF1, ORF52, and LANA to inhibit the cGAS-STING pathway [203–205], while KSHV lytic replication leads to the degradation of IFI16 though the mechanism remains unclear [207].

To facilitate viral evasion of cytotoxic reaction, KSHV induces a Th2-polarized rather than a Th1-polarized response. Three KSHV-encoded CC-chemokine ligands (vCCL), the homologs of cellular chemokines, compete with cellular chemokines to prevent activation of chemokine receptors [192]. KSHV complement control protein (KCP/ORF4) is a functional homolog of the complement regulatory protein which inhibits the activation of the complement system [208]. This mechanism is likely to protect both KSHV-infected cells and free virions from complement-mediated neutralization during acute viral infection. In contrast, KSHV activates the alternative complement system by downregulating the complement regulatory proteins CD55 and CD59 during latency, which is essential for cell survival and persistent infection [209].

KSHV has evolved strategies to evade the natural killer (NK) cells. ORF-K5 decreases cell surface expression of NK-activating ligands including MICA, MICB, and AICL as well as the costimulatory molecules ICAM and B7.2 [192, 210]. Similarly, miR-K12-7 targets MICB mRNA [211] while ORF54 decreases the expression of another NK ligand, NKp44L [212].

2.4.2 KSHV and Adaptive Immunity

Both KS patients and asymptomatic individuals develop T cell responses against several KSHV lytic and latent proteins [213]. Importantly, reconstitution of the immune system through antiretroviral therapy can lead to KS tumor regression [214], suggesting an important role of the KSHV-specific T cell response, particularly the CD8+ T cell response, in the development of KSHV-associated malignancies [213]. KSHV also induces strong humoral responses as antibodies against various viral antigens are present in KS and KSHV-infected patients [72–74, 215, 216].

B cell activation and differentiation into antibody-producing plasma cells or memory B cells are critical aspects of the adaptive immune response. Several studies suggest that KSHV targets both aspects of the B cell biology to evade the humoral immune response. ORF-K1 reduces the expression of bone marrow stromal antigen 2 (BST-2), which is constitutively expressed in mature B cells [196], and downregulates B cell receptor on the cell surface [217] while ORF-K15 blocks BCR transduction signal, contributing to decreased B cell activation [217].

Evading the cell-mediated immune response is an important strategy for KSHV persistent infection. ORF-K3 and -K5 enhance the internalization and lysosomal degradation of MHC-I molecules through ubiquitination of the cytosolic tails [218–221]. vIRF1 and vFLIP mediate MHC-I downregulation [222]. LANA evades immune surveillance by inhibiting MHC class I peptide presentation [76]. KSHV induces cellular suppressor of cytokine signaling 3 (SOCS3), and together with vIRF3, interferes with MHC class II antigen presentation to evade KSHV-specific CD4+ T helper cell immune response [223, 224]. Besides impairing antigen presentation, KSHV interferes with the function of antigen presenting cells by inhibiting differentiation of monocytes into dendritic cells [225] and downregulating costimulatory molecules required for efficient activation of CD8+ T cells [218, 226].

2.5 KSHV and Tumorigenesis

In KS tumors, most of the tumor cells are latently infected by KSHV, suggesting the importance of latent infection and latent genes in the development of KS tumors [65]. A small number of tumor cells undergo spontaneous lytic replication in early stage of tumors, which is essential for the spread and progression of this stage of tumors. However, there is no lytic cell in late stage of KS tumors [65]. Spontaneous lytic replication is also present in small number of cells in PEL and MCD [65]. Numerous KSHV latent and lytic genes have been shown to have oncogenic and tumor-promoting functions [192]. The recent development of a model of KSHV-induced cellular transformation and tumorigenesis of primary cells has allowed the delineation of the cellular pathways and viral genes that promote tumorigenesis in the context of viral infection [14].

2.5.1 Models of KSHV-Induced Cellular Transformation and Tumorigenesis

The origin of KS tumor cells remains controversial. KSHV infects both vascular and lymphatic endothelial cells and reprograms them to acquire KS-like cell surface markers [227, 228]. KSHV efficiently infects primary endothelial cells and prolongs their life span but cellular transformation remains elusive [229, 230]. Transfection of mouse bone marrow endothelial-lineage cells with recombinant KSHV BAC36 genomes results in immortalization of a subset of cells, which induces tumors in nude mice [231]. However, the exact target cells are unclear and the efficiency is low. In contrast, KSHV efficiently infects and transforms primary rat embryonic metanephric mesenchymal precursor (MM) cells [14]. KSHV-transformed MM cells (KMM) lose contact inhibition and form colonies in soft agar. KMM cells efficiently induce tumors in nude mice with virological and pathological features reminiscent of KS [14]. While KSHV can also infect and transform human mesenchymal stem cells of diverse origins, the efficiency of cellular transformation is much lower [232]. KSHV infection of rat and human mesenchymal stem cells reprograms them to acquire KS-like phenotypes including cell surface markers, and enhances their angiogenic, invasive, and transforming phenotypes [14, 232, 233].

2.5.2 KSHV Viral Genes and Tumorigenesis

The roles of the KSHV latent genes LANA, vFLIP and vCyclin, and miRNAs in tumorigenesis have been extensively studied. LANA promotes cell proliferation and survival by inhibiting tumor suppressor genes p53, p73, pRb, and TGF-β signaling [234–237], and activating c-Myc, emmprin, and survivin [238–241]. LANA promotes tumorigenesis by upregulating BMP-p-Smad1-Id1 pathway in KMM cells [242]. LANA upregulates Par3, SNAIL, and MMP9 while downregulates E-cadherin to promote cell proliferation in B cells [243]. KSHV-encoded miRNAs and vFLIP activate the NF-κB pathway and are essential for KSHV-induced cellular transformation and tumorigenesis by regulating cell proliferation, survival, homeostasis, and metabolic pathways [244–246]. vCyclin alone can interact with numerous CDKs to promote cell cycle progression and tumorigenesis, and antagonizes the senescence/G1 arrest response triggered by NF-κB hyper-activation [247–250]. In the context of KSHV infection, vCyclin only promotes cellular transformation and tumorigenesis by overriding cell contact inhibition [251].

Numerous KSHV lytic genes possess cellular transforming and growth-promoting activities. vIRF1 is the first KSHV oncogene identified [190]. It targets type I interferons, p53, and TGF–β pathways [190, 252, 253]. vGPCR is unique in that it is constitutive active without the need of a ligand. It has robust oncogenic activity [254]. vGPCR transgenic mice develop KS-like lesions [255]. Mechanistically, vGPCR activates Akt and mTOR pathways, and promotes genomic instability through miR-34a [256–258]. Unlike human IL-6, vIL6 only signals through gp130 to activate several downstream pathways, such as JAK/STAT, MAPK, and Akt, driving cellular proliferation, inflammation, and apoptosis inhibition [259, 260]. Of interest, vIL6 can induce intracellular signaling and interacts with splice variant 2 of vitamin K epoxide reductase complex subunit 1 (VKORC1) to promote PEL cell proliferation and survival [261, 262]. ORF-K1 also possesses oncogenic activity [263]. It activates Akt and AMPK pathway to promote cell proliferation and survival [264, 265]. Whether KSHV lytic genes contribute to KSHV-induced tumorigenesis remains to be tested in the context of viral infection.

2.5.3 Cellular Genes/Pathways in KSHV-Associated Malignancies

Extensive studies have identified cellular genes/pathways required for KSHV-associated malignancies. Transcriptional factors, such as c-Myc and STAT3, are required for cell survival, and the inhibition of c-Myc and STAT3 induced apoptosis in PEL cells [266, 267]. Epigenetic factors, including class I and II HDACs, as well as class III HDAC SIRT1 are essential for cell proliferation and survival in PEL cells, and their inhibitors SAHA and tennovin-6 significantly induced cell cycle arrest and apoptosis in vitro and in vivo [268–270]. In addition to the NF-κB and BMP-Smad1-Id pathways that are essential for cell proliferation and cellular transformation [242, 244–246], Akt and mTOR are also essential for cell proliferation in KS cells [271] while hepatocyte growth factor (HGF)/c-MET pathway is essential for cell cycle progression and cell survival in PEL cells [272, 273].

2.6 KSHV and Inflammation

2.6.1 Kaposi's Sarcoma: A Tumor Associated with Inflammation

Inflammation can have a double role in the development of cancer. Acute inflammatory response is considered as a physiological process required for the control of microbial infections and tumor growth. However, by stimulating cell proliferation and inhibiting apoptosis, chronic inflammation becomes a pathologic process participating in the modifications of the microenvironment, enhancing uncontrolled tissue regeneration, angiogenesis, and tumorigenesis [274]. It is estimated that more than 25% of cancers are associated with inflammation [275].

Chronic inflammation, a hallmark of KSHV-associated malignancies, participates in KS progression through the complex interplay between viral and cellular factors. By interfering with the intracellular signaling pathways during lytic and latent phases of infection, KSHV induces an inflammatory neoplastic network in the tumor microenvironment, which is mainly associated with the abnormal lympho-endothelial proliferations and the recruitment of activated myeloid and lymphoid immune cells [276]. Indeed, at the early stage of tumors, the KS microenvironment has a high level of pro- and anti-inflammatory cytokines (IL-6, TNF-α, and IL-10, respectively), chemokines (CXCL12, CXCR4, CXCR7), interferon (IFN-γ), as well as growth factors (VEGF) [277–279]. These cytokines can be released by different cell types including monocytes, endothelial cells, and KS tumors. During tumor growth, these mediators stimulate resting and non-proliferative lympho-endothelial cells to enhance inflammation, and therefore promote angiogenesis [280, 281].

2.6.2 Latent Viral Factors Involved in Inflammation

LANA upregulates emmprin expression, which induces the secretion of IL-6, VEGF, and MMPs, and enhances inflammation and angiogenesis [240, 282]. By stabilizing the Notch effector Hey-1, LANA also represses the expression of Prox-1 to modulate the differentiation of lymphatic endothelial cells [283]. Moreover, LANA activation of the Notch pathway enhances the invasiveness of KS tumors by activating PDGFRβ [284]. As stated in the previous section, LANA participates in the suppression of specific T cells immune response by inhibiting MHC-I antigen presentation through its acidic central repeat domain [285], and by downregulating MHC-II gene expression on APCs through the interaction with RFX proteins to inhibit the recruitment of CIITA to the MHC-II promoter [286].

vFLIP activates the classical and alternative NF-κB signaling pathways and participates in the upregulation of pro-inflammatory cytokines [287, 288]. Particularly, vFLIP promotes tumorigenesis through the induction of COX-2 and its inflammatory metabolite PGE2 in an NF-κB-dependent manner [287, 289].

Most of the KSHV-encoded miRNAs are expressed during latency and play significant roles in tumor growth, inflammation, and angiogenesis. Numerous KSHV miRNAs induce inflammation by activating the NF-κB pathway [112, 246]. Ectopic expression of the miRNA cluster in endothelial cells induces the expression of pro-inflammatory and pro-angiogenic cytokines MMP1, MMP13, and VEGFA [290]. VEGF is important for the recruitment of stem cells and macrophages at the site of infection, and therefore participates in the inflammatory microenvironment of KS tumors [291]. By inducing CXCR2 and activating Akt signaling pathway through targeting GRK2 stimulation, miR-K3 promotes angiogenesis, migration, and invasion of endothelial cells [292, 293]. The miRNAs derived from miR-K6, miR-K6-3p, and miR-K6-5p promote cell migration, invasion, and angiogenesis by targeting SH3BGR to activate the STAT3 pathway and CD8 to activate the c-Met pathway, respectively [294, 295]. miR-K12 promotes cell survival and proliferation by targeting the angiogenesis inhibitor THBS1 and SMAD5 to downregulate TGF-β signaling [296, 297].

In the context of inflammation, Kaposin B participates in the lymphatic reprogramming of vascular endothelial cells [298]. Kaposin B activates the p38/MK2 pathway leading to the stabilization of targeted gene transcripts including pro-inflammatory cytokines IL-6 and GM-CSF, as well as the lympho-endothelial differentiation factor PROX1 [298, 299]. By cooperating with c-Myc, Kaposin B triggers angiogenesis by mediating the expression of cellular miRNAs in endothelial cells [300].

In the latent phase, KSHV also expresses vIRF3 (LANA-2) in PEL cells [71]. vIRF3 plays a major role in PEL pathogenesis by promoting viral latency and inhibiting the host innate responses. By stabilizing HIF-1α, vIRF3 induces its accumulation and activation in the nucleus contributing to the uncontrolled expression of VEGF in KSHV-infected cells [301].

2.6.3 Viral Lytic Genes Involved in Inflammation

ORF-K15, predominantly expressed during the lytic cycle, mediates inflammation by activating JNK, and NF-κB pathways as well as NFAT/AP1 activities [302, 303]. These signaling pathways induce the expression of cytokines and chemokines such as IL-6, IL-1β, IL-8, CCL20, CXCL3, and COX-2. Depletion of ORF-K15 dramatically impairs KSHV-induced angiogenesis mediated by the recruitment of PLCɣ1 and the activation of NFAT1-dependant RCAN1 expression in endothelial cells [304].

As stated earlier, ORF-K1 participates in KSHV-induced tumorigenesis by performing multiple functions. Among them, ORF-K1 induces the secretion of VEGF, IL-6, GM-CSF, IL-1β, IL-8, and IL-10 in endothelial cells [305], and stimulates the expression of MMP-9, a matrix metalloproteinase involved in the angiogenic switch during tumor progression [306, 307]. In AIDS-related KS, ORF-K1 can synergize with HIV-1 proteins such as Tat to promote inflammation by activating NF-κB signaling [308] and NEF to promote cellular proliferation, vascular tube formation, and angiogenesis by regulating the PTEN/AKT/mTOR pathway [309]. In addition to inhibition of innate and adaptive immune responses, ORF-K5 enhances angiogenesis by disrupting VE-cadherin/β-catenin signaling, promoting the remodeling of cellular tight junctions [310]. vIL6 promotes angiogenesis and hematopoiesis by stimulating the secretion of VEGF [311]. By inducing several signaling pathways such as PKC, MAPK, mTOR, NF-κB, AP1, HIF-1α, and NFAT, vGPCR mediates the upregulation of pro-inflammatory and pro-angiogenic mediators (IL-2, IL-4, IL-6, IL-8, TNF-α, and VEGF) [312, 313]. vGPCR activation of NF-κB induces the expression of RANTES, IL-8, and GM-CSF as well as adhesion molecules VCAM-1, ICAM-1 and E-selectin [314].

KSHV encodes three homologues of cellular chemokines: vCCL1, vCCL2, and vCCL3. These viral chemokines activate their respective G-coupled protein receptors CCR8, CCR3 and CCR4 expressed on Th2 lymphocytes [315]. These viral chemokines inhibit T cells immune response by inducing Th2 polarization and attracting Th2 lymphocytes to the site of infection, and promoting angiogenesis by inducing the expression of VEGF [316, 317]. Moreover, vCCL2 antagonize CCR1 and CCR5 to inhibit host immune responses of Th1 lymphocytes [318].

2.7 KSHV and Metabolism

During latent infection and cellular transformation, KSHV reprograms cellular metabolic pathways to provide biosynthetic and bioenergetic precursors to support the fast anabolic cellular proliferation. During viral lytic replication, KSHV also reprograms specific metabolic pathways to support the production of infectious virions.

2.7.1 KSHV Reprograms Glucose Metabolism

A hallmark of tumorigenesis involves the switch of energy metabolism from oxidative phosphorylation to aerobic glycolysis. In untransformed telomerase-immortalized microvascular endothelial cells (TIME cells) and primary dermal microvascular endothelial cells (DMVECs), KSHV infection increases aerobic glycolysis by upregulating hexokinase 2 (HK2) and glucose transporter 3 (GLUT3) [319]. Thus, oxygen consumption and oxidative phosphorylation are decreased, and lactate production is increased. Inhibition of glycolysis leads to apoptosis in KSHV-infected TIME cells but not in uninfected cells, demonstrating the critical role of the aerobic glycolysis on cell survival in untransformed KSHV-infected cells [319]. A similar study in KSHV-infected primary dermal microvascular lymphatic endothelial cells (KLEC) also demonstrated increased aerobic glycolysis [320]. Mechanistically, KSHV miRNAs stabilize HIF-1α and inhibit mitochondrial biogenesis by downregulating EGLN2 and HSPA9. Moreover, HIF-1α is stabilized in KSHV-infected telomerase-immortalized HUVEC (TIVE) cells, which results in the upregulation of glycolytic effector-isoform 2 of pyruvate kinase (PKM2) and increased aerobic glycolysis [320].

However, KSHV-induced glycolysis does not occur in HFF cells [319], which implies cell type specificity in KSHV-induced metabolic reprogramming. In contrast to untransformed KSHV-infected cells, KSHV-transformed KMM cells have reduced glucose and oxygen consumption, lactate production, and intracellular ATP [244]. Mechanistically, vFLIP and the miRNA cluster inhibit the aerobic glycolysis in KMM cells by downregulating glucose transporters GLUT1 and GLUT3 through NF-κB activation. The decreased glycolytic flux confers a survival advantage to KMM cells in a nutrient deficient tumor microenvironment [244].

2.7.2 KSHV Reprogramming of Glutamine Metabolism for Host Cell Proliferation and Survival

Glutamine is required for cancer cell proliferation and survival [321, 322]. KSHV infection increases both the intracellular glutamine levels and glutamine uptake in TIME cells. KSHV-infected TIME cells rely on glutamine for their survival and glutamine deprivation-induced apoptosis in KSHV-infected TIME cells with a lesser effect on TIME cells [323]. Suppressing glutaminolytic enzymes in the presence of glutamine causes cell death at the similar levels to those deprived of glutamine in KSHV-infected TIME cells with little effect on TIME cells. The sensitivity to the absence of glutamine can be restored by the addition of TCA cycle intermediates, indicating that in untransformed cells, glutaminolysis is required for cell survivals by feeding the TCA cycle through anaplerosis [323].

KSHV-transformed KMM cells also rely on glutamine for their proliferation and transformation. The expression of glutaminolytic enzymes is upregulated in KMM cells compared to MM cells, and inhibition of any of those enzymes reduces KMM cell proliferation, implying glutaminolysis is required for KMM cell survival [324].

Interestingly, the addition of carbon sources, such as TCA intermediates, only partially rescues the proliferation of KMM cells following glutamine depletion. In contrast, nonessential amino acid asparagine fully rescues the effects of glutamine deprivation, indicating that glutamine and asparagine provide not only carbon source but also nitrogen source [324]. Specifically, glutamine provides the γ-nitrogen for nucleotide synthesis in KSHV-transformed cells. Overall, KSHV reprograms glutaminolysis to supply the building blocks for synthesizing nucleotides, nonessential amino acids, and TCA cycle intermediates to support KSHV-infected cell proliferation and transformation [324].

2.7.3 KSHV Infection Induces Lipogenesis

KSHV infection of TIME cells induces lipogenesis with an increase of metabolites involved in de novo fatty acid synthesis (FAS) and formation of lipid droplets [325]. Inhibitors of FAS induce a dose-dependent cell death in KSHV-infected TIME, which can be partially rescued by supplying cells with fatty acid precursors, indicating FAS is necessary for the survival of untransformed KSHV-infected TIME cells [325]. A separate study shows that KSHV infection increases peroxisomes in TIME cells [326]. A major function of peroxisomes is to break down the long-chain fatty acids through β-oxidation. Inhibition of enzymes involved in the peroxisomal β-oxidation leads to increased cell death in KSHV-infected TIME cells. Together, these observations suggest that KSHV-induced FAS and peroxisomal lipid metabolism are required for KSHV-infected TIME cell survival [326]. Additionally, PEL cells also have highly upregulated FAS compared to primary B cells and are sensitive to FAS inhibitors [327].

2.7.4 KSHV Depends on Glycolysis, Glutaminolysis, and FAS for Lytic Replication

Reprogramming of metabolic pathways is expected to be important for supporting KSHV lytic replication. However, there is so far limited work on metabolic rewiring during KSHV primary infection and reactivation. Inhibitors of glycolysis, glutaminolysis, and FAS significantly reduce the production of virions in both endothelial and SLK cells [328]. Inhibition of glycolysis and glutaminolysis suppresses KSHV replication by stalling early gene transcription and translation, respectively [328]. While inhibition of FAS decreases the production of extracellular virions, it does not affect intracellular viral genome levels, suggesting that FAS is required for virion assembly and maturation [328]. However, some of these inhibitors are not entirely specific and the mechanisms underlying the support of viral lytic replication by glycolysis and glutaminolysis remain unclear. Nevertheless, these results indicate that different stages of viral lytic replication might require different metabolites within the host cells.

2.8 Conclusion and Perspectives

Rapid progresses have been made in the KSHV field in the last decade, providing insights into the biology of virus and the scientific basis for developing novel therapeutic approaches for its associated malignancies. KSHV has evolved to hijack cellular machinery for completing its life cycle, which often results in the dysregulation of cellular functions. It is now clear that KSHV-induced uncontrolled cellular proliferation, cell survival, abnormal immune responses, and reprogrammed metabolism promote malignant tumor growth, angiogenesis, and inflammation, which are the hallmarks of KS.

The standard KS chemotherapy with liposomal doxorubicin, daunorubicin, or taxol is highly toxic and ineffective despite effective antiretroviral therapy in some cases [329]. Both PEL and MCD also do not any have effective therapy [330]. Therefore, alternative treatments and new therapeutic targets, particularly those targeting malignant proliferation, angiogenesis, inflammation, and dysregulated immune responses, are needed for KSHV-associated malignancies. Laboratory studies have so far identified numerous new targets and agents. These include sirtuin inhibitors (Tenovin-6 and nicotinamide), HDACs inhibitor, AMPK inhibitor, mTOR inhibitor Rapamycin (sirolimus), and p53 activator Nutlin-3 [268–270, 331–335]. Numerous potential therapeutic targets, particularly those targeting KSHV-specific epigenetics and metabolism, are attractive. Nevertheless, rigorous clinical trials are required to evaluate the efficacies of the inhibitors before their extensive usages in the patients. In fact, new drugs bevacizumab and imatinib for KS, and siltuximab for KSHV-MCD have been examined in clinical trials [336–338]. Ongoing clinical trials are testing the efficacies of Tocilizumab (NCT01441063) for MCD, and lenalidomide (NCT01057121) and pomalidomide (NCT02659930) for KS [339]. Because cellular pathways often interact with one another, it would be interesting to evaluate the interaction effects of multiple pathways and inhibitors. For example, while Rapamycin inhibits the mTOR pathway, it also activates the Akt pathway. Hence, the combination of inhibitors of both pathways would be predicted to be favorable, which has been demonstrated for both KS and PEL cells [271, 340].

Since KSHV is not a ubiquitous herpesvirus and immunosuppression is required for the development of the KSHV-associated malignancies, it would be essential to develop effective strategies to prevent its person-to-person transmission and manage immunosuppression in the affected populations. Development of KSHV vaccines should be one of the focuses of future research.

References

1. Chang Y, Cesarman E, Pessin MS, Lee F, Culpepper J, Knowles DM, Moore PS (1994) Identification of herpesvirus-like DNA sequences in AIDS-associated Kaposi's sarcoma. Science 266:1865–1869

2. Cesarman E, Chang Y, Moore PS, Said JW, Knowles DM (1995) Kaposi's sarcoma-associated herpesvirus-like DNA sequences in AIDS-related body-cavity-based lymphomas. N Engl J Med 332:1186–1191
3. Soulier J, Grollet L, Oksenhendler E, Cacoub P, Cazals-Hatem D, Babinet P, d'Agay MF, Clauvel JP, Raphael M, Degos L et al (1995) Kaposi's sarcoma-associated herpesvirus-like DNA sequences in multicentric Castleman's disease. Blood 86:1276–1280
4. Uldrick TS, Wang V, O'Mahony D, Aleman K, Wyvill KM, Marshall V, Steinberg SM, Pittaluga S, Maric I, Whitby D, Tosato G, Little RF, Yarchoan R (2010) An interleukin-6-related systemic inflammatory syndrome in patients co-infected with Kaposi sarcoma-associated herpesvirus and HIV but without Multicentric Castleman disease. Clin Infect Dis 51:350–358
5. Greene W, Kuhne K, Ye F, Chen J, Zhou F, Lei X, Gao SJ (2007) Molecular biology of KSHV in relation to AIDS-associated oncogenesis. Cancer Treat Res 133:69–127
6. Robey RC, Bower M (2015) Facing up to the ongoing challenge of Kaposi's sarcoma. Curr Opin Infect Dis 28:31–40
7. Okada S, Goto H, Yotsumoto M (2014) Current status of treatment for primary effusion lymphoma. Intractable Rare Dis Res 3:65–74
8. Carbone A, De Paoli P, Gloghini A, Vaccher E (2015) KSHV-associated multicentric Castleman disease: a tangle of different entities requiring multitarget treatment strategies. Int J Cancer 137:251–261
9. Polizzotto MN, Uldrick TS, Hu D, Yarchoan R (2012) Clinical Manifestations of Kaposi Sarcoma Herpesvirus Lytic Activation: Multicentric Castleman Disease (KSHV-MCD) and the KSHV Inflammatory Cytokine Syndrome. Front Microbiol 3:73
10. Tso FY, Sawyer A, Kwon EH, Mudenda V, Langford D, Zhou Y, West J, Wood C (2016) Kaposi's sarcoma-associated herpesvirus infection of neurons in HIV positive patients. J Infect Dis
11. Bechtel JT, Liang Y, Hvidding J, Ganem D (2003) Host range of Kaposi's sarcoma-associated herpesvirus in cultured cells. J Virol 77:6474–6481
12. Hahn AS, Kaufmann JK, Wies E, Naschberger E, Panteleev-Ivlev J, Schmidt K, Holzer A, Schmidt M, Chen J, Konig S, Ensser A, Myoung J, Brockmeyer NH, Sturzl M, Fleckenstein B, Neipel F (2012) The ephrin receptor tyrosine kinase A2 is a cellular receptor for Kaposi's sarcoma-associated herpesvirus. Nat Med 18:961–966
13. Akula SM, Pramod NP, Wang FZ, Chandran B (2002) Integrin alpha3beta1 (CD 49c/29) is a cellular receptor for Kaposi's sarcoma-associated herpesvirus (KSHV/HHV-8) entry into the target cells. Cell 108:407–419
14. Jones T, Ye F, Bedolla R, Huang Y, Meng J, Qian L, Pan H, Zhou F, Moody R, Wagner B, Arar M, Gao SJ (2012) Direct and efficient cellular transformation of primary rat mesenchymal precursor cells by KSHV. J Clin Invest 122:1076–1081
15. Foglieni C, Scabini S, Belloni D, Broccolo F, Lusso P, Malnati MS, Ferrero E (2005) Productive infection of HUVEC by HHV-8 is associated with changes compatible with angiogenic transformations. Eur J Histochem 49:273–284
16. Gao SJ, Deng JH, Zhou FC (2003) Productive lytic replication of a recombinant Kaposi's sarcoma-associated herpesvirus in efficient primary infection of primary human endothelial cells. J Virol 77:9738–9749
17. Kumar B, Chandran B (2016) KSHV entry and trafficking in target cells-hijacking of cell signal pathways, actin and membrane dynamics. Viruses 8
18. Zhang W, Gao SJ (2012) Exploitation of cellular cytoskeletons and signaling pathways for cell entry by Kaposi's sarcoma-associated herpesvirus and the closely related rhesus rhadinovirus. Pathogens 1:102–127
19. Akula SM, Wang FZ, Vieira J, Chandran B (2001) Human herpesvirus 8 interaction with target cells involves heparan sulfate. Virology 282:245–255

20. Hahn A, Birkmann A, Wies E, Dorer D, Mahr K, Sturzl M, Titgemeyer F, Neipel F (2009) Kaposi's sarcoma-associated herpesvirus gH/gL: glycoprotein export and interaction with cellular receptors. J Virol 83:396–407
21. Wang FZ, Akula SM, Pramod NP, Zeng L, Chandran B (2001) Human herpesvirus 8 envelope glycoprotein K8.1A interaction with the target cells involves heparan sulfate. J Virol 75:7517–7527
22. Birkmann A, Mahr K, Ensser A, Yaguboglu S, Titgemeyer F, Fleckenstein B, Neipel F (2001) Cell surface heparan sulfate is a receptor for human herpesvirus 8 and interacts with envelope glycoprotein K8.1. J Virol 75:11583–11593
23. Kerur N, Veettil MV, Sharma-Walia N, Sadagopan S, Bottero V, Paul AG, Chandran B (2010) Characterization of entry and infection of monocytic THP-1 cells by Kaposi's sarcoma associated herpesvirus (KSHV): role of heparan sulfate, DC-SIGN, integrins and signaling. Virology 406:103–116
24. Veettil MV, Sadagopan S, Sharma-Walia N, Wang FZ, Raghu H, Varga L, Chandran B (2008) Kaposi's sarcoma-associated herpesvirus forms a multimolecular complex of integrins (alphaVbeta5, alphaVbeta3, and alpha3beta1) and CD98-xCT during infection of human dermal microvascular endothelial cells, and CD98-xCT is essential for the postentry stage of infection. J Virol 82:12126–12144
25. Garrigues HJ, DeMaster LK, Rubinchikova YE, Rose TM (2014) KSHV attachment and entry are dependent on alphaVbeta3 integrin localized to specific cell surface microdomains and do not correlate with the presence of heparan sulfate. Virology 464–465:118–133
26. Rappocciolo G, Hensler HR, Jais M, Reinhart TA, Pegu A, Jenkins FJ, Rinaldo CR (2008) Human herpesvirus 8 infects and replicates in primary cultures of activated B lymphocytes through DC-SIGN. J Virol 82:4793–4806
27. Rappocciolo G, Jenkins FJ, Hensler HR, Piazza P, Jais M, Borowski L, Watkins SC, Rinaldo CR Jr (2006) DC-SIGN is a receptor for human herpesvirus 8 on dendritic cells and macrophages. J Immunol 176:1741–1749
28. Lewerenz J, Hewett SJ, Huang Y, Lambros M, Gout PW, Kalivas PW, Massie A, Smolders I, Methner A, Pergande M, Smith SB, Ganapathy V, Maher P (2013) The cystine/glutamate antiporter system x(c)(-) in health and disease: from molecular mechanisms to novel therapeutic opportunities. Antioxid Redox Signal 18:522–555
29. Kaleeba JA, Berger EA (2006) Kaposi's sarcoma-associated herpesvirus fusion-entry receptor: cystine transporter xCT. Science 311:1921–1924
30. Hahn AS, Desrosiers RC (2014) Binding of the Kaposi's sarcoma-associated herpesvirus to the ephrin binding surface of the EphA2 receptor and its inhibition by a small molecule. J Virol 88:8724–8734
31. Chakraborty S, Veettil MV, Bottero V, Chandran B (2012) Kaposi's sarcoma-associated herpesvirus interacts with EphrinA2 receptor to amplify signaling essential for productive infection. Proc Natl Acad Sci U S A 109:E1163–72
32. Dutta D, Chakraborty S, Bandyopadhyay C, Valiya Veettil M, Ansari MA, Singh VV, Chandran B (2013) EphrinA2 regulates clathrin mediated KSHV endocytosis in fibroblast cells by coordinating integrin-associated signaling and c-Cbl directed polyubiquitination. PLoS Pathog 9:e1003510
33. Akula SM, Naranatt PP, Walia NS, Wang FZ, Fegley B, Chandran B (2003) Kaposi's sarcoma-associated herpesvirus (human herpesvirus 8) infection of human fibroblast cells occurs through endocytosis. J Virol 77:7978–7990
34. Greene W, Gao SJ (2009) Actin dynamics regulate multiple endosomal steps during Kaposi's sarcoma-associated herpesvirus entry and trafficking in endothelial cells. PLoS Pathog 5:e1000512
35. Inoue N, Winter J, Lal RB, Offermann MK, Koyano S (2003) Characterization of entry mechanisms of human herpesvirus 8 by using an Rta-dependent reporter cell line. J Virol 77:8147–8152

36. Raghu H, Sharma-Walia N, Veettil MV, Sadagopan S, Chandran B (2009) Kaposi's sarcoma-associated herpesvirus utilizes an actin polymerization-dependent macropinocytic pathway to enter human dermal microvascular endothelial and human umbilical vein endothelial cells. J Virol 83:4895–4911
37. Valiya Veettil M, Sadagopan S, Kerur N, Chakraborty S, Chandran B (2010) Interaction of c-Cbl with myosin IIA regulates Bleb associated macropinocytosis of Kaposi's sarcoma-associated herpesvirus. PLoS Pathog 6:e1001238
38. Veettil MV, Kumar B, Ansari MA, Dutta D, Iqbal J, Gjyshi O, Bottero V, Chandran B (2016) ESCRT-0 component Hrs Promotes Macropinocytosis of Kaposi's Sarcoma-associated herpesvirus in human dermal microvascular endothelial cells. J Virol 90:3860–3872
39. Kumar B, Dutta D, Iqbal J, Ansari MA, Roy A, Chikoti L, Pisano G, Veettil MV, Chandran B (2016) ESCRT-I protein Tsg101 plays a role in the post-macropinocytic trafficking and infection of endothelial cells by Kaposi's sarcoma-associated herpesvirus. PLoS Pathog 12:e1005960
40. Naranatt PP, Krishnan HH, Smith MS, Chandran B (2005) Kaposi's sarcoma-associated herpesvirus modulates microtubule dynamics via RhoA-GTP-diaphanous 2 signaling and utilizes the dynein motors to deliver its DNA to the nucleus. J Virol 79:1191–1206
41. Sharma-Walia N, Naranatt PP, Krishnan HH, Zeng L, Chandran B (2004) Kaposi's sarcoma-associated herpesvirus/human herpesvirus 8 envelope glycoprotein gB induces the integrin-dependent focal adhesion kinase-Src-phosphatidylinositol 3-kinase-rho GTPase signal pathways and cytoskeletal rearrangements. J Virol 78:4207–4223
42. Krishnan HH, Sharma-Walia N, Streblow DN, Naranatt PP, Chandran B (2006) Focal adhesion kinase is critical for entry of Kaposi's sarcoma-associated herpesvirus into target cells. J Virol 80:1167–1180
43. Naik MU, Naik UP (2003) Calcium-and integrin-binding protein regulates focal adhesion kinase activity during platelet spreading on immobilized fibrinogen. Blood 102:3629–3636
44. Naik MU, Naik UP (2011) Contra-regulation of calcium- and integrin-binding protein 1-induced cell migration on fibronectin by PAK1 and MAP kinase signaling. J Cell Biochem 112:3289–3299
45. Bandyopadhyay C, Valiya-Veettil M, Dutta D, Chakraborty S, Chandran B (2014) CIB1 synergizes with EphrinA2 to regulate Kaposi's sarcoma-associated herpesvirus macropinocytic entry in human microvascular dermal endothelial cells. PLoS Pathog 10:e1003941
46. Chakraborty S, ValiyaVeettil M, Sadagopan S, Paudel N, Chandran B (2011) c-Cbl-mediated selective virus-receptor translocations into lipid rafts regulate productive Kaposi's sarcoma-associated herpesvirus infection in endothelial cells. J Virol 85:12410–12430
47. Greene W, Zhang W, He M, Witt C, Ye F, Gao SJ (2012) The ubiquitin/proteasome system mediates entry and endosomal trafficking of Kaposi's sarcoma-associated herpesvirus in endothelial cells. PLoS Pathog 8:e1002703
48. Bottero V, Chakraborty S, Chandran B (2013) Reactive oxygen species are induced by Kaposi's sarcoma-associated herpesvirus early during primary infection of endothelial cells to promote virus entry. J Virol 87:1733–1749
49. Pan H, Xie J, Ye F, Gao SJ (2006) Modulation of Kaposi's sarcoma-associated herpesvirus infection and replication by MEK/ERK, JNK, and p38 multiple mitogen-activated protein kinase pathways during primary infection. J Virol 80:5371–5382
50. Sharma-Walia N, Krishnan HH, Naranatt PP, Zeng L, Smith MS, Chandran B (2005) ERK1/2 and MEK1/2 induced by Kaposi's sarcoma-associated herpesvirus (human herpesvirus 8) early during infection of target cells are essential for expression of viral genes and for establishment of infection. J Virol 79:10308–10329
51. Xie J, Pan H, Yoo S, Gao SJ (2005) Kaposi's sarcoma-associated herpesvirus induction of AP-1 and interleukin 6 during primary infection mediated by multiple mitogen-activated protein kinase pathways. J Virol 79:15027–15037

52. Qin Z, Dai L, Defee M, Findlay VJ, Watson DK, Toole BP, Cameron J, Peruzzi F, Kirkwood K, Parsons C (2013) Kaposi's sarcoma-associated herpesvirus suppression of DUSP1 facilitates cellular pathogenesis following de novo infection. J Virol 87:621–635
53. Cheng F, Sawant TV, Lan K, Lu C, Jung JU, Gao SJ (2015) Screening of the human kinome identifies MSK1/2-CREB1 as an essential pathway mediating Kaposi's sarcoma-associated herpesvirus lytic replication during primary infection. J Virol 89:9262–9280
54. Cheng F, He M, Jung JU, Lu C, Gao SJ (2016) Suppression of Kaposi's sarcoma-associated herpesvirus infection and replication by 5'-AMP-activated protein kinase. J Virol 90:6515–6525
55. Sadagopan S, Sharma-Walia N, Veettil MV, Raghu H, Sivakumar R, Bottero V, Chandran B (2007) Kaposi's sarcoma-associated herpesvirus induces sustained NF-kappaB activation during de novo infection of primary human dermal microvascular endothelial cells that is essential for viral gene expression. J Virol 81:3949–3968
56. Gjyshi O, Bottero V, Veettil MV, Dutta S, Singh VV, Chikoti L, Chandran B (2014) Kaposi's sarcoma-associated herpesvirus induces Nrf2 during de novo infection of endothelial cells to create a microenvironment conducive to infection. PLoS Pathog 10: e1004460
57. Yoo SM, Zhou FC, Ye FC, Pan HY, Gao SJ (2005) Early and sustained expression of latent and host modulating genes in coordinated transcriptional program of KSHV productive primary infection of human primary endothelial cells. Virology 343:47–64
58. Krishnan HH, Naranatt PP, Smith MS, Zeng L, Bloomer C, Chandran B (2004) Concurrent expression of latent and a limited number of lytic genes with immune modulation and antiapoptotic function by Kaposi's sarcoma-associated herpesvirus early during infection of primary endothelial and fibroblast cells and subsequent decline of lytic gene expression. J Virol 78:3601–3620
59. Purushothaman P, Thakker S, Verma SC (2015) Transcriptome analysis of Kaposi's sarcoma-associated herpesvirus during de novo primary infection of human B and endothelial cells. J Virol 89:3093–3111
60. Toth Z, Brulois K, Lee HR, Izumiya Y, Tepper C, Kung HJ, Jung JU (2013) Biphasic euchromatin-to-heterochromatin transition on the KSHV genome following de novo infection. PLoS Pathog 9:e1003813
61. Gunther T, Grundhoff A (2010) The epigenetic landscape of latent Kaposi's sarcoma-associated herpesvirus genomes. PLoS Pathog 6:e1000935
62. Toth Z, Maglinte DT, Lee SH, Lee HR, Wong LY, Brulois KF, Lee S, Buckley JD, Laird PW, Marquez VE, Jung JU (2010) Epigenetic analysis of KSHV latent and lytic genomes. PLoS Pathog 6:e1001013
63. Toth Z, Papp B, Brulois K, Choi YJ, Gao SJ, Jung JU (2016) LANA-mediated recruitment of host polycomb repressive complexes onto the KSHV genome during de novo infection. PLoS Pathog 12:e1005878
64. Singh VV, Dutta D, Ansari MA, Dutta S, Chandran B (2014) Kaposi's sarcoma-associated herpesvirus induces the ATM and H2AX DNA damage response early during de novo infection of primary endothelial cells, which play roles in latency establishment. J Virol 88:2821–2834
65. Ye F, Lei X, Gao SJ (2011) Mechanisms of Kaposi's sarcoma-associated herpesvirus latency and reactivation. Adv Virol
66. Dittmer D, Lagunoff M, Renne R, Staskus K, Haase A, Ganem D (1998) A cluster of latently expressed genes in Kaposi's sarcoma-associated herpesvirus. J Virol 72:8309–8315
67. Pearce M, Matsumura S, Wilson AC (2005) Transcripts encoding K12, v-FLIP, v-cyclin, and the microRNA cluster of Kaposi's sarcoma-associated herpesvirus originate from a common promoter. J Virol 79:14457–14464
68. Sarid R, Wiezorek JS, Moore PS, Chang Y (1999) Characterization and cell cycle regulation of the major Kaposi's sarcoma-associated herpesvirus (human herpesvirus 8) latent genes and their promoter. J Virol 73:1438–1446

69. Staudt MR, Dittmer DP (2006) Promoter switching allows simultaneous transcription of LANA and K14/vGPCR of Kaposi's sarcoma-associated herpesvirus. Virology 350:192–205
70. Samols MA, Hu J, Skalsky RL, Renne R (2005) Cloning and identification of a microRNA cluster within the latency-associated region of Kaposi's sarcoma-associated herpesvirus. J Virol 79:9301–9305
71. Cunningham C, Barnard S, Blackbourn DJ, Davison AJ (2003) Transcription mapping of human herpesvirus 8 genes encoding viral interferon regulatory factors. J Gen Virol 84:1471–1483
72. Gao S-J, Kingsley L, Hoover DR, Spira TJ, Rinaldo CR, Saah A, Phair J, Detels R, Parry P, Chang Y, Moore PS (1996) Seroconversion to antibodies against Kaposi's sarcoma-associated herpesvirus-related latent nuclear antigens before the development of Kaposi's sarcoma. N Engl J Med 335:233–241
73. Gao SJ, Kingsley L, Li M, Zheng W, Parravicini C, Ziegler J, Newton R, Rinaldo CR, Saah A, Phair J, Detels R, Chang Y, Moore PS (1996) KSHV antibodies among Americans, Italians and Ugandans with and without Kaposi's sarcoma. Nat Med 2:925–928
74. Kedes DH, Operskalski E, Busch M, Kohn R, Flood J, Ganem D (1996) The seroepidemiology of human herpesvirus 8 (Kaposi's sarcoma-associated herpesvirus): distribution of infection in KS risk groups and evidence for sexual transmission. Nat Med 2:918–924
75. Zhang YJ, Deng JH, Rabkin C, Gao SJ (2000) Hot-spot variations of Kaposi's sarcoma-associated herpesvirus latent nuclear antigen and application in genotyping by PCR-RFLP. J Gen Virol 81:2049–2058
76. Kwun HJ, da Silva SR, Qin H, Ferris RL, Tan R, Chang Y, Moore PS (2011) The central repeat domain 1 of Kaposi's sarcoma-associated herpesvirus (KSHV) latency associated-nuclear antigen 1 (LANA1) prevents cis MHC class I peptide presentation. Virology 412:357–365
77. Kwun HJ, da Silva SR, Shah IM, Blake N, Moore PS, Chang Y (2007) Kaposi's sarcoma-associated herpesvirus latency-associated nuclear antigen 1 mimics Epstein-Barr virus EBNA1 immune evasion through central repeat domain effects on protein processing. J Virol 81:8225–8235
78. Canham M, Talbot SJ (2004) A naturally occurring C-terminal truncated isoform of the latent nuclear antigen of Kaposi's sarcoma-associated herpesvirus does not associate with viral episomal DNA. J Gen Virol 85:1363–1369
79. Kwun HJ, Toptan T, Ramos da Silva S, Atkins JF, Moore PS, Chang Y (2014) Human DNA tumor viruses generate alternative reading frame proteins through repeat sequence recoding. Proc Natl Acad Sci U S A 111:E4342–9
80. Ballestas ME, Chatis PA, Kaye KM (1999) Efficient persistence of extrachromosomal KSHV DNA mediated by latency-associated nuclear antigen. Science 284:641–644
81. Ueda K, Sakakibara S, Ohsaki E, Yada K (2006) Lack of a mechanism for faithful partition and maintenance of the KSHV genome. Virus Res 122:85–94
82. Gao SJ, Zhang YJ, Deng JH, Rabkin CS, Flore O, Jenson HB (1999) Molecular polymorphism of Kaposi's sarcoma-associated herpesvirus (Human herpesvirus 8) latent nuclear antigen: evidence for a large repertoire of viral genotypes and dual infection with different viral genotypes. J Infect Dis 180:1466–1476
83. Ye FC, Zhou FC, Yoo SM, Xie JP, Browning PJ, Gao SJ (2004) Disruption of Kaposi's sarcoma-associated herpesvirus latent nuclear antigen leads to abortive episome persistence. J Virol 78:11121–11129
84. Cherezova L, Burnside KL, Rose TM (2011) Conservation of complex nuclear localization signals utilizing classical and non-classical nuclear import pathways in LANA homologs of KSHV and RFHV. PLoS ONE 6:e18920

85. Ballestas ME, Kaye KM (2001) Kaposi's sarcoma-associated herpesvirus latency-associated nuclear antigen 1 mediates episome persistence through cis-acting terminal repeat (TR) sequence and specifically binds TR DNA. J Virol 75:3250–3258
86. Barbera AJ, Chodaparambil JV, Kelley-Clarke B, Luger K, Kaye KM (2006) Kaposi's sarcoma-associated herpesvirus LANA hitches a ride on the chromosome. Cell Cycle 5:1048–1052
87. Barbera AJ, Chodaparambil JV, Kelley-Clarke B, Joukov V, Walter JC, Luger K, Kaye KM (2006) The nucleosomal surface as a docking station for Kaposi's sarcoma herpesvirus LANA. Science 311:856–861
88. Kelley-Clarke B, De Leon-Vazquez E, Slain K, Barbera AJ, Kaye KM (2009) Role of Kaposi's sarcoma-associated herpesvirus C-terminal LANA chromosome binding in episome persistence. J Virol 83:4326–4337
89. Lim C, Lee D, Seo T, Choi C, Choe J (2003) Latency-associated nuclear antigen of Kaposi's sarcoma-associated herpesvirus functionally interacts with heterochromatin protein 1. J Biol Chem 278:7397–7405
90. Pan HY, Zhang YJ, Wang XP, Deng JH, Zhou FC, Gao SJ (2003) Identification of a novel cellular transcriptional repressor interacting with the latent nuclear antigen of Kaposi's sarcoma-associated herpesvirus. J Virol 77:9758–9768
91. Matsumura S, Persson LM, Wong L, Wilson AC (2010) The latency-associated nuclear antigen interacts with MeCP2 and nucleosomes through separate domains. J Virol 84:2318–2330
92. Ottinger M, Christalla T, Nathan K, Brinkmann MM, Viejo-Borbolla A, Schulz TF (2006) Kaposi's sarcoma-associated herpesvirus LANA-1 interacts with the short variant of BRD4 and releases cells from a BRD4- and BRD2/RING3-induced G1 cell cycle arrest. J Virol 80:10772–10786
93. Xiao B, Verma SC, Cai Q, Kaul R, Lu J, Saha A, Robertson ES (2010) Bub1 and CENP-F can contribute to Kaposi's sarcoma-associated herpesvirus genome persistence by targeting LANA to kinetochores. J Virol 84:9718–9732
94. Si H, Verma SC, Lampson MA, Cai Q, Robertson ES (2008) Kaposi's sarcoma-associated herpesvirus-encoded LANA can interact with the nuclear mitotic apparatus protein to regulate genome maintenance and segregation. J Virol 82:6734–6746
95. Verma SC, Choudhuri T, Kaul R, Robertson ES (2006) Latency-associated nuclear antigen (LANA) of Kaposi's sarcoma-associated herpesvirus interacts with origin recognition complexes at the LANA binding sequence within the terminal repeats. J Virol 80:2243–2256
96. Lan K, Kuppers DA, Verma SC, Robertson ES (2004) Kaposi's sarcoma-associated herpesvirus-encoded latency-associated nuclear antigen inhibits lytic replication by targeting Rta: a potential mechanism for virus-mediated control of latency. J Virol 78:6585–6594
97. Li Q, Zhou F, Ye F, Gao SJ (2008) Genetic disruption of KSHV major latent nuclear antigen LANA enhances viral lytic transcriptional program. Virology 379:234–244
98. Li Q, He M, Zhou F, Ye F, Gao SJ (2014) Activation of Kaposi's sarcoma-associated herpesvirus (KSHV) by inhibitors of class III histone deacetylases: identification of sirtuin 1 as a regulator of the KSHV life cycle. J Virol 88:6355–6367
99. He M, Gao SJ (2014) A novel role of SIRT1 in gammaherpesvirus latency and replication. Cell Cycle 13:3328–3330
100. Stedman W, Deng Z, Lu F, Lieberman PM (2004) ORC, MCM, and histone hyperacetylation at the Kaposi's sarcoma-associated herpesvirus latent replication origin. J Virol 78:12566–12575
101. Lu F, Zhou J, Wiedmer A, Madden K, Yuan Y, Lieberman PM (2003) Chromatin remodeling of the Kaposi's sarcoma-associated herpesvirus ORF50 promoter correlates with reactivation from latency. J Virol 77:11425–11435
102. Chen J, Ueda K, Sakakibara S, Okuno T, Parravicini C, Corbellino M, Yamanishi K (2001) Activation of latent Kaposi's sarcoma-associated herpesvirus by demethylation of the promoter of the lytic transactivator. Proc Natl Acad Sci U S A 98:4119–4124

103. Ohsaki E, Ueda K, Sakakibara S, Do E, Yada K, Yamanishi K (2004) Poly(ADP-ribose) polymerase 1 binds to Kaposi's sarcoma-associated herpesvirus (KSHV) terminal repeat sequence and modulates KSHV replication in latency. J Virol 78:9936–9946
104. Hyun TS, Subramanian C, Cotter MA 2nd, Thomas RA, Robertson ES (2001) Latency-associated nuclear antigen encoded by Kaposi's sarcoma-associated herpesvirus interacts with Tat and activates the long terminal repeat of human immunodeficiency virus type 1 in human cells. J Virol 75:8761–8771
105. Garber AC, Shu MA, Hu J, Renne R (2001) DNA binding and modulation of gene expression by the latency-associated nuclear antigen of Kaposi's sarcoma-associated herpesvirus. J Virol 75:7882–7892
106. Stedman W, Kang H, Lin S, Kissil JL, Bartolomei MS, Lieberman PM (2008) Cohesins localize with CTCF at the KSHV latency control region and at cellular c-myc and H19/Igf2 insulators. EMBO J 27:654–666
107. Kang H, Lieberman PM (2009) Cell cycle control of Kaposi's sarcoma-associated herpesvirus latency transcription by CTCF-cohesin interactions. J Virol 83:6199–6210
108. Kang H, Wiedmer A, Yuan Y, Robertson E, Lieberman PM (2011) Coordination of KSHV latent and lytic gene control by CTCF-cohesin mediated chromosome conformation. PLoS Pathog 7:e1002140
109. Chen HS, Wikramasinghe P, Showe L, Lieberman PM (2012) Cohesins repress Kaposi's sarcoma-associated herpesvirus immediate early gene transcription during latency. J Virol 86:9454–9464
110. Kang H, Cho H, Sung GH, Lieberman PM (2013) CTCF regulates Kaposi's sarcoma-associated herpesvirus latency transcription by nucleosome displacement and RNA polymerase programming. J Virol 87:1789–1799
111. Ye FC, Zhou FC, Xie JP, Kang T, Greene W, Kuhne K, Lei XF, Li QH, Gao SJ (2008) Kaposi's sarcoma-associated herpesvirus latent gene vFLIP inhibits viral lytic replication through NF-kappaB-mediated suppression of the AP-1 pathway: a novel mechanism of virus control of latency. J Virol 82:4235–4249
112. Lei X, Bai Z, Ye F, Xie J, Kim CG, Huang Y, Gao SJ (2010) Regulation of NF-kappaB inhibitor IkappaBalpha and viral replication by a KSHV microRNA. Nat Cell Biol 12:193–199
113. Bellare P, Ganem D (2009) Regulation of KSHV lytic switch protein expression by a virus-encoded microRNA: an evolutionary adaptation that fine-tunes lytic reactivation. Cell Host Microbe 6:570–575
114. Lu CC, Li Z, Chu CY, Feng J, Sun R, Rana TM (2010) MicroRNAs encoded by Kaposi's sarcoma-associated herpesvirus regulate viral life cycle. EMBO Rep 11:784–790
115. Lu F, Stedman W, Yousef M, Renne R, Lieberman PM (2010) Epigenetic regulation of Kaposi's sarcoma-associated herpesvirus latency by virus-encoded microRNAs that target Rta and the cellular Rbl2-DNMT pathway. J Virol 84:2697–2706
116. Liang D, Gao Y, Lin X, He Z, Zhao Q, Deng Q, Lan K (2011) A human herpesvirus miRNA attenuates interferon signaling and contributes to maintenance of viral latency by targeting IKKepsilon. Cell Res 21:793–806
117. Lin X, Liang D, He Z, Deng Q, Robertson ES, Lan K (2011) miR-K12-7-5p encoded by Kaposi's sarcoma-associated herpesvirus stabilizes the latent state by targeting viral ORF50/RTA. PLoS ONE 6:e16224
118. Bai Z, Huang Y, Li W, Zhu Y, Jung JU, Lu C, Gao SJ (2014) Genomewide mapping and screening of Kaposi's sarcoma-associated herpesvirus (KSHV) 3′ untranslated regions identify bicistronic and polycistronic viral transcripts as frequent targets of KSHV microRNAs. J Virol 88:377–392
119. Arias C, Weisburd B, Stern-Ginossar N, Mercier A, Madrid AS, Bellare P, Holdorf M, Weissman JS, Ganem D (2014) KSHV 2.0: a comprehensive annotation of the Kaposi's sarcoma-associated herpesvirus genome using next-generation sequencing reveals novel genomic and functional features. PLoS Pathog 10:e1003847

120. Lukac DM, Renne R, Kirshner JR, Ganem D (1998) Reactivation of Kaposi's sarcoma-associated herpesvirus infection from latency by expression of the ORF 50 transactivator, a homolog of the EBV R protein. Virology 252:304–312
121. Sun R, Lin SF, Gradoville L, Yuan Y, Zhu F, Miller G (1998) A viral gene that activates lytic cycle expression of Kaposi's sarcoma-associated herpesvirus. Proc Natl Acad Sci U S A 95:10866–10871
122. Gradoville L, Gerlach J, Grogan E, Shedd D, Nikiforow S, Metroka C, Miller G (2000) Kaposi's sarcoma-associated herpesvirus open reading frame 50/Rta protein activates the entire viral lytic cycle in the HH-B2 primary effusion lymphoma cell line. J Virol 74:6207–6212
123. Song MJ, Brown HJ, Wu TT, Sun R (2001) Transcription activation of polyadenylated nuclear rna by rta in human herpesvirus 8/Kaposi's sarcoma-associated herpesvirus. J Virol 75:3129–3140
124. Bu W, Palmeri D, Krishnan R, Marin R, Aris VM, Soteropoulos P, Lukac DM (2008) Identification of direct transcriptional targets of the Kaposi's sarcoma-associated herpesvirus Rta lytic switch protein by conditional nuclear localization. J Virol 82:10709–10723
125. Chen J, Ye F, Xie J, Kuhne K, Gao SJ (2009) Genome-wide identification of binding sites for Kaposi's sarcoma-associated herpesvirus lytic switch protein, RTA. Virology 386:290–302
126. Ziegelbauer J, Grundhoff A, Ganem D (2006) Exploring the DNA binding interactions of the Kaposi's sarcoma-associated herpesvirus lytic switch protein by selective amplification of bound sequences in vitro. J Virol 80:2958–2967
127. Ye J, Shedd D, Miller G (2005) An Sp1 response element in the Kaposi's sarcoma-associated herpesvirus open reading frame 50 promoter mediates lytic cycle induction by butyrate. J Virol 79:1397–1408
128. Carroll KD, Khadim F, Spadavecchia S, Palmeri D, Lukac DM (2007) Direct interactions of Kaposi's sarcoma-associated herpesvirus/human herpesvirus 8 ORF50/Rta protein with the cellular protein octamer-1 and DNA are critical for specifying transactivation of a delayed-early promoter and stimulating viral reactivation. J Virol 81:8451–8467
129. Wilson SJ, Tsao EH, Webb BL, Ye H, Dalton-Griffin L, Tsantoulas C, Gale CV, Du MQ, Whitehouse A, Kellam P (2007) X box binding protein XBP-1 s transactivates the Kaposi's sarcoma-associated herpesvirus (KSHV) ORF50 promoter, linking plasma cell differentiation to KSHV reactivation from latency. J Virol 81:13578–13586
130. Chang PJ, Boonsiri J, Wang SS, Chen LY, Miller G (2010) Binding of RBP-Jkappa (CSL) protein to the promoter of the Kaposi's sarcoma-associated herpesvirus ORF47 (gL) gene is a critical but not sufficient determinant of transactivation by ORF50 protein. Virology 398:38–48
131. Wang SE, Wu FY, Fujimuro M, Zong J, Hayward SD, Hayward GS (2003) Role of CCAAT/enhancer-binding protein alpha (C/EBPalpha) in activation of the Kaposi's sarcoma-associated herpesvirus (KSHV) lytic-cycle replication-associated protein (RAP) promoter in cooperation with the KSHV replication and transcription activator (RTA) and RAP. J Virol 77:600–623
132. Guito J, Lukac DM (2012) KSHV Rta promoter specification and viral reactivation. Front Microbiol 3:30
133. Sun Z, Jha HC, Pei YG, Robertson ES (2016) Major histocompatibility complex class II HLA-DRalpha is downregulated by Kaposi's Sarcoma-associated herpesvirus-encoded lytic transactivator RTA and MARCH8. J Virol 90:8047–8058
134. Chmura JC, Herold K, Ruffin A, Atuobi T, Fabiyi Y, Mitchell AE, Choi YB, Ehrlich ES (2017) The Itch ubiquitin ligase is required for KSHV RTA induced vFLIP degradation. Virology 501:119–126
135. Ehrlich ES, Chmura JC, Smith JC, Kalu NN, Hayward GS (2014) KSHV RTA abolishes NFkappaB responsive gene expression during lytic reactivation by targeting vFLIP for degradation via the proteasome. PLoS ONE 9:e91359

136. AuCoin DP, Colletti KS, Cei SA, Papouskova I, Tarrant M, Pari GS (2004) Amplification of the Kaposi's sarcoma-associated herpesvirus/human herpesvirus 8 lytic origin of DNA replication is dependent upon a cis-acting AT-rich region and an ORF50 response element and the trans-acting factors ORF50 (K-Rta) and K8 (K-bZIP). Virology 318:542–555
137. Wang Y, Li H, Chan MY, Zhu FX, Lukac DM, Yuan Y (2004) Kaposi's sarcoma-associated herpesvirus ori-Lyt-dependent DNA replication: cis-acting requirements for replication and ori-Lyt-associated RNA transcription. J Virol 78:8615–8629
138. Wakeman BS, Izumiya Y, Speck SH (2016) Identification of novel KSHV Orf50 transcripts: discovery of new RTA isoforms with variable transactivation potential. J Virol
139. Han Z, Swaminathan S (2006) Kaposi's sarcoma-associated herpesvirus lytic gene ORF57 is essential for infectious virion production. J Virol 80:5251–5260
140. Majerciak V, Yamanegi K, Allemand E, Kruhlak M, Krainer AR, Zheng ZM (2008) Kaposi's sarcoma-associated herpesvirus ORF57 functions as a viral splicing factor and promotes expression of intron-containing viral lytic genes in spliceosome-mediated RNA splicing. J Virol 82:2792–2801
141. Malik P, Blackbourn DJ, Cheng MF, Hayward GS, Clements JB (2004) Functional co-operation between the Kaposi's sarcoma-associated herpesvirus ORF57 and ORF50 regulatory proteins. J Gen Virol 85:2155–2166
142. Majerciak V, Zheng ZM (2015) KSHV ORF57, a protein of many faces. Viruses 7:604–633
143. Pilkington GR, Majerciak V, Bear J, Uranishi H, Zheng ZM, Felber BK (2012) Kaposi's sarcoma-associated herpesvirus ORF57 is not a bona fide export factor. J Virol 86:13089–13094
144. Massimelli MJ, Kang JG, Majerciak V, Le SY, Liewehr DJ, Steinberg SM, Zheng ZM (2011) Stability of a long noncoding viral RNA depends on a 9-nt core element at the RNA 5′ end to interact with viral ORF57 and cellular PABPC1. Int J Biol Sci 7:1145–1160
145. Sei E, Conrad NK (2011) Delineation of a core RNA element required for Kaposi's sarcoma-associated herpesvirus ORF57 binding and activity. Virology 419:107–116
146. Kang JG, Pripuzova N, Majerciak V, Kruhlak M, Le SY, Zheng ZM (2011) Kaposi's sarcoma-associated herpesvirus ORF57 promotes escape of viral and human interleukin-6 from microRNA-mediated suppression. J Virol 85:2620–2630
147. Boyne JR, Jackson BR, Taylor A, Macnab SA, Whitehouse A (2010) Kaposi's sarcoma-associated herpesvirus ORF57 protein interacts with PYM to enhance translation of viral intronless mRNAs. EMBO J 29:1851–1864
148. Lin SF, Robinson DR, Miller G, Kung HJ (1999) Kaposi's sarcoma-associated herpesvirus encodes a bZIP protein with homology to BZLF1 of Epstein-Barr virus. J Virol 73:1909–1917
149. Purushothaman P, Uppal T, Verma SC (2015) Molecular biology of KSHV lytic reactivation. Viruses 7:116–153
150. Martinez FP, Tang Q (2012) Leucine zipper domain is required for Kaposi sarcoma-associated herpesvirus (KSHV) K-bZIP protein to interact with histone deacetylase and is important for KSHV replication. J Biol Chem 287:15622–15634
151. Izumiya Y, Ellison TJ, Yeh ET, Jung JU, Luciw PA, Kung HJ (2005) Kaposi's sarcoma-associated herpesvirus K-bZIP represses gene transcription via SUMO modification. J Virol 79:9912–9925
152. Chang PC, Izumiya Y, Wu CY, Fitzgerald LD, Campbell M, Ellison TJ, Lam KS, Luciw PA, Kung HJ (2010) Kaposi's sarcoma-associated herpesvirus (KSHV) encodes a SUMO E3 ligase that is SIM-dependent and SUMO-2/3-specific. J Biol Chem 285:5266–5273
153. Yang WS, Hsu HW, Campbell M, Cheng CY, Chang PC (2015) K-bZIP mediated SUMO-2/3 specific modification on the KSHV genome negatively regulates lytic gene expression and viral reactivation. PLoS Pathog 11:e1005051

154. Yang WS, Campbell M, Chang PC (2017) SUMO modification of a heterochromatin histone demethylase JMJD2A enables viral gene transactivation and viral replication. PLoS Pathog 13:e1006216
155. Rossetto C, Yamboliev I, Pari GS (2009) Kaposi's sarcoma-associated herpesvirus/human herpesvirus 8 K-bZIP modulates latency-associated nuclear protein-mediated suppression of lytic origin-dependent DNA synthesis. J Virol 83:8492–8501
156. Izumiya Y, Izumiya C, Van Geelen A, Wang DH, Lam KS, Luciw PA, Kung HJ (2007) Kaposi's sarcoma-associated herpesvirus-encoded protein kinase and its interaction with K-bZIP. J Virol 81:1072–1082
157. Toth Z, Brulois K, Jung JU (2013) The chromatin landscape of Kaposi's sarcoma-associated herpesvirus. Viruses 5:1346–1373
158. Gwack Y, Baek HJ, Nakamura H, Lee SH, Meisterernst M, Roeder RG, Jung JU (2003) Principal role of TRAP/mediator and SWI/SNF complexes in Kaposi's sarcoma-associated herpesvirus RTA-mediated lytic reactivation. Mol Cell Biol 23:2055–2067
159. Ye F, Zeng Y, Sha J, Jones T, Kuhne K, Wood C, Gao SJ (2016) High glucose induces reactivation of latent Kaposi's sarcoma-associated herpesvirus. J Virol Aug 17. pii: JVI.01049-16. [Epub ahead of print]
160. Thakker S, Verma SC (2016) Co-infections and pathogenesis of KSHV-associated malignancies. Front Microbiol 7:151
161. Merat R, Amara A, Lebbe C, de The H, Morel P, Saib A (2002) HIV-1 infection of primary effusion lymphoma cell line triggers Kaposi's sarcoma-associated herpesvirus (KSHV) reactivation. Int J Cancer 97:791–795
162. Zeng Y, Zhang X, Huang Z, Cheng L, Yao S, Qin D, Chen X, Tang Q, Lv Z, Zhang L, Lu C (2007) Intracellular Tat of human immunodeficiency virus type 1 activates lytic cycle replication of Kaposi's sarcoma-associated herpesvirus: role of JAK/STAT signaling. J Virol 81:2401–2417
163. Aoki Y, Tosato G (2004) HIV-1 Tat enhances Kaposi sarcoma-associated herpesvirus (KSHV) infectivity. Blood 104:810–814
164. Spadavecchia S, Gonzalez-Lopez O, Carroll KD, Palmeri D, Lukac DM (2010) Convergence of Kaposi's sarcoma-associated herpesvirus reactivation with Epstein-Barr virus latency and cellular growth mediated by the notch signaling pathway in coinfected cells. J Virol 84:10488–10500
165. Jiang Y, Xu D, Zhao Y, Zhang L (2008) Mutual inhibition between Kaposi's sarcoma-associated herpesvirus and Epstein-Barr virus lytic replication initiators in dually-infected primary effusion lymphoma. PLoS ONE 3:e1569
166. Lu C, Zeng Y, Huang Z, Huang L, Qian C, Tang G, Qin D (2005) Human herpesvirus 6 activates lytic cycle replication of Kaposi's sarcoma-associated herpesvirus. Am J Pathol 166:173–183
167. Tang Q, Qin D, Lv Z, Zhu X, Ma X, Yan Q, Zeng Y, Guo Y, Feng N, Lu C (2012) Herpes simplex virus type 2 triggers reactivation of Kaposi's sarcoma-associated herpesvirus from latency and collaborates with HIV-1 Tat. PLoS ONE 7:e31652
168. Blauvelt A (2001) Skin diseases associated with human herpesvirus 6, 7, and 8 infection. J Investig Dermatol Symp Proc 6:197–202
169. Roupelieva M, Griffiths SJ, Kremmer E, Meisterernst M, Viejo-Borbolla A, Schulz T, Haas J (2010) Kaposi's sarcoma-associated herpesvirus Lana-1 is a major activator of the serum response element and mitogen-activated protein kinase pathways via interactions with the mediator complex. J Gen Virol 91:1138–1149
170. Vieira J, O'Hearn P, Kimball L, Chandran B, Corey L (2001) Activation of Kaposi's sarcoma-associated herpesvirus (human herpesvirus 8) lytic replication by human cytomegalovirus. J Virol 75:1378–1386
171. Wells R, Stensland L, Vieira J (2009) The human cytomegalovirus UL112-113 locus can activate the full Kaposi's sarcoma-associated herpesvirus lytic replication cycle. J Virol 83:4695–4699

172. Dai L, DeFee MR, Cao Y, Wen J, Wen X, Noverr MC, Qin Z (2014) Lipoteichoic acid (LTA) and lipopolysaccharides (LPS) from periodontal pathogenic bacteria facilitate oncogenic herpesvirus infection within primary oral cells. PLoS ONE 9:e101326
173. Morris TL, Arnold RR, Webster-Cyriaque J (2007) Signaling cascades triggered by bacterial metabolic end products during reactivation of Kaposi's sarcoma-associated herpesvirus. J Virol 81:6032–6042
174. Yu X, Shahir AM, Sha J, Feng Z, Eapen B, Nithianantham S, Das B, Karn J, Weinberg A, Bissada NF, Ye F (2014) Short-chain fatty acids from periodontal pathogens suppress histone deacetylases, EZH2, and SUV39H1 to promote Kaposi's sarcoma-associated herpesvirus replication. J Virol 88:4466–4479
175. Davis DA, Rinderknecht AS, Zoeteweij JP, Aoki Y, Read-Connole EL, Tosato G, Blauvelt A, Yarchoan R (2001) Hypoxia induces lytic replication of Kaposi sarcoma-associated herpesvirus. Blood 97:3244–3250
176. Haque M, Wang V, Davis DA, Zheng ZM, Yarchoan R (2006) Genetic organization and hypoxic activation of the Kaposi's sarcoma-associated herpesvirus ORF34-37 gene cluster. J Virol 80:7037–7051
177. Haque M, Davis DA, Wang V, Widmer I, Yarchoan R (2003) Kaposi's sarcoma-associated herpesvirus (human herpesvirus 8) contains hypoxia response elements: relevance to lytic induction by hypoxia. J Virol 77:6761–6768
178. Veeranna RP, Haque M, Davis DA, Yang M, Yarchoan R (2012) Kaposi's sarcoma-associated herpesvirus latency-associated nuclear antigen induction by hypoxia and hypoxia-inducible factors. J Virol 86:1097–1108
179. Cai Q, Lan K, Verma SC, Si H, Lin D, Robertson ES (2006) Kaposi's sarcoma-associated herpesvirus latent protein LANA interacts with HIF-1 alpha to upregulate RTA expression during hypoxia: Latency control under low oxygen conditions. J Virol 80:7965–7975
180. Dalton-Griffin L, Wilson SJ, Kellam P (2009) X-box binding protein 1 contributes to induction of the Kaposi's sarcoma-associated herpesvirus lytic cycle under hypoxic conditions. J Virol 83:7202–7209
181. Yu F, Feng J, Harada JN, Chanda SK, Kenney SC, Sun R (2007) B cell terminal differentiation factor XBP-1 induces reactivation of Kaposi's sarcoma-associated herpesvirus. FEBS Lett 581:3485–3488
182. Zhang L, Zhu C, Guo Y, Wei F, Lu J, Qin J, Banerjee S, Wang J, Shang H, Verma SC, Yuan Z, Robertson ES, Cai Q (2014) Inhibition of KAP1 enhances hypoxia-induced Kaposi's sarcoma-associated herpesvirus reactivation through RBP-Jkappa. J Virol 88:6873–6884
183. Sun R, Liang D, Gao Y, Lan K (2014) Kaposi's sarcoma-associated herpesvirus-encoded LANA interacts with host KAP1 to facilitate establishment of viral latency. J Virol 88:7331–7344
184. Ye F, Zhou F, Bedolla RG, Jones T, Lei X, Kang T, Guadalupe M, Gao SJ (2011) Reactive oxygen species hydrogen peroxide mediates Kaposi's sarcoma-associated herpesvirus reactivation from latency. PLoS Pathog 7:e1002054
185. Li X, Feng J, Sun R (2011) Oxidative stress induces reactivation of Kaposi's sarcoma-associated herpesvirus and death of primary effusion lymphoma cells. J Virol 85:715–724
186. Hesser CR, Karijolich J, Dominissini D, He C, Glaunsinger BA (2018) N6-methyladenosine modification and the YTHDF2 reader protein play cell type specific roles in lytic viral gene expression during Kaposi's sarcoma-associated herpesvirus infection. PLoS Pathog 14(4): e1006995
187. Tan B, Liu H, Zhang S, et al (2018) Viral and cellular N(6)-methyladenosine and N(6),2'-O-dimethyladenosine epitranscriptomes in the KSHV life cycle. Nat microbiol 3(1):108–120
188. Du H, Zhao Y, He J, et al (2016) YTHDF2 destabilizes m(6)A-containing RNA through direct recruitment of the CCR4-NOT deadenylase complex. Nat commun 7:12626

189. Wang X, Lu Z, Gomez A, et al (2014) N6-methyladenosine-dependent regulation of messenger RNA stability. Nat 505(7481):117–120
190. Gao SJ, Boshoff C, Jayachandra S, Weiss RA, Chang Y, Moore PS (1997) KSHV ORF K9 (vIRF) is an oncogene which inhibits the interferon signaling pathway. Oncogene 15:1979–1985
191. Hwang SW, Kim D, Jung JU, Lee HR (2017) KSHV-encoded viral interferon regulatory factor 4 (vIRF4) interacts with IRF7 and inhibits interferon alpha production. Biochem Biophys Res Commun 486:700–705
192. Giffin L, Damania B (2014) KSHV: pathways to tumorigenesis and persistent infection. Adv Virus Res 88:111–159
193. Liang Q, Fu B, Wu F, Li X, Yuan Y, Zhu F (2012) ORF45 of Kaposi's sarcoma-associated herpesvirus inhibits phosphorylation of interferon regulatory factor 7 by IKKepsilon and TBK1 as an alternative substrate. J Virol 86:10162–10172
194. Yu Y, Wang SE, Hayward GS (2005) The KSHV immediate-early transcription factor RTA encodes ubiquitin E3 ligase activity that targets IRF7 for proteosome-mediated degradation. Immunity 22:59–70
195. Lefort S, Soucy-Faulkner A, Grandvaux N, Flamand L (2007) Binding of Kaposi's sarcoma-associated herpesvirus K-bZIP to interferon-responsive factor 3 elements modulates antiviral gene expression. J Virol 81:10950–10960
196. Bartee E, McCormack A, Fruh K (2006) Quantitative membrane proteomics reveals new cellular targets of viral immune modulators. PLoS Pathog 2:e107
197. Li Q, Means R, Lang S, Jung JU (2007) Downregulation of gamma interferon receptor 1 by Kaposi's sarcoma-associated herpesvirus K3 and K5. J Virol 81:2117–2127
198. Lagos D, Vart RJ, Gratrix F, Westrop SJ, Emuss V, Wong PP, Robey R, Imami N, Bower M, Gotch F, Boshoff C (2008) Toll-like receptor 4 mediates innate immunity to Kaposi sarcoma herpesvirus. Cell Host Microbe 4:470–483
199. Jacobs SR, Gregory SM, West JA, Wollish AC, Bennett CL, Blackbourn DJ, Heise MT, Damania B (2013) The viral interferon regulatory factors of kaposi's sarcoma-associated herpesvirus differ in their inhibition of interferon activation mediated by toll-like receptor 3. J Virol 87:798–806
200. Jacobs SR, Stopford CM, West JA, Bennett CL, Giffin L, Damania B (2015) Kaposi's Sarcoma-associated herpesvirus viral interferon regulatory factor 1 interacts with a member of the interferon-stimulated gene 15 pathway. J Virol 89:11572–11583
201. Bussey KA, Reimer E, Todt H, Denker B, Gallo A, Konrad A, Ottinger M, Adler H, Sturzl M, Brune W, Brinkmann MM (2014) The gammaherpesviruses Kaposi's sarcoma-associated herpesvirus and murine gammaherpesvirus 68 modulate the Toll-like receptor-induced proinflammatory cytokine response. J Virol 88:9245–9259
202. Gregory SM, Davis BK, West JA, Taxman DJ, Matsuzawa S, Reed JC, Ting JP, Damania B (2011) Discovery of a viral NLR homolog that inhibits the inflammasome. Science 331:330–334
203. Zhang G, Chan B, Samarina N, Abere B, Weidner-Glunde M, Buch A, Pich A, Brinkmann MM, Schulz TF (2016) Cytoplasmic isoforms of Kaposi sarcoma herpesvirus LANA recruit and antagonize the innate immune DNA sensor cGAS. Proc Natl Acad Sci U S A 113:E1034–43
204. Wu JJ, Li W, Shao Y, Avey D, Fu B, Gillen J, Hand T, Ma S, Liu X, Miley W, Konrad A, Neipel F, Sturzl M, Whitby D, Li H, Zhu F (2015) Inhibition of cGAS DNA Sensing by a Herpesvirus Virion Protein. Cell Host Microbe 18:333–344
205. Ma Z, Jacobs SR, West JA, Stopford C, Zhang Z, Davis Z, Barber GN, Glaunsinger BA, Dittmer DP, Damania B (2015) Modulation of the cGAS-STING DNA sensing pathway by gammaherpesviruses. Proc Natl Acad Sci U S A 112:E4306–15
206. Kerur N, Veettil MV, Sharma-Walia N, Bottero V, Sadagopan S, Otageri P, Chandran B (2011) IFI16 acts as a nuclear pathogen sensor to induce the inflammasome in response to Kaposi Sarcoma-associated herpesvirus infection. Cell Host Microbe 9:363–375

207. Roy A, Dutta D, Iqbal J, Pisano G, Gjyshi O, Ansari MA, Kumar B, Chandran B (2016) Nuclear innate immune DNA sensor IFI16 is degraded during lytic reactivation of Kaposi's Sarcoma-Associated Herpesvirus (KSHV): role of IFI16 in maintenance of KSHV Latency. J Virol 90:8822–8841
208. Areste C, Blackbourn DJ (2009) Modulation of the immune system by Kaposi's sarcoma-associated herpesvirus. Trends Microbiol 17:119–129
209. Lee MS, Jones T, Song DY, Jang JH, Jung JU, Gao SJ (2014) Exploitation of the complement system by oncogenic Kaposi's sarcoma-associated herpesvirus for cell survival and persistent infection. PLoS Pathog 10:e1004412
210. Thomas M, Wills M, Lehner PJ (2008) Natural killer cell evasion by an E3 ubiquitin ligase from Kaposi's sarcoma-associated herpesvirus. Biochem Soc Trans 36:459–463
211. Nachmani D, Stern-Ginossar N, Sarid R, Mandelboim O (2009) Diverse herpesvirus microRNAs target the stress-induced immune ligand MICB to escape recognition by natural killer cells. Cell Host Microbe 5:376–385
212. Madrid AS, Ganem D (2012) Kaposi's sarcoma-associated herpesvirus ORF54/dUTPase downregulates a ligand for the NK activating receptor NKp44. J Virol 86:8693–8704
213. Robey RC, Mletzko S, Bower M, Meys R, Boffito M, Nelson M, Bunker CB, Gotch FM (2011) Ex-vivo recognition of late-lytic CD8 epitopes specific for Kaposi's sarcoma-associated herpesvirus (KSHV) by HIV/KSHV-coinfected individuals. Viral Immunol 24:211–220
214. Bihl F, Mosam A, Henry LN, Chisholm JV 3rd, Dollard S, Gumbi P, Cassol E, Page T, Mueller N, Kiepiela P, Martin JN, Coovadia HM, Scadden DT, Brander C (2007) Kaposi's sarcoma-associated herpesvirus-specific immune reconstitution and antiviral effect of combined HAART/chemotherapy in HIV clade C-infected individuals with Kaposi's sarcoma. Aids 21:1245–1252
215. Miller G, Rigsby MO, Heston L, Grogan E, Sun R, Metroka C, Levy JA, Gao SJ, Chang Y, Moore P (1996) Antibodies to butyrate-inducible antigens of Kaposi's sarcoma-associated herpesvirus in patients with HIV-1 infection. N Engl J Med 334:1292–1297
216. Simpson GR, Schulz TF, Whitby D, Cook PM, Boshoff C, Rainbow L, Howard MR, Gao SJ, Bohenzky RA, Simmonds P, Lee C, de Ruiter A, Hatzakis A, Tedder RS, Weller IV, Weiss RA, Moore PS (1996) Prevalence of Kaposi's sarcoma associated herpesvirus infection measured by antibodies to recombinant capsid protein and latent immunofluorescence antigen. Lancet 348:1133–1138
217. Choi JK, Lee BS, Shim SN, Li M, Jung JU (2000) Identification of the novel K15 gene at the rightmost end of the Kaposi's sarcoma-associated herpesvirus genome. J Virol 74:436–446
218. Coscoy L, Ganem D (2001) A viral protein that selectively downregulates ICAM-1 and B7-2 and modulates T cell costimulation. J Clin Invest 107:1599–1606
219. Ishido S, Wang C, Lee BS, Cohen GB, Jung JU (2000) Downregulation of major histocompatibility complex class I molecules by Kaposi's sarcoma-associated herpesvirus K3 and K5 proteins. J Virol 74:5300–5309
220. Brulois K, Toth Z, Wong LY, Feng P, Gao SJ, Ensser A, Jung JU (2014) Kaposi's sarcoma-associated herpesvirus K3 and K5 ubiquitin E3 ligases have stage-specific immune evasion roles during lytic replication. J Virol 88:9335–9349
221. Brulois K, Jung JU (2014) Interplay between Kaposi's sarcoma-associated herpesvirus and the innate immune system. Cytokine Growth Factor Rev 25:597–609
222. Lagos D, Trotter MW, Vart RJ, Wang HW, Matthews NC, Hansen A, Flore O, Gotch F, Boshoff C (2007) Kaposi sarcoma herpesvirus-encoded vFLIP and vIRF1 regulate antigen presentation in lymphatic endothelial cells. Blood 109:1550–1558
223. Butler LM, Jeffery HC, Wheat RL, Long HM, Rae PC, Nash GB, Blackbourn DJ (2012) Kaposi's sarcoma-associated herpesvirus inhibits expression and function of endothelial cell major histocompatibility complex class II via suppressor of cytokine signaling 3. J Virol 86:7158–7166

224. Zuo J, Hislop AD, Leung CS, Sabbah S, Rowe M (2013) Kaposi's sarcoma-associated herpesvirus-encoded viral IRF3 modulates major histocompatibility complex class II (MHC-II) antigen presentation through MHC-II transactivator-dependent and -independent mechanisms: implications for oncogenesis. J Virol 87:5340–5350
225. Cirone M, Lucania G, Bergamo P, Trivedi P, Frati L, Faggioni A (2007) Human herpesvirus 8 (HHV-8) inhibits monocyte differentiation into dendritic cells and impairs their immunostimulatory activity. Immunol Lett 113:40–46
226. Gregory SM, Wang L, West JA, Dittmer DP, Damania B (2012) Latent Kaposi's sarcoma-associated herpesvirus infection of monocytes downregulates expression of adaptive immune response costimulatory receptors and proinflammatory cytokines. J Virol 86:3916–3923
227. Hong YK, Foreman K, Shin JW, Hirakawa S, Curry CL, Sage DR, Libermann T, Dezube BJ, Fingeroth JD, Detmar M (2004) Lymphatic reprogramming of blood vascular endothelium by Kaposi sarcoma-associated herpesvirus. Nat Genet 36:683–685
228. Wang HW, Trotter MW, Lagos D, Bourboulia D, Henderson S, Makinen T, Elliman S, Flanagan AM, Alitalo K, Boshoff C (2004) Kaposi sarcoma herpesvirus-induced cellular reprogramming contributes to the lymphatic endothelial gene expression in Kaposi's sarcoma. Nat Genet 36:687–693
229. Flore O, Rafii S, Ely S, O'Leary JJ, Hyjek EM, Cesarman E (1998) Transformation of primary human endothelial cells by Kaposi's sarcoma-associated herpesvirus. Nature 394:588–592
230. Wang L, Damania B (2008) Kaposi's sarcoma-associated herpesvirus confers a survival advantage to endothelial cells. Cancer Res 68:4640–4648
231. Mutlu AD, Cavallin LE, Vincent L, Chiozzini C, Eroles P, Duran EM, Asgari Z, Hooper AT, La Perle KM, Hilsher C, Gao SJ, Dittmer DP, Rafii S, Mesri EA (2007) In vivo-restricted and reversible malignancy induced by human herpesvirus-8 KSHV: a cell and animal model of virally induced Kaposi's sarcoma. Cancer Cell 11:245–258
232. Lee MS, Yuan H, Jeon H, Zhu Y, Yoo S, Shi S, Krueger B, Renne R, Lu C, Jung JU, Gao SJ (2016) Human mesenchymal stem cells of diverse origins support persistent infection with Kaposi's sarcoma-associated herpesvirus and manifest distinct angiogenic, invasive, and transforming phenotypes. MBio 7:e02109–15
233. Wu W, Vieira J, Fiore N, Banerjee P, Sieburg M, Rochford R, Harrington W Jr, Feuer G (2006) KSHV/HHV-8 infection of human hematopoietic progenitor (CD34+) cells: persistence of infection during hematopoiesis in vitro and in vivo. Blood 108:141–151
234. Di Bartolo DL, Cannon M, Liu YF, Renne R, Chadburn A, Boshoff C, Cesarman E (2008) KSHV LANA inhibits TGF-beta signaling through epigenetic silencing of the TGF-beta type II receptor. Blood 111:4731–4740
235. Santag S, Jager W, Karsten CB, Kati S, Pietrek M, Steinemann D, Sarek G, Ojala PM, Schulz TF (2013) Recruitment of the tumour suppressor protein p73 by Kaposi's sarcoma herpesvirus latent nuclear antigen contributes to the survival of primary effusion lymphoma cells. Oncogene 32:3676–3685
236. Friborg J Jr, Kong W, Hottiger MO, Nabel GJ (1999) p53 inhibition by the LANA protein of KSHV protects against cell death. Nature 402:889–894
237. Radkov SA, Kellam P, Boshoff C (2000) The latent nuclear antigen of Kaposi sarcoma-associated herpesvirus targets the retinoblastoma-E2F pathway and with the oncogene Hras transforms primary rat cells. Nat Med 6:1121–1127
238. Lu J, Verma SC, Murakami M, Cai Q, Kumar P, Xiao B, Robertson ES (2009) Latency-associated nuclear antigen of Kaposi's sarcoma-associated herpesvirus (KSHV) upregulates survivin expression in KSHV-Associated B-lymphoma cells and contributes to their proliferation. J Virol 83:7129–7141
239. Bubman D, Guasparri I, Cesarman E (2007) Deregulation of c-Myc in primary effusion lymphoma by Kaposi's sarcoma herpesvirus latency-associated nuclear antigen. Oncogene 26:4979–4986

240. Qin Z, Dai L, Slomiany MG, Toole BP, Parsons C (2010) Direct activation of emmprin and associated pathogenesis by an oncogenic herpesvirus. Cancer Res 70:3884–3889
241. Liu J, Martin HJ, Liao G, Hayward SD (2007) The Kaposi's sarcoma-associated herpesvirus LANA protein stabilizes and activates c-Myc. J Virol 81:10451–10459
242. Liang D, Hu H, Li S, Dong J, Wang X, Wang Y, He L, He Z, Gao Y, Gao SJ, Lan K (2014) Oncogenic herpesvirus KSHV Hijacks BMP-Smad1-Id signaling to promote tumorigenesis. PLoS Pathog 10:e1004253
243. Jha HC, Sun Z, Upadhyay SK, El-Naccache DW, Singh RK, Sahu SK, Robertson ES (2016) KSHV-Mediated Regulation of Par3 and SNAIL Contributes to B-Cell Proliferation. PLoS Pathog 12:e1005801
244. Zhu Y, Ramos da Silva S, He M, Liang Q, Lu C, Feng P, Jung JU, Gao SJ (2016) An oncogenic virus promotes cell survival and cellular transformation by suppressing glycolysis. PLoS Pathog 12:e1005648
245. Ballon G, Chen K, Perez R, Tam W, Cesarman E (2011) Kaposi's sarcoma herpesvirus (KSHV) vFLIP oncoprotein induces B cell transdifferentiation and tumorigenesis in mice. J Clin Invest 121:1141–1153
246. Moody R, Zhu Y, Huang Y, Cui X, Jones T, Bedolla R, Lei X, Bai Z, Gao SJ (2013) KSHV microRNAs mediate cellular transformation and tumorigenesis by redundantly targeting cell growth and survival pathways. PLoS Pathog 9:e1003857
247. Zhi H, Zahoor MA, Shudofsky AM, Giam CZ (2015) KSHV vCyclin counters the senescence/G1 arrest response triggered by NF-kappaB hyperactivation. Oncogene 34:496–505
248. Godden-Kent D, Talbot SJ, Boshoff C, Chang Y, Moore P, Weiss RA, Mittnacht S (1997) The cyclin encoded by Kaposi's sarcoma-associated herpesvirus stimulates cdk6 to phosphorylate the retinoblastoma protein and histone H1. J Virol 71:4193–4198
249. Swanton C, Mann DJ, Fleckenstein B, Neipel F, Peters G, Jones N (1997) Herpes viral cyclin/Cdk6 complexes evade inhibition by CDK inhibitor proteins. Nature 390:184–187
250. Verschuren EW, Klefstrom J, Evan GI, Jones N (2002) The oncogenic potential of Kaposi's sarcoma-associated herpesvirus cyclin is exposed by p53 loss in vitro and in vivo. Cancer Cell 2:229–241
251. Jones T, Ramos da Silva S, Bedolla R, Ye F, Zhou F, Gao SJ (2014) Viral cyclin promotes KSHV-induced cellular transformation and tumorigenesis by overriding contact inhibition. Cell Cycle 13:845–858
252. Seo T, Park J, Lee D, Hwang SG, Choe J (2001) Viral interferon regulatory factor 1 of Kaposi's sarcoma-associated herpesvirus binds to p53 and represses p53-dependent transcription and apoptosis. J Virol 75:6193–6198
253. Seo T, Park J, Choe J (2005) Kaposi's sarcoma-associated herpesvirus viral IFN regulatory factor 1 inhibits transforming growth factor-beta signaling. Cancer Res 65:1738–1747
254. Bais C, Santomasso B, Coso O, Arvanitakis L, Raaka EG, Gutkind JS, Asch AS, Cesarman E, Gershengorn MC, Mesri EA (1998) G-protein-coupled receptor of Kaposi's sarcoma-associated herpesvirus is a viral oncogene and angiogenesis activator. Nature 391:86–89
255. Yang TY, Chen SC, Leach MW, Manfra D, Homey B, Wiekowski M, Sullivan L, Jenh CH, Narula SK, Chensue SW, Lira SA (2000) Transgenic expression of the chemokine receptor encoded by human herpesvirus 8 induces an angioproliferative disease resembling Kaposi's sarcoma. J Exp Med 191:445–454
256. Krause CJ, Popp O, Thirunarayanan N, Dittmar G, Lipp M, Muller G (2016) MicroRNA-34a promotes genomic instability by a broad suppression of genome maintenance mechanisms downstream of the oncogene KSHV-vGPCR. Oncotarget 7:10414–10432
257. Martin D, Galisteo R, Molinolo AA, Wetzker R, Hirsch E, Gutkind JS (2011) PI3Kgamma mediates kaposi's sarcoma-associated herpesvirus vGPCR-induced sarcomagenesis. Cancer Cell 19:805–813

258. Martin D, Nguyen Q, Molinolo A, Gutkind JS (2014) Accumulation of dephosphorylated 4EBP after mTOR inhibition with rapamycin is sufficient to disrupt paracrine transformation by the KSHV vGPCR oncogene. Oncogene 33:2405–2412
259. Wu J, Xu Y, Mo D, Huang P, Sun R, Huang L, Pan S, Xu J (2014) Kaposi's sarcoma-associated herpesvirus (KSHV) vIL-6 promotes cell proliferation and migration by upregulating DNMT1 via STAT3 activation. PLoS ONE 9:e93478
260. Hideshima T, Chauhan D, Teoh G, Raje N, Treon SP, Tai YT, Shima Y, Anderson KC (2000) Characterization of signaling cascades triggered by human interleukin-6 versus Kaposi's sarcoma-associated herpes virus-encoded viral interleukin 6. Clin Cancer Res 6:1180–1189
261. Chen D, Choi YB, Sandford G, Nicholas J (2009) Determinants of secretion and intracellular localization of human herpesvirus 8 interleukin-6. J Virol 83:6874–6882
262. Chen D, Cousins E, Sandford G, Nicholas J (2012) Human herpesvirus 8 viral interleukin-6 interacts with splice variant 2 of vitamin K epoxide reductase complex subunit 1. J Virol 86:1577–1588
263. Lee H, Veazey R, Williams K, Li M, Guo J, Neipel F, Fleckenstein B, Lackner A, Desrosiers RC, Jung JU (1998) Deregulation of cell growth by the K1 gene of Kaposi's sarcoma-associated herpesvirus. Nat Med 4:435–440
264. Tomlinson CC, Damania B (2004) The K1 protein of Kaposi's sarcoma-associated herpesvirus activates the Akt signaling pathway. J Virol 78:1918–1927
265. Anders PM, Zhang Z, Bhende PM, Giffin L, Damania B (2016) The KSHV K1 protein modulates AMPK function to enhance cell survival. PLoS Pathog 12:e1005985
266. Tolani B, Gopalakrishnan R, Punj V, Matta H, Chaudhary PM (2014) Targeting Myc in KSHV-associated primary effusion lymphoma with BET bromodomain inhibitors. Oncogene 33:2928–2937
267. Aoki Y, Feldman GM, Tosato G (2003) Inhibition of STAT3 signaling induces apoptosis and decreases survivin expression in primary effusion lymphoma. Blood 101:1535–1542
268. He M, Tan B, Vasan K, Yuan H, Cheng F, Ramos da Silva S, Lu C, Gao SJ (2017) SIRT1 and AMPK pathways are essential for the proliferation and survival of primary effusion lymphoma cells. J Pathol
269. He M, Yuan H, Tan B, Bai R, Kim HS, Bae S, Che L, Kim JS, Gao SJ (2016) SIRT1-mediated downregulation of p27Kip1 is essential for overcoming contact inhibition of Kaposi's sarcoma-associated herpesvirus transformed cells. Oncotarget 7:75698–75711
270. Bhatt S, Ashlock BM, Toomey NL, Diaz LA, Mesri EA, Lossos IS, Ramos JC (2013) Efficacious proteasome/HDAC inhibitor combination therapy for primary effusion lymphoma. J Clin Invest 123:2616–2628
271. Chaisuparat R, Hu J, Jham BC, Knight ZA, Shokat KM, Montaner S (2008) Dual inhibition of PI3Kalpha and mTOR as an alternative treatment for Kaposi's sarcoma. Cancer Res 68:8361–8368
272. Dai L, Trillo-Tinoco J, Cao Y, Bonstaff K, Doyle L, Del Valle L, Whitby D, Parsons C, Reiss K, Zabaleta J, Qin Z (2015) Targeting HGF/c-MET induces cell cycle arrest, DNA damage, and apoptosis for primary effusion lymphoma. Blood 126:2821–2831
273. Lam BQ, Dai L, Li L, Qiao J, Lin Z, Qin Z (2017) Molecular mechanisms of activating c-MET in KSHV+ primary effusion lymphoma. Oncotarget 8:18373–18380
274. Korniluk A, Koper O, Kemona H, Dymicka-Piekarska V (2017) From inflammation to cancer. Ir J Med Sci 186:57–62
275. Balkwill FR, Mantovani A (2012) Cancer-related inflammation: common themes and therapeutic opportunities. Semin Cancer Biol 22:33–40
276. Riva G, Barozzi P, Torelli G, Luppi M (2010) Immunological and inflammatory features of Kaposi's sarcoma and other Kaposi's sarcoma-associated herpesvirus/human herpesvirus 8-associated neoplasias. AIDS Rev 12:40–51
277. Guedes F, de Andrade HF Jr, Fernandes ER, Tuon FF, Brasil RA, Pagliari C, Duarte MI (2008) The effects of human herpesvirus 8 infection and interferon-gamma response in

cutaneous lesions of Kaposi sarcoma differ among human immunodeficiency virus-infected and uninfected individuals. Br J Dermatol 159:839–846
278. Breuer-McHam JN, Ledbetter LS, Sarris AH, Duvic M (2000) Cytokine expression patterns distinguish HIV associated skin diseases. Exp Dermatol 9:341–350
279. Desnoyer A, Dupin N, Assoumou L, Carlotti A, Gaudin F, Deback C, Peytavin G, Marcelin AG, Boue F, Balabanian K, Pourcher V, group A. L. t. (2016) Expression pattern of the CXCL12/CXCR4-CXCR7 trio in Kaposi sarcoma skin lesions. Br J Dermatol 175:1251–1262
280. DiMaio TA, Gutierrez KD, Lagunoff M (2014) Kaposi's sarcoma-associated herpesvirus downregulates transforming growth factor beta2 to promote enhanced stability of capillary-like tube formation. J Virol 88:14301–14309
281. Douglas JL, Gustin JK, Moses AV, Dezube BJ, Pantanowitz L (2010) Kaposi's sarcoma pathogenesis: a triad of viral infection, oncogenesis and chronic inflammation. Transl Biomed 1. pii: 172
282. Dai L, Bratoeva M, Toole BP, Qin Z, Parsons C (2012) KSHV activation of VEGF secretion and invasion for endothelial cells is mediated through viral upregulation of emmprin-induced signal transduction. Int J Cancer 131:834–843
283. Wang X, He Z, Xia T, Li X, Liang D, Lin X, Wen H, Lan K (2014) Latency-associated nuclear antigen of Kaposi sarcoma-associated herpesvirus promotes angiogenesis through targeting notch signaling effector Hey1. Cancer Res 74:2026–2037
284. Cheng F, Pekkonen P, Laurinavicius S, Sugiyama N, Henderson S, Gunther T, Rantanen V, Kaivanto E, Aavikko M, Sarek G, Hautaniemi S, Biberfeld P, Aaltonen L, Grundhoff A, Boshoff C, Alitalo K, Lehti K, Ojala PM (2011) KSHV-initiated notch activation leads to membrane-type-1 matrix metalloproteinase-dependent lymphatic endothelial-to-mesenchymal transition. Cell Host Microbe 10:577–590
285. Zaldumbide A, Ossevoort M, Wiertz EJ, Hoeben RC (2007) In cis inhibition of antigen processing by the latency-associated nuclear antigen I of Kaposi sarcoma herpes virus. Mol Immunol 44:1352–1360
286. Thakker S, Purushothaman P, Gupta N, Challa S, Cai Q, Verma SC (2015) Kaposi's sarcoma-associated herpesvirus latency-associated nuclear antigen inhibits major histocompatibility complex class II expression by disrupting enhanceosome assembly through binding with the regulatory factor X complex. J Virol 89:5536–5556
287. Liu L, Eby MT, Rathore N, Sinha SK, Kumar A, Chaudhary PM (2002) The human herpes virus 8-encoded viral FLICE inhibitory protein physically associates with and persistently activates the Ikappa B kinase complex. J Biol Chem 277:13745–13751
288. Matta H, Chaudhary PM (2004) Activation of alternative NF-kappa B pathway by human herpes virus 8-encoded Fas-associated death domain-like IL-1 beta-converting enzyme inhibitory protein (vFLIP). Proc Natl Acad Sci U S A 101:9399–9404
289. Sharma-Walia N, Patel K, Chandran K, Marginean A, Bottero V, Kerur N, Paul AG (2012) COX-2/PGE2: molecular ambassadors of Kaposi's sarcoma-associated herpes virus oncoprotein-v-FLIP. Oncogenesis 1:e5
290. Guo Y, Li W, Qin J, Lu C, Fan W (2017) Kaposi's sarcoma-associated herpesvirus (KSHV)-encoded microRNAs promote matrix metalloproteinases (MMPs) expression and pro-angiogenic cytokine secretion in endothelial cells. J Med Virol 89:1274–1280
291. Breen EC (2007) VEGF in biological control. J Cell Biochem 102:1358–1367
292. Hu M, Wang C, Li W, Lu W, Bai Z, Qin D, Yan Q, Zhu J, Krueger BJ, Renne R, Gao SJ, Lu C (2015) A KSHV microRNA directly targets G protein-coupled receptor Kinase 2 to promote the migration and invasion of endothelial cells by inducing CXCR2 and activating AKT signaling. PLoS Pathog 11:e1005171
293. Li W, Jia X, Shen C, Zhang M, Xu J, Shang Y, Zhu K, Hu M, Yan Q, Qin D, Lee MS, Zhu J, Lu H, Krueger BJ, Renne R, Gao SJ, Lu C (2016) A KSHV microRNA enhances viral latency and induces angiogenesis by targeting GRK2 to activate the CXCR2/AKT pathway. Oncotarget 7:32286–32305

294. Li W, Yan Q, Ding X, Shen C, Hu M, Zhu Y, Qin D, Lu H, Krueger BJ, Renne R, Gao SJ, Lu C (2016) The SH3BGR/STAT3 pathway regulates cell migration and angiogenesis induced by a gammaherpesvirus microRNA. PLoS Pathog 12:e1005605
295. Li W, Hu M, Wang C, Lu H, Chen F, Xu J, Shang Y, Wang F, Qin J, Yan Q, Krueger BJ, Renne R, Gao SJ, Lu C (2017) A viral microRNA downregulates metastasis suppressor CD82 and induces cell invasion and angiogenesis by activating the c-Met signaling. Oncogene. May 22. https://doi.org/10.1038/onc.2017.139. [Epub ahead of print]
296. Liu Y, Sun R, Lin X, Liang D, Deng Q, Lan K (2012) Kaposi's sarcoma-associated herpesvirus-encoded microRNA miR-K12-11 attenuates transforming growth factor beta signaling through suppression of SMAD5. J Virol 86:1372–1381
297. Samols MA, Skalsky RL, Maldonado AM, Riva A, Lopez MC, Baker HV, Renne R (2007) Identification of cellular genes targeted by KSHV-encoded microRNAs. PLoS Pathog 3:e65
298. Yoo J, Kang J, Lee HN, Aguilar B, Kafka D, Lee S, Choi I, Lee J, Ramu S, Haas J, Koh CJ, Hong YK (2010) Kaposin-B enhances the PROX1 mRNA stability during lymphatic reprogramming of vascular endothelial cells by Kaposi's sarcoma herpes virus. PLoS Pathog 6:e1001046
299. McCormick C, Ganem D (2005) The kaposin B protein of KSHV activates the p38/MK2 pathway and stabilizes cytokine mRNAs. Science 307:739–741
300. Chang HC, Hsieh TH, Lee YW, Tsai CF, Tsai YN, Cheng CC, Wang HW (2016) c-Myc and viral cofactor Kaposin B co-operate to elicit angiogenesis through modulating miRNome traits of endothelial cells. BMC Syst Biol 10(Suppl 1):1
301. Shin YC, Joo CH, Gack MU, Lee HR, Jung JU (2008) Kaposi's sarcoma-associated herpesvirus viral IFN regulatory factor 3 stabilizes hypoxia-inducible factor-1 alpha to induce vascular endothelial growth factor expression. Cancer Res 68:1751–1759
302. Brinkmann MM, Glenn M, Rainbow L, Kieser A, Henke-Gendo C, Schulz TF (2003) Activation of mitogen-activated protein kinase and NF-kappaB pathways by a Kaposi's sarcoma-associated herpesvirus K15 membrane protein. J Virol 77:9346–9358
303. Cho NH, Choi YK, Choi JK (2008) Multi-transmembrane protein K15 of Kaposi's sarcoma-associated herpesvirus targets Lyn kinase in the membrane raft and induces NFAT/AP1 activities. Exp Mol Med 40:565–573
304. Bala K, Bosco R, Gramolelli S, Haas DA, Kati S, Pietrek M, Havemeier A, Yakushko Y, Singh VV, Dittrich-Breiholz O, Kracht M, Schulz TF (2012) Kaposi's sarcoma herpesvirus K15 protein contributes to virus-induced angiogenesis by recruiting PLCgamma1 and activating NFAT1-dependent RCAN1 expression. PLoS Pathog 8:e1002927
305. Lee BS, Lee SH, Feng P, Chang H, Cho NH, Jung JU (2005) Characterization of the Kaposi's sarcoma-associated herpesvirus K1 signalosome. J Virol 79:12173–12184
306. Wang L, Wakisaka N, Tomlinson CC, DeWire SM, Krall S, Pagano JS, Damania B (2004) The Kaposi's sarcoma-associated herpesvirus (KSHV/HHV-8) K1 protein induces expression of angiogenic and invasion factors. Cancer Res 64:2774–2781
307. Bergers G, Brekken R, McMahon G, Vu TH, Itoh T, Tamaki K, Tanzawa K, Thorpe P, Itohara S, Werb Z, Hanahan D (2000) Matrix metalloproteinase-9 triggers the angiogenic switch during carcinogenesis. Nat Cell Biol 2:737–744
308. Yao S, Hu M, Hao T, Li W, Xue X, Xue M, Zhu X, Zhou F, Qin D, Yan Q, Zhu J, Gao SJ, Lu C (2015) MiRNA-891a-5p mediates HIV-1 Tat and KSHV Orf-K1 synergistic induction of angiogenesis by activating NF-kappaB signaling. Nucleic Acids Res 43:9362–9378
309. Xue M, Yao S, Hu M, Li W, Hao T, Zhou F, Zhu X, Lu H, Qin D, Yan Q, Zhu J, Gao SJ, Lu C (2014) HIV-1 Nef and KSHV oncogene K1 synergistically promote angiogenesis by inducing cellular miR-718 to regulate the PTEN/AKT/mTOR signaling pathway. Nucleic Acids Res 42:9862–9879
310. Mansouri M, Rose PP, Moses AV, Fruh K (2008) Remodeling of endothelial adherens junctions by Kaposi's sarcoma-associated herpesvirus. J Virol 82:9615–9628

311. Aoki Y, Jaffe ES, Chang Y, Jones K, Teruya-Feldstein J, Moore PS, Tosato G (1999) Angiogenesis and hematopoiesis induced by Kaposi's sarcoma-associated herpesvirus-encoded interleukin-6. Blood 93:4034–4043
312. Cannon M (2007) The KSHV and other human herpesviral G protein-coupled receptors. Curr Top Microbiol Immunol 312:137–156
313. de Munnik SM, Smit MJ, Leurs R, Vischer HF (2015) Modulation of cellular signaling by herpesvirus-encoded G protein-coupled receptors. Front Pharmacol 6:40
314. Pati S, Cavrois M, Guo HG, Foulke JS Jr, Kim J, Feldman RA, Reitz M (2001) Activation of NF-kappaB by the human herpesvirus 8 chemokine receptor ORF74: evidence for a paracrine model of Kaposi's sarcoma pathogenesis. J Virol 75:8660–8673
315. Choi YB, Nicholas J (2008) Autocrine and paracrine promotion of cell survival and virus replication by human herpesvirus 8 chemokines. J Virol 82:6501–6513
316. Stine JT, Wood C, Hill M, Epp A, Raport CJ, Schweickart VL, Endo Y, Sasaki T, Simmons G, Boshoff C, Clapham P, Chang Y, Moore P, Gray PW, Chantry D (2000) KSHV-encoded CC chemokine vMIP-III is a CCR4 agonist, stimulates angiogenesis, and selectively chemoattracts TH2 cells. Blood 95:1151–1157
317. Nicholas J (2010) Human herpesvirus 8-encoded cytokines. Future Virol 5:197–206
318. Szpakowska M, Chevigne A (2016) vCCL2/vMIP-II, the viral master KEYmokine. J Leukoc Biol 99:893–900
319. Delgado T, Carroll PA, Punjabi AS, Margineantu D, Hockenbery DM, Lagunoff M (2010) Induction of the Warburg effect by Kaposi's sarcoma herpesvirus is required for the maintenance of latently infected endothelial cells. Proc Natl Acad Sci U S A 107:10696–10701
320. Ma T, Patel H, Babapoor-Farrokhran S, Franklin R, Semenza GL, Sodhi A, Montaner S (2015) KSHV induces aerobic glycolysis and angiogenesis through HIF-1-dependent upregulation of pyruvate kinase 2 in Kaposi's sarcoma. Angiogenesis 18:477–488
321. Daye D, Wellen KE (2012) Metabolic reprogramming in cancer: unraveling the role of glutamine in tumorigenesis. Semin Cell Dev Biol 23:362–369
322. Jain M, Nilsson R, Sharma S, Madhusudhan N, Kitami T, Souza AL, Kafri R, Kirschner MW, Clish CB, Mootha VK (2012) Metabolite profiling identifies a key role for glycine in rapid cancer cell proliferation. Science 336:1040–1044
323. Sanchez EL, Carroll PA, Thalhofer AB, Lagunoff M (2015) Latent KSHV infected endothelial cells are glutamine addicted and require glutaminolysis for survival. PLoS Pathog 11:e1005052
324. Zhu Y, Li TT, Ramos da Silva S, Lee JJ, Lu C, Eoh HJ, Jung JU, Gao SJ (2017) A critical role of glutamine γ-nitrogen for nucleotide biosynthesis in cancer cells hijacked by an oncogenic virus mBio, submitted
325. Delgado T, Sanchez EL, Camarda R, Lagunoff M (2012) Global metabolic profiling of infection by an oncogenic virus: KSHV induces and requires lipogenesis for survival of latent infection. PLoS Pathog 8:e1002866
326. Sychev ZE, Hu A, DiMaio TA, Gitter A, Camp ND, Noble WS, Wolf-Yadlin A, Lagunoff M (2017) Integrated systems biology analysis of KSHV latent infection reveals viral induction and reliance on peroxisome mediated lipid metabolism. PLoS Pathog 13:e1006256
327. Bhatt AP, Jacobs SR, Freemerman AJ, Makowski L, Rathmell JC, Dittmer DP, Damania B (2012) Dysregulation of fatty acid synthesis and glycolysis in non-Hodgkin lymphoma. Proc Natl Acad Sci U S A 109:11818–11823
328. Sanchez EL, Pulliam TH, Dimaio TA, Thalhofer AB, Delgado T, Lagunoff M (2017) Glycolysis, glutaminolysis, and fatty acid synthesis are required for distinct stages of Kaposi's sarcoma-associated herpesvirus lytic replication. J Virol 91: Apr 28;91(10). pii: e02237-16. https://doi.org/10.1128/jvi.02237-16. Print 2017 May 15
329. Cianfrocca M, Lee S, Von Roenn J, Tulpule A, Dezube BJ, Aboulafia DM, Ambinder RF, Lee JY, Krown SE, Sparano JA (2010) Randomized trial of paclitaxel versus pegylated

liposomal doxorubicin for advanced human immunodeficiency virus-associated Kaposi sarcoma: evidence of symptom palliation from chemotherapy. Cancer 116:3969–3977

330. Pinzone MR, Berretta M, Cacopardo B, Nunnari G (2015) Epstein-barr virus- and Kaposi sarcoma-associated herpesvirus-related malignancies in the setting of human immunodeficiency virus infection. Semin Oncol 42:258–271
331. Roy D, Sin SH, Lucas A, Venkataramanan R, Wang L, Eason A, Chavakula V, Hilton IB, Tamburro KM, Damania B, Dittmer DP (2013) mTOR inhibitors block Kaposi sarcoma growth by inhibiting essential autocrine growth factors and tumor angiogenesis. Cancer Res 73:2235–2246
332. Sin SH, Roy D, Wang L, Staudt MR, Fakhari FD, Patel DD, Henry D, Harrington WJ Jr, Damania BA, Dittmer DP (2007) Rapamycin is efficacious against primary effusion lymphoma (PEL) cell lines in vivo by inhibiting autocrine signaling. Blood 109:2165–2173
333. Petre CE, Sin SH, Dittmer DP (2007) Functional p53 signaling in Kaposi's sarcoma-associated herpesvirus lymphomas: implications for therapy. J Virol 81:1912–1922
334. Sarek G, Kurki S, Enback J, Iotzova G, Haas J, Laakkonen P, Laiho M, Ojala PM (2007) Reactivation of the p53 pathway as a treatment modality for KSHV-induced lymphomas. J Clin Invest 117:1019–1028
335. Ye F, Lattif AA, Xie J, Weinberg A, Gao S (2012) Nutlin-3 induces apoptosis, disrupts viral latency and inhibits expression of angiopoietin-2 in Kaposi sarcoma tumor cells. Cell Cycle 11:1393–1399
336. Uldrick TS, Wyvill KM, Kumar P, O'Mahony D, Bernstein W, Aleman K, Polizzotto MN, Steinberg SM, Pittaluga S, Marshall V, Whitby D, Little RF, Yarchoan R (2012) Phase II study of bevacizumab in patients with HIV-associated Kaposi's sarcoma receiving antiretroviral therapy. J Clin Oncol 30:1476–1483
337. Koon HB, Krown SE, Lee JY, Honda K, Rapisuwon S, Wang Z, Aboulafia D, Reid EG, Rudek MA, Dezube BJ, Noy A (2014) Phase II trial of imatinib in AIDS-associated Kaposi's sarcoma: AIDS Malignancy Consortium Protocol 042. J Clin Oncol 32:402–408
338. van Rhee F, Wong RS, Munshi N, Rossi JF, Ke XY, Fossa A, Simpson D, Capra M, Liu T, Hsieh RK, Goh YT, Zhu J, Cho SG, Ren H, Cavet J, Bandekar R, Rothman M, Puchalski TA, Reddy M, van de Velde H, Vermeulen J, Casper C (2014) Siltuximab for multicentric Castleman's disease: a randomised, double-blind, placebo-controlled trial. Lancet Oncol 15:966–974
339. Dittmer DP, Damania B (2016) Kaposi sarcoma-associated herpesvirus: immunobiology, oncogenesis, and therapy. J Clin Invest 126:3165–3175
340. Bhatt AP, Bhende PM, Sin SH, Roy D, Dittmer DP, Damania B (2010) Dual inhibition of PI3 K and mTOR inhibits autocrine and paracrine proliferative loops in PI3 K/Akt/mTOR-addicted lymphomas. Blood 115:4455–4463

Kaposi's Sarcoma-Associated Herpesvirus (KSHV)-Associated Disease in the AIDS Patient: An Update

3

Dirk P. Dittmer and Blossom Damania

Contents

Abstract

In this book chapter, we review the current knowledge of the biology and pathogenesis of Kaposi's sarcomaassociated herpesvirus (KSHV). We describe the lifecycle of KSHV, the cancers associated with this virus, as well as current treatment modalities.

D. P. Dittmer · B. Damania (✉)
Department of Microbiology & Immunology, Lineberger Comprehensive Cancer Center, University of North Carolina, CB #7295, NC 27599 Chapel Hill, USA
e-mail: damania@med.unc.edu

C. Meyers (ed.), *HIV/AIDS-Associated Viral Oncogenesis*, Cancer Treatment and Research177, https://doi.org/10.1007/978-3-030-03502-0_3

3.1 Introduction

Approximately, 25 % of all human cancers are etiologically linked to an infectious agent including viruses and bacteria. These pathogens are usually controlled by the host immune system. In individuals that are immunodeficient, such as acquired immunodeficiency syndrome (**AIDS**) patients or patients receiving immunosuppressive therapies following organ transplantation, this checkpoint fails and there is a significantly higher risk for the development of cancers associated with infectious agents. It is important to remember, though, that temporal immune deficiency is a normal physiological process, e.g. during aging and infant development. Viruses contribute to cancer development either cell autonomously through the activities of viral oncogenes acting within a cell, or through paracrine mechanisms that modulate the transformed cell and the tumor microenvironment [27].

Kaposi's sarcoma (**KS**) was described in 1872 by Moritz Kaposi, the head of the Vienna dermatology clinic, as *"idiopathisches multiples Pigmentsarkom"* a rare angiosarcoma in elderly men of Mediterranean descent [49]. In the mid-1980s, the human immunodeficiency virus (**HIV**) epidemic lead to a significant increase in the incidence of KS in high-risk populations. Today, over 30 years later, the number of new HIV infections has declined due to combination Anti Retroviral Therapy (**cART**). Yet, because of cART the number of persons living with HIV is increasing and the mean age of the cohort of HIV-infected persons is also increasing. Many HIV-positive individuals are now entering the age bracket, in which Moritiz Kaposi initially described classic KS in the elderly. As a result, KS remains the single most common neoplasm seen in individuals living with HIV today [88].

Chang and Moore identified KSHV (also known as human herpesvirus 8) in KS lesions of AIDS patients in 1994 [13] using representational difference analysis. KSHV has since been found in HIV+ and HIV− negative KS patients as well as in a number of B-cell hyperplasias and frank lymphomas. Ninety-nine per cent of all KS lesions, regardless of clinical type or HIV status, contain KSHV viral DNA and express a least one viral protein, the latency-associated nuclear antigen (**LANA**), as well as all viral micro RNAs, thereby linking KS to KSHV infection [27].

3.2 KSHV and the Development of KS

KS is divided into four subtypes delineated by clinical manifestations: classic, endemic, AIDS-associated, and iatrogenic. Classic KS is a disease of elderly Mediterranean and Eastern European men, while endemic KS is found in parts of equatorial Africa such as Uganda, Zambia, Malawi, Kenya, and South Africa in the elderly as well as in children [59]. KS represents the most common cancer in countries with high, coincident HIV and KSHV prevalence [45]. In endemic regions, transmission of KSHV is thought to occur early in childhood [32]. Endemic KS tends to be more aggressive than classic KS of the elderly, and occurs at almost equal proportions in men and women, the elderly and children [27].

Widespread HIV infection has given rise to an epidemic of KS. KSHV antibodies prevalent in black South African HIV patients, and KS has become the most common neoplasm in regions of sub-Saharan Africa that are ravaged by HIV infection. In the U.S., KSHV antibody prevalence also exceeds 30% in cities with high HIV burden and in high-risk populations [54]. This is most likely, because among adults, HIV and KSHV are transmitted by similar routes, though the efficiency of KSHV transmission (or basic reproductive ratio, which is a function of viral load among other factors) is less that that of acute HIV-1 infection.

In 1981, KS was recognized as a defining pathology for HIV diagnosis but the introduction of cART has led to a substantial decline of AIDS-related KS in the United States. The Centers for Disease control (**CDC**) estimated in 2016 that the average American had a 1 in 99 chance of being diagnosed with HIV at some point in his or her life. Even in the cART era, standardized incidence rates for KS are higher than that of any other AIDS-defining or non-AIDS-defining cancers [61]. This suggests that KS will remain a permanent health problem for years to come. As HIV-positive men in the U.S. age, it is speculated that the incidence of AIDS-KS may rise again.

Iatrogenic KS occurs after solid organ transplantation in patients receiving immunosuppressive therapy [16]. KS comprises an estimated 3% of all tumors associated with transplantation [63]. Iatrogenic KS is observed in regions of high KSHV prevalence, such as Southern Italy, Saudi Arabia and Turkey. KSHV may already be present in the recipient prior to organ transplantation, and may be acquired during induced immunosuppression after transplantation, or may even be acquired through the graft itself [5]. The frequency of KS in AIDS patients is 20,000 times higher than in the general population [6] and the frequency of KS in transplant recipients is 500 times higher than in healthy individuals [91].

In the mid-1980s, incidence rates for KS displayed an exponential increase. Back then, KS was primarily observed in AIDS patients with a history of men who had sex with men, but not in individuals who became HIV-infected through blood transfusion [37]. In AIDS-associated KS, there was a correlation between incidence rates and the lifetime number of male sexual partners [59]. This established KSHV as a sexually transmitted agent responsible for the development of this cancer. Today, more women are becoming infected with HIV and consequently AIDS-KS is also seen in this group. Interestingly, African KS affects both genders; while classic (Mediterranean) KS affects predominantly elder men. The reason for the gender bias in classic KS is unknown. In the U.S., KS incidence rates follow a bimodal distribution that peaks at ages 30–36 and again at ages >70.

KS lesions are classified as plaque, patched, or nodular. As the KS tumor clinically advances, the KSHV-infected cells increase in number along with the endothelial cell population in the lesion. There is evidence for both polyclonality and monoclonality of the lesions [47, 76]. It is thought that KS likely initiates as a polyclonal hyperplasia and develops into a clonal neoplasia. Kaposi's sarcoma not only affects the skin but can also involve multiple organs such as the liver, lung, spleen, and gastrointestinal tract. In some forms of KS, only lymphoid and internal organs are affected. Oral KS in the setting of AIDS is associated with advanced disease and visceral development. However, in the setting of cART-controlled HIV infection, it may occur in isolation and represent limited disease. Edema is common in KS patients. Aggressive types of KS can lead to foci formation in the visceral organs and ultimately result in hemorrhage and death.

KSHV viral load in PBMC rise up to 6 months prior to lesion formation [101]. A rise in viral load predicts the imminent appearance of KS [72]. However, systemic viral load in plasma varies widely across KS patients and does not correlate with the number of skin lesions [44]. Inhibitors of the viral polymerase reduce overall risk of future KS, but do not lead to regression of established KS lesions. KSHV is found in circulating B cells as well as monocytes, macrophages, endothelial cells, and epithelial cells [21, 77, 92]. The presence of the most common anti-KSHV antibodies, which are directed against the LANA protein, documents prior exposure but does not allow a prediction of KS development, since in HIV-positive individuals the median time from seroconversion to disease is seven years or greater [37, 59].

The KS lesion is highly angiogenic and is comprised of spindle-shaped cells, slit-like endothelium-lined vasculature and infiltrating blood cells. The spindle cells appear to arise from lymphatic endothelial cells and form the majority of the neoplasm [31]. In fact, experimental KSHV infection can reprogram the blood endothelial gene expression profile into that of the lymphatic endothelium and vice versa [42, 43, 98, 100], though the profile also shows the presence of mesenchymal markers including various Notch isoforms [15, 58] consistent with dedifferentiation into a progenitor stage.

The primary receptor for KSHV infection of endothelial cells is ephrin receptor tyrosine kinase A2 [41]. Ephrins and their corresponding kinases are differentially expressed across different cell lineages. Hence, the expression pattern of EphA2 may express the tropism of KSHV. It may also become a target of novel, directed KS therapy [14, 85]. KS tumor explants lose the virus after serial passage in tissue culture over time. KSHV-infected endothelial cell preparations in culture generally also lose the virus over time [40, 55].

3.3 KSHV and the Development of Lymphomas

KSHV is also found in B lymphoproliferative diseases; primary effusion lymphoma (**PEL**) and the plasmablastic variant of multicentric Castleman's disease (**MCD**). In fact, the first association of KS and a B-cell lymphoproliferative disorder, MCD, was reported in a patient who presented with both diseases [81]. Greater than 50% of KSHV-positive transplant recipients develop lymphoproliferative disease [35]. KSHV is most certainly the causal agent of both MCD and PEL [12, 90]. MCD is a B-cell lymphoproliferative disorder. Patients usually present with diffuse lymphadenopathys. In addition to B cell proliferation, MCD displays vascular proliferation of the germinal centers of the lymph node. There are two forms of MCD: (i) a plasmablastic variant form that is associated with lymphadenopathy and immune dysregulation and (ii) a hyaline vascular form, which presents as a solid mass. Close to 100% of AIDS-associated MCD is associated with KSHV. AIDS-associated MCD is usually accompanied by the development of KS in the affected individual, often in the same lymph node.

MCD is a polyclonal tumor and is highly dependent on cytokines such as human interleukin 6 (IL-6) (reviewed in [103]). KSHV itself encodes a viral IL-6 that is also expressed in these lesions [71, 73, 94]. Expression of either human IL-6 or viral IL-6 in transgenic mice causes B-cell hyperplasia and lymphoma. Viral antigens can be detected in the immunoblastic B cells in the mantle zone of the lymph node. The plasmablasts in MCD express monotypic IgM light chains [29] and MCD patients frequently develop cytopenia, autoimmune disease and other malignancies such as KS and non-Hodgkin's lymphoma [1]. Anti-IL-6 or anti-IL-6R antibodies show efficacy in KSHV-negative Castleman's disease and there is every reason to believe that siltuximab or tocilizumab (also known as atlizumab) will also be active in KSHV-positive, HIV-associated MCD and perhaps even PEL.

PEL, sometimes referred to as body cavity-based lymphoma (BCBLs), represent a specific subset of non-Hodgkin's B-cell lymphoma (NHL) that involve body cavities (peritoneal, pleural or pericardial cavities) and form a distinct clinico-pathologic group from other NHL [67]. All PEL are KSHV-positive, and are often coinfected with EBV as well. These tumors are typically large-cell immunoblastic or anaplastic large-cell lymphomas that express CD45, but not CD19, carry clonal immunoglobulin gene rearrangements, and lack mutations in c-myc, bcl-2, ras, and p53 [1, 67].

PEL display the characteristics of a preterminal stage of B-cell differentiation. Since PEL have mutations in their immunoglobulin genes, they are thought to arise from post-germinal center B cells. However, PEL do not express immunoglobulins. Most PEL express CD138/syndecan-1 antigen, which is normally also expressed by a subset of plasma cells. Most PEL also express high levels of human IL-6 and IL10.

Although KSHV is linked to PEL and MCD in HIV patients, there are cases of KSHV-positive lymphomas that do not fit the classic PEL phenotypes. There appears to be a high incidence of KSHV infection in solid HIV-associated immunoblastic/plasmablastic non-Hodgkin's lymphomas that developed in patients lacking PEL and MCD [22] and yet others have found KSHV associated with solid lymphomas, which resemble PEL cell morphology but do not present as effusions [10]. KSHV has also been linked to cases of germinotropic lymphoproliferative disease (GLD) [28]. This disease also involves plasmablasts but unlike plasmablastic lymphomas, the GLD lymphomas contain polyclonal immunoglobulin receptors. This suggests a model in which KSHV infects an early germinal center B cell that can still differentiate into multiple lymphoma phenotypes dependent on secondary mutations to the cellular genome.

Finally, KSHV infection can also lead to KS-immune reconstitution syndrome (KS-IRIS) [8, 18] and KSHV-inflammatory cytokine syndrome (KICS) [74]. Patients with KICS have high KSHV viral loads and levels of viral IL-6, human IL-6, human IL-10 as well as C-reactive protein.

The evidence linking KSHV to KS, PEL, MCD and KICS, is overwhelming and has been confirmed by multiple laboratories and indepent methods such as the presence of viral DNA in the lesions, viral protein expression and anti-KSHV antibodies (directed against LANA/orf73, orf K8.1 and others). KSHV DNA has also been detected in multiple myeloma, primary pulmonary hypertension, angiosarcomas, as well as malignant skin tumors in posttransplant patients such as Bowen's disease, squamous cell carcinomas, actinic keratosis, and extramammary Paget's disease. However, these disease associations were never substantiated and have largely been discarded [1, 27].

3.4 Prevalence of Viral Infection

Several serology studies have suggested that KSHV infection is widespread in Africa with 30–60% of people being KSHV-positive, but is uncommon in the United States and Western Europe with seropositivity ranging from 3 to 10% in the general population [50]. KSHV seropositivity is considerably higher in high-risk populations reaching 38% in participants seen at AIDS clinical trials centers [54]. Regions such as Italy, Greece, Turkey, and Saudi Arabia show a higher prevalence of KSHV at about 4–35% [102], which correlates with correspondingly higher incidence rates for classical or transplant-associated KS. Transmission routes include sexual transmission, mother-to-child transmission, but probably all forms involve salivary transmission [9, 59, 96]. There is no evidence that transmission rates decline, as most KSHV transmission, similar to other herpesviruses, appears during episodes of asymptomatic shedding.

3.5 The KSHV Genome

A hallmark of herpesviruses including KSHV is their ability to establish a latent infection for the lifetime of their host. Pathogenesis caused by these viruses is usually seen in the context of host immunesuppression. All herpesviruses share a common evolutionary origin, which is evident from the homology seen among a substantial number of herpesviral genes (reviewed in [25]). Based on biological characteristics and genomic organization, herpesviruses are classified into three subfamilies: alpha, beta, and gamma. The gamma herpesviruses are lymphotropic and some are capable of undergoing lytic replication in epithelial, endothelial, or fibroblast cells. The gammaherpesvirinae are grouped into two classes: lymphocryptoviruses (gamma-1) and rhadinoviruses (gamma-2). Epstein–Barr virus (EBV) or human herpesvirus 4 (HHV4) is a lymphocryptovirus while KSHV or human herpesvirus 8 (HHV8) is a rhadinovirus.

During latent infection, viral gene expression is highly attenuated and the viral genome remains stably associated with the cell. In the lytic phase of infection, viral gene expression and DNA replication ensue, leading to the production of progeny virions and eventual lysis of the infected cell. The KSHV viral genome is comprised of a ~140 kb long unique region flanked by multiple terminal repeat sequences with the total genomic size being ~160–170 kb. KSHV encodes for more than 80 open reading frames (ORFs) that encode for proteins greater than 100 amino acids [83]. The viral genes encoded by KSHV can be divided into three classes—(i) genes common to all herpesviruses (ii) genes unique to KSHV (these are generally given a "K" designation followed by the number of the open reading frame (ORF), and (iii) KSHV encoded genes that are homologous to cellular genes (these may be unique to KSHV or shared with other herpesviruses), and are likely to have been usurped from the host genome during the course of evolution. It is likely that several viral genes contribute to the neoplastic process [19].

While there exist distinct clades of KSHV, most of the variation is concentrated in a few proteins, such as the extracellular regions of the K1 and K15 proteins, which are exposed to the host immune system, or in extended repeat regions, where the genome is inherently unstable, such as in the two origins of replication and the central protein coding region of LANA. Whole genome sequencing has shown that all other regions are conserved across strains with just a few single nucleotide variations inside protein coding regions [70]. At this point, none of the genomic variation seen within KSHV has been associated with overt clinical or cellular phenotypes, though specific point mutations in the viral micro RNA precursors lead to the absence of certain mature miRNAs in PEL or KS lesions.

3.6 Molecular Biology of KSHV-Associated Disease

KSHV gene expression in human KS, PEL and MCD disease has involved the use of microarrays to profile viral gene expression. Since the KSHV genome is orders of magnitude smaller than the human genome, it has been feasible to develop whole genome arrays based upon real-time quantitative RT-PCR for all individual viral genes and to analyze primary KS biopsy samples and KSHV-infected lymphomas [24, 33]. Conventional microarray-based viral gene expression in KSHV-infected lymphomas as well as RNAseq studies has also been performed. These techniques generate a viral signature for each disease state and offer a chance to classify KS beyond Moritz Kaposi's observational diagnosis. High-throughput genomic profiling offers the chance to accelerate our investigations into KSHV-associated cancers as much as it has benefited research into nonviral cancers. Microarray analyses of host cell transcription [34, 46, 51] proved that KSHV-positive PEL differ from other types of B-cell lymphomas. This is consistent with the idea that KSHV reprograms the tumor cell.

It has been shown that KSHV infection reprograms endothelial cells. Blood endothelial cells are reprogrammed toward lympathic endothelium and conversely, lymphatic endothelium is reprogrammed toward blood endothelium [42, 43, 98, 100]. Several studies have ascertained the host transcription profile in tissue culture models of KSHV infection [66, 68, 75, 79]. KS has a cellular transcription signature that is distinct from other cancers and tied to the unique pathology of this disease, as an angioproliferative, cytokine driven disease. For instance, c-Kit and other growth factor receptors in microarray studies of KSHV-infected endothelial cells led to a successful pilot study using the kinase inhibitor gleevec (Imatinib) [52]. Other studies found response rates of KS to a matrix metalloproteinase inhibitor [23] or anti-VEGF antibodies such as bevacicumab [95].

Every KS tumor transcribes high levels of the canonical KSHV latency transcripts encoding LANA, vFLIP, vCyclin, the viral micro RNAs, and Kaposin. These genes are under control of the same promoter and are expressed in every KS tumor cell [26, 30]. Kaposin is located immediately downstream of these three genes and in addition to the common promoter can be regulated by a promoter located between LANA and cyclin [56] and during lytic reactivation yet another, ORF-proximal promoter [84]. Like LANA, Kaposin too is expressed in every tumor cell [92] and has been shown to stabilize cellular cytokine mRNAs [62]. In addition to these latent proteins, many KS tumors as well as PEL engrafts [93, 97] express an extended set of proteins that were initially classified as lytic viral genes, but in the context of the tumor may be the result of abortive or incomplete viral reactivation. These include the KSHV interferon regulatory factor (vIRF-1) and G-coupled receptor (vGPCR) homologs [24] and the K1 constitutive signal protein [3, 97, 99, 104], as well as K15, a constitutive signaling protein located at opposite end of K1 [39]. This suggests that a subset of KS phenotypes may be attributable to these genes and the paracrine mechanisms that they invoke [4, 64, 65]. The vIRF-3, a duplicated KSHV IRF homolog, is constitutively transcribed in KSHV-infected PEL [80]. Thus, we speculate that KSHV has to interfere with the host cell's innate interferon response in every infected cell regardless of cell lineage or mode of infection and has thus placed multiple copies of the vIRFs, all of which interfere with normal interferon signaling, under different control elements, e.g., vIRF-3 is specific for B cells while vIRF-1 is specific for endothelial cells. Thus, both latent and select lytic genes can be considered tumor-specific therapy targets for KS.

3.7 Therapies to Treat KS, PEL, and MCD

Treatment modalities for KS include observation, local therapy, or systemic chemotherapy specifically paclitaxel and anthracyclines, such as doxorubicin/adriamycin [69], depending on the severity of the disease. Response rates approach 70% depending on comorbidities. KS is know to reapear and to require repeated treatment; a complete cure is seldom achieved as none of the anti-cancer treatments erradicate the latent virus. A key development was the demonstration that liposomal formulation of the peggylated-anthracyclins were as efficactious as the initial drug, but had significant fewer side effects. No new theraphies against KS have been introduced since the liposomal anthracyclines such as liposomal Doxorubicin or liposomal Daunorubicin. Whether a protein-bound formulation of paclitaxel (Abraxane) has activity with reduced toxicity is unknown. Interferon alpha was initially approved to treat KS, but is no longer in use. KS is a highly angiogenic tumor but clinical trials targeting the angiogenic nature of KS have shown limited efficacy as single agent [95]. This is expected, since most of these agents, such as the humanized anti-VEGF antibody bevacicumab are tumorstatic and do not kill the tumor cell directly.

A clinical trial involving daily doses of Imatinib mesylate (Gleevec), which targets c-kit and platelet-derived growth factor receptor (PDGFR) signaling, resulted in clinical and histologic regression of cutaneous KS [52], as did a trial of a matrix metalloproteinase inhibitor [23]. As more receptor tyrosine kinase (RTK)-targeting molecules become available, targeting PDGFR, VEGFR, and related mediators of paracrine tumor promoters, offer promise for KS.

Organ transplants, who developed KS due to immunosuppressive therapy, benefited from treatment with rapamycin [91]. This observation has been repeated in multiple settings and switching from cyclosporine A or FK506, which suppress T cell activation, but not B cell or endothelial cell activation to rapamycin, which suppresses proliferation in all three cell types, has emerged as the informal standard of care of iatrogenic KS. Rapamycin/Sirolimus and its derivatives Temsirolimus and Everolimus are allosteric inhibitors of the mTOR pathway and display both immunosuppressive and antineoplastic properties. The clinical effect of rapamycin could be reproduced in animal models [82, 89]. Of note, rapamycin was active against doxorubicin-resistant PEL. Rapamycin acted via an antiangiogenic mechanism ultimately reducing the levels of VEGF and of VEGF receptor on endothelial cells. Again, as single agent rapamycin was tumorstatic, rather than tumortoxic. Newer, competitive inhibitors of the mTOR pathway are likely to produce superior results. Additional inhibitors targeting the active site of PI3K and mTOR have also proved effective in animal models [2, 7].

A series of clinical trials is exploring the efficacy of "imids", i.e., thalidomide, lenalidomide, and pomalidomide in KS that develops in HIV-suppressed indiviudals. These compounds have an as yet ill-defined mechanism of action that affects the immunesystem as well as potential KS tumor cells directly, through modulating gene expression [20]. In 2018 Pomalidomide received orphan drug designation for KS by the FDA of the US.

The risk for KS and virally associated lymphomas increases rapidly as the CD4+ cell counts of HIV-infected individuals diminish [17], and the risk of developing AIDS-associated cancers is lower for individuals who are less severely immune suppressed. Since the prevalence of KS in AIDS patients is very high, and HIV coinfection is thought to be an important factor in the development of KS, attempts to control KS by improving the immune system of HIV-infected individuals through cART are recommended. Indeed, the incidence of KS has declined considerably following the introduction of cART therapy and often cART alone will lead to KS regression in AIDS patients. However, it is important to note that even in the face of cART therapy, the likelihood of an HIV-positive individual developing KS is still 20 times higher than uninfected individuals [17] and that by now one-fourth of KS develops in individuals who are HIV-suppressed [53].

Current treatments for MCD, PEL, and other AIDS lymphomas include standard chemotherapy such as CHOP, which contains four drugs; prednisone, vincristine, cyclophosphamide, and doxorubicin, or EPOCH, which in addition contains etoposide. These can be given coincidentally with cART [78, 86]. Case reports in the literature also suggest that Rituximab (rituxan) is effective against PEL. Rituximab is an anti-CD20 antibody, but because Rituximab targets normal B cells as well, it can be associated with an increased risk of infection when used in AIDS patients [48]. Scott et al. have reported on two MCD patients that went into sustained remission with just oral etoposide [86], but a more modern approach would be neutralizing human IL-6 using anti-IL-6 antibodies or anti-IL-6 receptor antibodies. Whether the concept of neutralizing paracrine factors can also be applied to viral IL-6 remains to be explored.

Another line of thinking has lead to exploratory studies using anti-herpesviral drugs that inhibit herpesviral replication such as ganciclovir or AZT [11, 38, 60, 94] in patients. There are two possible mechanisms of action. First, these inhibitors suppress viral dissemination and thus the pool of infected cells rather than acting directly on the tumor. Second, there is the observation that AZT as well as ganciclovir has direct cytotoxicity on the infected cell, and selectivity for infected cells, as only those cells express the viral kinases that convert these prodrugs into their active forms. The later can be enhanced by inducing viral reactivation using histone-deactylase inhibitors such as vorinostat, butyrate, or valproic acid. Cidofovir, another herpesvirus polymerase inhibitor, did not show a clinical benefit [57].

cART therapy has resulted in varying degrees of success with respect to decline in the incidence of non-Hodgkin lymphoma. It is estimated that cART therapy decreases the incidence of non-Hodgkin lymphoma anywhere in the range of 40–76%. Moreover, there is emerging evidence that protease inhibitors such as indinavir or nelfinavir, which also inhibit matrix metalloproteinase may have direct anti-KS activity [36] in addition to HAART-associated reconstitution of the immune system [87]. More information on current trials that are underway to treat KS, PEL and MCD can be gleaned by visiting the National Cancer Institute (NCI) website: http://www-dcs.nci.nih.gov/branches/aidstrials/adlist.html.

3.8 Conclusions

As a consequence of cART, the life expectancy of HIV-infected individuals now equals that of other persons with chronically managed diseases such as diabetis or heart disease. As these HIV-infected patients continue to age, there will be a corresponding increase in the incidence of AIDS-defining, since HIV+ are disproportionally exposed to KSHV, human papilloma virus, and Epstein–Barr Virus, as well as cancers not associated with infectious causes. Most of the current therapies with the exception of anti-herpesviral drugs do not take advantage of the unique viral etiology of KSHV-associated cancers, and anti-herpesviral drugs themselves are not effective against latent virus. Thus, it will be important to show that "traditional" anticancer therapies are safe in the context of cART and HIV infection, and to develop future therapies that directly impact upon, and obliterate, the function of viral genes.

Acknowledgements Due to space restrictions, we regret that we had to omit many important references. BD is a Leukemia & Lymphoma Society Scholar. DPD is a member of the AIDS malignancies clinical trials consortium (AMC). Please note that this chapter is an updated version of the chapter we wrote for the previous version of this book (AIDS-Associated Viral Oncogenesis, 2007. Springer Press. Editor: Craig Meyers).

References

1. Ablashi DV, Chatlynne LG, Whitman Jr JE, Cesarman E Jr (2002) Spectrum of Kaposi's sarcoma-associated herpesvirus, or human herpesvirus 8, diseases. Clin Microbiol Rev 15(3):439–464
2. Anders P, Bhende PM, Foote M, Dittmer DP, Park SI, Damania B (2015) Dual inhibition of phosphatidylinositol 3-kinase/mammalian target of rapamycin and mitogen activated protein kinase pathways in non-Hodgkin lymphoma. Leuk Lymphoma 56(1):263–266. https://doi.org/10.3109/10428194.2014.917639
3. Anders PM, Zhang Z, Bhende PM, Giffin L, Damania B (2016) The KSHV K1 protein modulates AMPK function to enhance cell survival. PLoS Pathog 12(11):e1005985. https://doi.org/10.1371/journal.ppat.1005985
4. Bais C, Van Geelen A, Eroles P, Mutlu A, Chiozzini C, Dias S, Silverstein RL, Rafii S, Mesri EA (2003) Kaposi's sarcoma associated herpesvirus G protein-coupled receptor immortalizes human endothelial cells by activation of the VEGF receptor-2/ KDR. Cancer Cell 3(2):131–143
5. Barozzi P, Luppi M, Facchetti F, Mecucci C, Alu M, Sarid R, Rasini V, Ravazzini L, Rossi E, Festa S, Crescenzi B, Wolf DG, Schulz TF, Torelli G (2003) Post-transplant Kaposi sarcoma originates from the seeding of donor-derived progenitors. Nat Med 9(5):554–561
6. Beral V, Peterman TA, Berkelman RL, Jaffe HW (1990) Kaposi's sarcoma among persons with AIDS: a sexually transmitted infection? Lancet 335(8682):123–128
7. Bhatt AP, Bhende PM, Sin SH, Roy D, Dittmer DP, Damania B (2010) Dual inhibition of PI3 K and mTOR inhibits autocrine and paracrine proliferative loops in PI3K/Akt/mTOR-addicted lymphomas. Blood 115(22):4455–4463. https://doi.org/10.1182/blood-2009-10-251082
8. Bower M, Nelson M, Young AM, Thirlwell C, Newsom-Davis T, Mandalia S, Dhillon T, Holmes P, Gazzard BG, Stebbing J (2005) Immune reconstitution inflammatory syndrome associated with Kaposi's sarcoma. J Clin Oncol 23(22):5224–5228. https://doi.org/10.1200/JCO.2005.14.597
9. Brayfield BP, Kankasa C, West JT, Muyanga J, Bhat G, Klaskala W, Mitchell CD, Wood C (2004) Distribution of Kaposi sarcoma-associated herpesvirus/human herpesvirus 8 in maternal saliva and breast milk in Zambia: implications for transmission. J Infect Dis 189(12):2260–2270
10. Carbone A, Gloghini A, Vaccher E, Cerri M, Gaidano G, Dalla-Favera R, Tirelli U (2005) Kaposi's sarcoma-associated herpesvirus/human herpesvirus type 8-positive solid lymphomas: a tissue-based variant of primary effusion lymphoma. J Mol Diagn 7(1):17–27

11. Casper C, Nichols WG, Huang ML, Corey L, Wald A (2004) Remission of HHV-8 and HIV-associated multicentric Castleman disease with ganciclovir treatment. Blood 103(5): 1632–1634
12. Cesarman E, Chang Y, Moore PS, Said JW, Knowles DM (1995) Kaposi's sarcoma-associated herpesvirus-like DNA sequences in AIDS- related body-cavity-based lymphomas. N Engl J Med 332(18):1186–1191
13. Chang Y, Cesarman E, Pessin MS, Lee F, Culpepper J, Knowles DM, Moore PS (1994) Identification of herpesvirus-like DNA sequences in AIDS-associated Kaposi's sarcoma. Science 266(5192):1865–1869
14. Chen W, Sin SH, Wen KW, Damania B, Dittmer DP (2012) Hsp90 inhibitors are efficacious against Kaposi Sarcoma by enhancing the degradation of the essential viral gene LANA, of the viral co-receptor EphA2 as well as other client proteins. PLoS Pathog 8(11):e1003048. https://doi.org/10.1371/journal.ppat.1003048
15. Cheng F, Pekkonen P, Laurinavicius S, Sugiyama N, Henderson S, Gunther T, Rantanen V, Kaivanto E, Aavikko M, Sarek G, Hautaniemi S, Biberfeld P, Aaltonen L, Grundhoff A, Boshoff C, Alitalo K, Lehti K, Ojala PM (2011) KSHV-initiated notch activation leads to membrane-type-1 matrix metalloproteinase-dependent lymphatic endothelial-to-mesenchymal transition. Cell Host Microbe 10(6):577–590. https://doi.org/10.1016/j.chom.2011.10.011
16. Civati G, Busnach G, Brando B, Broggi ML, Brunati C, Casadei GP, Minetti L (1988) Occurrence of Kaposi's sarcoma in renal transplant recipients treated with low doses of cyclosporine. Transplant Proc 20(Suppl 3):924–928
17. Clifford GM, Polesel J, Rickenbach M, Dal Maso L, Keiser O, Kofler A, Rapiti E, Levi F, Jundt G, Fisch T, Bordoni A, De Weck D, Franceschi S (2005) Cancer risk in the Swiss HIV cohort study: associations with immunodeficiency, smoking, and highly active antiretroviral therapy. J Natl Cancer Inst 97(6):425–432
18. Connick E, Kane MA, White IE, Ryder J, Campbell TB (2004) Immune reconstitution inflammatory syndrome associated with Kaposi sarcoma during potent antiretroviral therapy. Clin Infect Dis 39(12):1852–1855. https://doi.org/10.1086/426078
19. Damania B (2004) Oncogenic gamma-herpesviruses: comparison of viral proteins involved in tumorigenesis. Nat Rev Microbiol 2(8):656–668
20. Davis DA, Mishra S, Anagho HA, Aisabor AI, Shrestha P, Wang V, Takamatsu Y, Maeda K, Mitsuya H, Zeldis JB, Yarchoan R (2017) Restoration of immune surface molecules in Kaposi sarcoma-associated herpes virus infected cells by lenalidomide and pomalidomide. Oncotarget 8(31):50342–50358, May 17
21. Decker LL, Shankar P, Khan G, Freeman RB, Dezube BJ, Lieberman J, Thorley-Lawson DA (1996) The Kaposi sarcoma-associated herpesvirus (KSHV) is present as an intact latent genome in KS tissue but replicates in the peripheral blood mononuclear cells of KS patients. J Exp Med 184(1):283–288
22. Deloose ST, Smit LA, Pals FT, Kersten MJ, van Noesel CJ, Pals ST (2005) High incidence of Kaposi sarcoma-associated herpesvirus infection in HIV-related solid immunoblastic/plasmablastic diffuse large B-cell lymphoma. Leukemia
23. Dezube BJ, Krown SE, Lee JY, Bauer KS, Aboulafia DM (2006) Randomized phase II trial of matrix metalloproteinase inhibitor COL-3 in AIDS-related Kaposi's sarcoma: an AIDS malignancy consortium study. J Clin Oncol 24(9):1389–1394
24. Dittmer DP (2003) Transcription profile of Kaposi's Sarcoma-associated herpesvirus in primary Kaposi's Sarcoma lesions as determined by real-time PCR arrays. Cancer Res 63(9): 2010–2015
25. Dittmer DP, Damania B (2016) Kaposi sarcoma-associated herpesvirus: immunobiology, oncogenesis, and therapy. J Clin Invest 126(9):3165–3175. https://doi.org/10.1172/JCI84418
26. Dittmer D, Lagunoff M, Renne R, Staskus K, Haase A, Ganem D (1998) A cluster of latently expressed genes in Kaposi's sarcoma-associated herpesvirus. J Virol 72(10):8309–8315

27. Dittmer DP, Damania B (2007) KSHV-associated disease in the AIDS patient. AIDS-associated viral oncogenesis. Editor: Craig Meyers, Springer
28. Du MQ, Diss TC, Liu H, Ye H, Hamoudi RA, Cabecadas J, Dong HY, Harris NL, Chan JK, Rees JW, Dogan A, Isaacson PG (2002) KSHV- and EBV-associated germinotropic lymphoproliferative disorder. Blood 100(9):3415–3418
29. Du MQ, Liu H, Diss TC, Ye H, Hamoudi RA, Dupin N, Meignin V, Oksenhendler E, Boshoff C, Isaacson PG (2001) Kaposi sarcoma-associated herpesvirus infects monotypic (IgM lambda) but polyclonal naive B cells in Castleman disease and associated lymphoproliferative disorders. Blood 97(7):2130–2136
30. Dupin N, Fisher C, Kellam P, Ariad S, Tulliez M, Franck N, van Marck E, Salmon D, Gorin I, Escande JP, Weiss RA, Alitalo K, Boshoff C (1999) Distribution of human herpesvirus-8 latently infected cells in Kaposi's sarcoma, multicentric Castleman's disease, and primary effusion lymphoma. Proc Natl Acad Sci U S A 96(8):4546–4551
31. Dupin N, Grandadam M, Calvez V, Gorin I, Aubin JT, Havard S, Lamy F, Leibowitch M, Huraux JM, Escande JP, Agut H (1995) Herpesvirus-like DNA sequences in patients with Mediterranean Kaposi's sarcoma. Lancet 345(8952):761–762
32. El-Mallawany NK, Kamiyango W, Slone JS, Villiera J, Kovarik CL, Cox CM, Dittmer DP, Ahmed S, Schutze GE, Scheurer ME, Kazembe PN, Mehta PS (2016) Clinical factors associated with long-term complete remission versus poor response to chemotherapy in HIV-infected children and adolescents with Kaposi sarcoma receiving bleomycin and vincristine: a retrospective observational study. PLoS One 11(4):e0153335. https://doi.org/10.1371/journal.pone.0153335
33. Fakhari FD (2002) Charting latency transcripts in Kaposi's sarcoma-associated herpesvirus by whole-genome real-time quantitative reverse transcription-PCR. J Virol 76(12)
34. Fan W, Bubman D, Chadburn A, Harrington WJ Jr, Cesarman E, Knowles DM (2005) Distinct subsets of primary effusion lymphoma can be identified based on their cellular gene expression profile and viral association. J Virol 79(2):1244–1251
35. Farge D, Lebbe C, Marjanovic Z, Tuppin P, Mouquet C, Peraldi MN, Lang P, Hiesse C, Antoine C, Legendre C, Bedrossian J, Gagnadoux MF, Loirat C, Pellet C, Sheldon J, Golmard JL, Agbalika F, Schulz TF (1999) Human herpes virus-8 and other risk factors for Kaposi's sarcoma in kidney transplant recipients. Groupe Cooperatif de Transplantation d' Ile de France (GCIF). Transplantation 67(9):1236–1242
36. Gantt S, Carlsson J, Ikoma M, Gachelet E, Gray M, Geballe AP, Corey L, Casper C, Lagunoff M, Vieira J (2011) The HIV protease inhibitor nelfinavir inhibits Kaposi's sarcoma-associated herpesvirus replication in vitro. Antimicrob Agents Chemother 55(6): 2696–2703. https://doi.org/10.1128/AAC.01295-10
37. Gao SJ, Kingsley L, Hoover DR, Spira TJ, Rinaldo CR, Saah A, Phair J, Detels R, Parry P, Chang Y, Moore PS (1996) Seroconversion to antibodies against Kaposi's sarcoma-associated herpesvirus-related latent nuclear antigens before the development of Kaposi's sarcoma. N Engl J Med 335(4):233–241
38. Ghosh SK, Wood C, Boise LH, Mian AM, Deyev VV, Feuer G, Toomey NL, Shank NC, Cabral L, Barber GN, Harrington WJ Jr (2003) Potentiation of TRAIL-induced apoptosis in primary effusion lymphoma through azidothymidine-mediated inhibition of NF-kappa B. Blood 101(6):2321–2327
39. Gramolelli S, Weidner-Glunde M, Abere B, Viejo-Borbolla A, Bala K, Ruckert J, Kremmer E, Schulz TF (2015) Inhibiting the recruitment of PLCgamma1 to Kaposi's sarcoma herpesvirus K15 protein reduces the invasiveness and angiogenesis of infected endothelial cells. PLoS Pathog 11(8):e1005105. https://doi.org/10.1371/journal.ppat.1005105
40. Grundhoff A, Ganem D (2004) Inefficient establishment of KSHV latency suggests an additional role for continued lytic replication in Kaposi sarcoma pathogenesis. J Clin Invest 113(1):124–136

41. Hahn AS, Kaufmann JK, Wies E, Naschberger E, Panteleev-Ivlev J, Schmidt K, Holzer A, Schmidt M, Chen J, Konig S, Ensser A, Myoung J, Brockmeyer NH, Sturzl M, Fleckenstein B, Neipel F (2012) The ephrin receptor tyrosine kinase A2 is a cellular receptor for Kaposi's sarcoma-associated herpesvirus. Nat Med 18(6):961–966. https://doi.org/10.1038/nm.2805
42. Hansen A, Henderson S, Lagos D, Nikitenko L, Coulter E, Roberts S, Gratrix F, Plaisance K, Renne R, Bower M, Kellam P, Boshoff C (2010) KSHV-encoded miRNAs target MAF to induce endothelial cell reprogramming. Genes Dev 24(2):195–205. https://doi.org/10.1101/gad.553410
43. Hong YK, Foreman K, Shin JW, Hirakawa S, Curry CL, Sage DR, Libermann T, Dezube BJ, Fingeroth JD, Detmar M (2004) Lymphatic reprogramming of blood vascular endothelium by Kaposi sarcoma-associated herpesvirus. Nat Genet 36(7):683–685
44. Hosseinipour MC, Sweet KM, Xiong J, Namarika D, Mwafongo A, Nyirenda M, Chiwoko L, Kamwendo D, Hoffman I, Lee J, Phiri S, Vahrson W, Damania B, Dittmer DP (2014) Viral profiling identifies multiple subtypes of Kaposi's sarcoma. MBio 5(5):e01614–e01633. https://doi.org/10.1128/mBio.01633-14
45. Host KM, Horner MJ, van der Gronde T, Moses A, Phiri S, Dittmer DP, Damania B, Gopal S (2017) Kaposi's sarcoma in Malawi: a continued problem for HIV-positive and HIV-negative individuals. AIDS 31(2):318–319. https://doi.org/10.1097/QAD.0000000000001341
46. Jenner RG, Maillard K, Cattini N, Weiss RA, Boshoff C, Wooster R, Kellam P (2003) Kaposi's sarcoma-associated herpesvirus-infected primary effusion lymphoma has a plasma cell gene expression profile. Proc Natl Acad Sci U S A 100(18):10399–10404
47. Kaaya SF, Leshabari MT, Mbwambo JK (1998) Risk behaviors and vulnerability to HIV infection among Tanzanian youth. J Health Popul Dev Ctries 1(2):51–60
48. Kaplan LD, Lee JY, Ambinder RF, Sparano JA, Cesarman E, Chadburn A, Levine AM, Scadden DT (2005) Rituximab does not improve clinical outcome in a randomized phase 3 trial of CHOP with or without rituximab in patients with HIV-associated non-Hodgkin lymphoma: AIDS-Malignancies Consortium Trial 010. Blood 106(5):1538–1543
49. Kaposi M (1872) Idiopatisches multiples Pigmentsarkom der Haut. Arch Dermatol Syphillis 4:265–273
50. Kedes DH, Operskalski E, Busch M, Kohn R, Flood J, Ganem D (1996) The seroepidemiology of human herpesvirus 8 (Kaposi's sarcoma- associated herpesvirus): distribution of infection in KS risk groups and evidence for sexual transmission. Nat Med 2(8):918–924
51. Klein U, Gloghini A, Gaidano G, Chadburn A, Cesarman E, Dalla-Favera R, Carbone A (2003) Gene expression profile analysis of AIDS-related primary effusion lymphoma (PEL) suggests a plasmablastic derivation and identifies PEL-specific transcripts. Blood 101 (10):4115–4121
52. Koon HB, Bubley GJ, Pantanowitz L, Masiello D, Smith B, Crosby K, Proper J, Weeden W, Miller TE, Chatis P, Egorin MJ, Tahan SR, Dezube BJ (2005) Imatinib-induced regression of AIDS-related Kaposi's sarcoma. J Clin Oncol 23(5):982–989
53. Krown SE, Lee JY, Dittmer DP, Consortium AM (2008) More on HIV-associated Kaposi's sarcoma. N Engl J Med 358(5):535–536; author reply 536. https://doi.org/10.1056/nejmc072994
54. Labo N, Miley W, Benson CA, Campbell TB, Whitby D (2015) Epidemiology of Kaposi's sarcoma-associated herpesvirus in HIV-1-infected US persons in the era of combination antiretroviral therapy. AIDS 29(10):1217–1225. https://doi.org/10.1097/QAD.0000000000000682
55. Lagunoff M, Bechtel J, Venetsanakos E, Roy AM, Abbey N, Herndier B, McMahon M, Ganem D (2002) De novo infection and serial transmission of Kaposi's sarcoma-associated herpesvirus in cultured endothelial cells. J Virol 76(5):2440–2448

56. Li H, Komatsu T, Dezube BJ, Kaye KM (2002) The Kaposi's sarcoma-associated herpesvirus K12 transcript from a primary effusion lymphoma contains complex repeat elements, is spliced, and initiates from a novel promoter. J Virol 76(23):11880–11888
57. Little RF, Merced-Galindez F, Staskus K, Whitby D, Aoki Y, Humphrey R, Pluda JM, Marshall V, Walters M, Welles L, Rodriguez-Chavez IR, Pittaluga S, Tosato G, Yarchoan R (2003) A pilot study of cidofovir in patients with kaposi sarcoma. J Infect Dis 187(1):149–153
58. Liu R, Li X, Tulpule A, Zhou Y, Scehnet JS, Zhang S, Lee JS, Chaudhary PM, Jung J, Gill PS (2010) KSHV-induced notch components render endothelial and mural cell characteristics and cell survival. Blood 115(4):887–895. https://doi.org/10.1182/blood-2009-08-236745
59. Martin JN, Ganem DE, Osmond DH, Page-Shafer KA, Macrae D, Kedes DH (1998) Sexual transmission and the natural history of human herpesvirus 8 infection. N Engl J Med 338 (14):948–954
60. Martin DF, Kuppermann BD, Wolitz RA, Palestine AG, Li H, Robinson CA (1999) Oral ganciclovir for patients with cytomegalovirus retinitis treated with a ganciclovir implant. Roche Ganciclovir Study Group. N Engl J Med 340(14):1063–1070
61. Mbulaiteye SM, Biggar RJ, Goedert JJ, Engels EA (2003) Immune deficiency and risk for malignancy among persons with AIDS. J Acquir Immune Defic Syndr 32(5):527–533
62. McCormick C, Ganem D (2005) The kaposin B protein of KSHV activates the p38/MK2 pathway and stabilizes cytokine mRNAs. Science 307(5710):739–741
63. Mendez JC, Procop GW, Espy MJ, Smith TF, McGregor CG, Paya CV (1999) Relationship of HHV8 replication and Kaposi's sarcoma after solid organ transplantation. Transplantation 67(8):1200–1201
64. Montaner S, Sodhi A, Molinolo A, Bugge TH, Sawai ET, He Y, Li Y, Ray PE, Gutkind JS (2003) Endothelial infection with KSHV genes in vivo reveals that vGPCR initiates Kaposi's sarcomagenesis and can promote the tumorigenic potential of viral latent genes. Cancer Cell 3(1):23–36
65. Moore PS, Kingsley LA, Holmberg SD, Spira T, Gupta P, Hoover DR, Parry JP, Conley LJ, Jaffe HW, Chang Y (1996) Kaposi's sarcoma-associated herpesvirus infection prior to onset of Kaposi's sarcoma. Aids 10(2):175–180
66. Moses AV, Jarvis MA, Raggo C, Bell YC, Ruhl R, Luukkonen BG, Griffith DJ, Wait CL, Druker BJ, Heinrich MC, Nelson JA, Fruh K (2002) Kaposi's sarcoma-associated herpesvirus-induced upregulation of the c- kit proto-oncogene, as identified by gene expression profiling, is essential for the transformation of endothelial cells. J Virol 76(16):8383–8399
67. Nador RG, Cesarman E, Chadburn A, Dawson DB, Ansari MQ, Sald J, Knowles DM (1996) Primary effusion lymphoma: a distinct clinicopathologic entity associated with the Kaposi's sarcoma-associated herpes virus. Blood 88(2):645–656
68. Naranatt PP, Krishnan HH, Svojanovsky SR, Bloomer C, Mathur S, Chandran B (2004) Host gene induction and transcriptional reprogramming in Kaposi's sarcoma-associated herpesvirus (KSHV/HHV-8)-infected endothelial, fibroblast, and B cells: insights into modulation events early during infection. Cancer Res 64(1):72–84
69. Northfelt DW, Dezube BJ, Thommes JA, Miller BJ, Fischl MA, Friedman-Kien A, Kaplan LD, Du Mond C, Mamelok RD, Henry DH (1998) Pegylated-liposomal doxorubicin versus doxorubicin, bleomycin, and vincristine in the treatment of AIDS-related Kaposi's sarcoma: results of a randomized phase III clinical trial. J Clin Oncol 16(7):2445–2451
70. Olp LN, Jeanniard A, Marimo C, West JT, Wood C (2015) Whole-genome sequencing of Kaposi's sarcoma-associated herpesvirus from zambian Kaposi's sarcoma biopsy specimens reveals unique viral diversity. J Virol 89(24):12299–12308. https://doi.org/10.1128/JVI.01712-15
71. Parravicini C, Chandran B, Corbellino M, Berti E, Paulli M, Moore PS, Chang Y (2000) Differential viral protein expression in Kaposi's sarcoma-associated herpesvirus-infected diseases: Kaposi's sarcoma, primary effusion lymphoma, and multicentric Castleman's disease. Am J Pathol 156(3):743–749

72. Pellet C, Kerob D, Dupuy A, Carmagnat MV, Mourah S, Podgorniak MP, Toledano C, Morel P, Verola O, Dosquet C, Hamel Y, Calvo F, Rabian C, Lebbe C (2006) Kaposi's sarcoma-associated herpesvirus viremia is associated with the progression of classic and endemic Kaposi's sarcoma. J Invest Dermatol 126(3):621–627
73. Polizzotto MN, Uldrick TS, Wang V, Aleman K, Wyvill KM, Marshall V, Pittaluga S, O'Mahony D, Whitby D, Tosato G, Steinberg SM, Little RF, Yarchoan R (2013) Human and viral interleukin-6 and other cytokines in Kaposi sarcoma herpesvirus-associated multicentric Castleman disease. Blood 122(26):4189–4198. https://doi.org/10.1182/blood-2013-08-519959
74. Polizzotto MN, Uldrick TS, Wyvill KM, Aleman K, Marshall V, Wang V, Whitby D, Pittaluga S, Jaffe ES, Millo C, Tosato G, Little RF, Steinberg SM, Sereti I, Yarchoan R (2016) Clinical features and outcomes of patients with symptomatic Kaposi sarcoma herpesvirus (KSHV)-associated Inflammation: prospective characterization of KSHV inflammatory cytokine syndrome (KICS). Clin Infect Dis 62(6):730–738. https://doi.org/10.1093/cid/civ996
75. Poole LJ, Yu Y, Kim PS, Zheng QZ, Pevsner J, Hayward GS (2002) Altered patterns of cellular gene expression in dermal microvascular endothelial cells infected with Kaposi's sarcoma-associated herpesvirus. J Virol 76(7):3395–3420
76. Rabkin CS, Janz S, Lash A, Coleman AE, Musaba E, Liotta L, Biggar RJ, Zhuang Z (1997) Monoclonal origin of multicentric Kaposi's sarcoma lesions. N Engl J Med 336(14):988–993
77. Rappocciolo G, Jenkins FJ, Hensler HR, Piazza P, Jais M, Borowski L, Watkins SC, Rinaldo CR Jr (2006) DC-SIGN is a receptor for human herpesvirus 8 on dendritic cells and macrophages. J Immunol 176(3):1741–1749
78. Ratner L, Lee J, Tang S, Redden D, Hamzeh F, Herndier B, Scadden D, Kaplan L, Ambinder R, Levine A, Harrington W, Grochow L, Flexner C, Tan B, Straus D (2001) Chemotherapy for human immunodeficiency virus-associated non-Hodgkin's lymphoma in combination with highly active antiretroviral therapy. J Clin Oncol 19(8):2171–2178
79. Renne R, Barry C, Dittmer D, Compitello N, Brown PO, Ganem D (2001) Modulation of cellular and viral gene expression by the latency-associated nuclear antigen of Kaposi's sarcoma-associated herpesvirus. J Virol 75(1):458–468
80. Rivas C, Thlick AE, Parravicini C, Moore PS, Chang Y (2001) Kaposi's sarcoma-associated herpesvirus LANA2 is a B-cell-specific latent viral protein that inhibits p53. J Virol 75(1):429–438
81. De Rosa G, Barra E, Guarino M, Gentile R (1989) Multicentric Castleman's disease in association with Kaposi's sarcoma. Appl Pathol 7(2):105–110
82. Roy D, Sin SH, Lucas A, Venkataramanan R, Wang L, Eason A, Chavakula V, Hilton IB, Tamburro KM, Damania B, Dittmer DP (2013) mTOR inhibitors block Kaposi sarcoma growth by inhibiting essential autocrine growth factors and tumor angiogenesis. Cancer Res 73(7):2235–2246. https://doi.org/10.1158/0008-5472.CAN-12-1851
83. Russo JJ, Bohenzky RA, Chien MC, Chen J, Yan M, Maddalena D, Parry JP, Peruzzi D, Edelman IS, Chang Y, Moore PS (1996) Nucleotide sequence of the Kaposi sarcoma-associated herpesvirus (HHV8). Proc Natl Acad Sci U S A 93(25):14862–14867
84. Sadler R, Wu L, Forghani B, Renne R, Zhong W, Herndier B, Ganem D (1999) A complex translational program generates multiple novel proteins from the latently expressed kaposin (K12) locus of Kaposi's sarcoma-associated herpesvirus. J Virol 73(7):5722–5730
85. Scehnet JS, Ley EJ, Krasnoperov V, Liu R, Manchanda PK, Sjoberg E, Kostecke AP, Gupta S, Kumar SR, Gill PS (2009) The role of Ephs, Ephrins, and growth factors in Kaposi sarcoma and implications of EphrinB2 blockade. Blood 113(1):254–263. https://doi.org/10.1182/blood-2008-02-140020
86. Scott D, Cabral L, Harrington WJ Jr (2001) Treatment of HIV-associated multicentric Castleman's disease with oral etoposide. Am J Hematol 66(2):148–150

87. Sgadari C, Barillari G, Toschi E, Carlei D, Bacigalupo I, Baccarini S, Palladino C, Leone P, Bugarini R, Malavasi L, Cafaro A, Falchi M, Valdembri D, Rezza G, Bussolino F, Monini P, Ensoli B (2002) HIV protease inhibitors are potent anti-angiogenic molecules and promote regression of Kaposi sarcoma. Nat Med 8(3):225–232
88. Silverberg MJ, Lau B, Achenbach CJ, Jing Y, Althoff KN, D'Souza G, Engels EA, Hessol NA, Brooks JT, Burchell AN, Gill MJ, Goedert JJ, Hogg R, Horberg MA, Kirk GD, Kitahata MM, Korthuis PT, Mathews WC, Mayor A, Modur SP, Napravnik S, Novak RM, Patel P, Rachlis AR, Sterling TR, Willig JH, Justice AC, Moore RD, Dubrow R, North American ACCoR, Design of the International Epidemiologic Databases to Evaluate A (2015) Cumulative incidence of cancer among persons with HIV in North America: a cohort study. Ann Intern Med 163(7):507–518. https://doi.org/10.7326/m14-2768
89. Sin SH, Roy D, Wang L, Staudt MR, Fakhari FD, Patel DD, Henry D, Harrington WJ Jr, Damania BA, Dittmer DP (2007) Rapamycin is efficacious against primary effusion lymphoma (PEL) cell lines in vivo by inhibiting autocrine signaling. Blood 109(5):2165–2173. https://doi.org/10.1182/blood-2006-06-028092
90. Soulier J, Grollet L, Oksenhendler E, Cacoub P, Cazals-Hatem D, Babinet P, d'Agay MF, Clauvel JP, Raphael M, Degos L et al (1995) Kaposi's sarcoma-associated herpesvirus-like DNA sequences in multicentric Castleman's disease. Blood 86(4):1276–1280
91. Stallone G, Schena A, Infante B, Di Paolo S, Loverre A, Maggio G, Ranieri E, Gesualdo L, Schena FP, Grandaliano G (2005) Sirolimus for Kaposi's sarcoma in renal-transplant recipients. N Engl J Med 352(13):1317–1323
92. Staskus KA, Zhong W, Gebhard K, Herndier B, Wang H, Renne R, Beneke J, Pudney J, Anderson DJ, Ganem D, Haase AT (1997) Kaposi's sarcoma-associated herpesvirus gene expression in endothelial (spindle) tumor cells. J Virol 71(1):715–719
93. Staudt MR, Kanan Y, Jeong JH, Papin JF, Hines-Boykin R, Dittmer DP (2004) The tumor microenvironment controls primary effusion lymphoma growth in vivo. Cancer Res 64 (14):4790–4799
94. Uldrick TS, Polizzotto MN, Aleman K, O'Mahony D, Wyvill KM, Wang V, Marshall V, Pittaluga S, Steinberg SM, Tosato G, Whitby D, Little RF, Yarchoan R (2011) High-dose zidovudine plus valganciclovir for Kaposi sarcoma herpesvirus-associated multicentric Castleman disease: a pilot study of virus-activated cytotoxic therapy. Blood 117(26):6977–6986. https://doi.org/10.1182/blood-2010-11-317610
95. Uldrick TS, Wyvill KM, Kumar P, O'Mahony D, Bernstein W, Aleman K, Polizzotto MN, Steinberg SM, Pittaluga S, Marshall V, Whitby D, Little RF, Yarchoan R (2012) Phase II study of bevacicumab in patients with HIV-associated Kaposi's sarcoma receiving antiretroviral therapy. J Clin Oncol 30(13):1476–1483. https://doi.org/10.1200/JCO.2011.39.6853
96. Vieira J, Huang ML, Koelle DM, Corey L (1997) Transmissible Kaposi's sarcoma-associated herpesvirus (human herpesvirus 8) in saliva of men with a history of Kaposi's sarcoma. J Virol 71(9):7083–7087
97. Wang L, Dittmer DP, Tomlinson CC, Fakhari FD, Damania B (2006) Immortalization of primary endothelial cells by the K1 protein of Kaposi's sarcoma-associated herpesvirus. Cancer Res 66(7):3658–3666
98. Wang HW, Trotter MW, Lagos D, Bourboulia D, Henderson S, Makinen T, Elliman S, Flanagan AM, Alitalo K, Boshoff C (2004) Kaposi sarcoma herpesvirus-induced cellular reprogramming contributes to the lymphatic endothelial gene expression in Kaposi sarcoma. Nat Genet 36(7):687–693
99. Wang L, Wakisaka N, Tomlinson CC, DeWire SM, Krall S, Pagano JS, Damania B (2004) The Kaposi's sarcoma-associated herpesvirus (KSHV/HHV-8) K1 protein induces expression of angiogenic and invasion factors. Cancer Research 64(8):2774–2781, Apr 15
100. Wang L, Wakisaka N, Tomlinson CC, DeWire S, Krall S, Pagano JS, Damania B (2004) The Kaposi's sarcoma-associated herpesvirus (KSHV/HHV8) K1 protein induces expression of angiogenic and invasion factors. Cancer Research

101. Whitby D, Howard MR, Tenant-Flowers M, Brink NS, Copas A, Boshoff C, Hatzioannou T, Suggett FE, Aldam DM, Denton AS et al (1995) Detection of Kaposi sarcoma associated herpesvirus in peripheral blood of HIV-infected individuals and progression to Kaposi's sarcoma. Lancet 346(8978):799–802
102. Whitby D, Luppi M, Barozzi P, Boshoff C, Weiss RA, Torelli G (1998) Human herpesvirus 8 seroprevalence in blood donors and lymphoma patients from different regions of Italy. J Natl Cancer Inst 90(5):395–397
103. Yu L, Tu M, Cortes J, Xu-Monette ZY, Miranda RN, Zhang J, Orlowski RZ, Neelapu S, Boddu PC, Akosile MA, Uldrick TS, Yarchoan R, Medeiros LJ, Li Y, Fajgenbaum DC, Young KH (2017) Clinical and pathological characteristics of HIV- and HHV-8-negative Castleman disease. Blood 129(12):1658–1668. https://doi.org/10.1182/blood-2016-11-748855
104. Zhang Z, Chen W, Sanders MK, Brulois KF, Dittmer DP, Damania B (2016) The K1 protein of Kaposi's sarcoma-associated herpesvirus augments viral lytic replication. J Virol 90(17): 7657–7666. https://doi.org/10.1128/JVI.03102-15

Molecular Biology of EBV in Relationship to HIV/AIDS-Associated Oncogenesis

4

Fengchao Lang, Yonggang Pei, Zachary L. Lamplugh and Erle S. Robertson

Contents

F. Lang · Y. Pei · Z. L. Lamplugh · E. S. Robertson
Department of Otorhinolaryngology–Head and Neck Surgery and Tumor Virology and Global Cancer Programs, Abramson Cancer Center, Philadelphia, USA

F. Lang · Y. Pei · Z. L. Lamplugh · E. S. Robertson
Perelman School of Medicine, University of Pennsylvania, Philadelphia, PA 19104, USA

E. S. Robertson (✉)
3610 Hamilton Walk, 201E Johnson Pavilion, Philadelphia, PA 19104, USA
e-mail: erle@upenn.edu

C. Meyers (ed.), *HIV/AIDS-Associated Viral Oncogenesis*, Cancer Treatment and Research177, https://doi.org/10.1007/978-3-030-03502-0_4

Abstract

Herpesvirus-induced disease is one of the most lethal factors which leads to high mortality in HIV/AIDS patients. EBV, also known as human herpesvirus 4, can transform naive B cells into immortalized cells in vitro through the regulation of cell cycle, cell proliferation, and apoptosis. EBV infection is associated with several lymphoma and epithelial cancers in humans, which occurs at a much higher rate in immune deficient individuals than in healthy people, demonstrating that the immune system plays a vital role in inhibiting EBV activities. EBV latency infection proteins can mimic suppression cytokines or upregulate PD-1 on B cells to repress the cytotoxic T cells response. Many malignancies, including Hodgkin Lymphoma and non-Hodgkin's lymphomas occur at a much higher frequency in EBV positive individuals than in EBV negative people during the development of HIV infection. Importantly, understanding EBV pathogenesis at the molecular level will aid the development of novel therapies for EBV-induced diseases in HIV/AIDS patients.

Keywords

EBV · HIV/AIDS · Oncogenesis · Latent infection · Lymphoma

4.1 Introduction

Epstein–Barr virus (EBV) was discovered and isolated from a Burkitt's lymphoma patient in 1964 [41]. It is known as the first tumor virus. EBV has 184 kb DNA base pairs, encoding 85 open reading frames (ORFs) and non-coding RNAs. There are two phases during EBV lifecycle, lytic infection producing new viral particles, and the followed long-term latent infection. EBV infects over 90% of the world's population. Typically, the infection is asymptomatic when humans are infected by EBV before adulthood but it will lead to mononucleosis when people are infected post-adolescence. EBV establishes extremely successful strategies to evade from host immune surveillance and contributes to about 0.5–2% of cancer occurrence [18], and is shown to be associated with Hodgkin Lymphoma (HL), Burkitt's Lymphoma (BL), Post-transplant Lymphoproliferative Disease (PTLD), Primary CNS Lymphoma (PCNSL), Nasopharyngeal Carcinoma (NPC), and gastric carcinoma (GC) [66].

In vivo, EBV primarily infects naive B cells of the tonsils followed by lytic viral replication, which spreads to the tonsil epithelial cells [142]. The replication and spreading process can also be suppressed by the host immune system. In latent infection, EBV expresses very limited proteins and non-coding RNAs in the memory B cells. There are now four distinct latency programs in the infected cells, latency type 0, I, II, and III [140]. They each exhibit a unique protein expression profile. In latency III, regarded as the first latency program established, following primary infection of resting naive B cells, EBV expresses six EBV nuclear antigens (EBNAs; EBNA1, EBNA2, EBNA3A, EBNA3B, EBNA3C, and EBNA-LP), three membrane proteins (latent membrane protein (LMP); LMP1, LMP2A, and LMP2B), two viral RNAs (EBERs), and the BamA rightward transcripts (BARTs) [157]. Viral protein expression pattern in latency III can potently drive B cells into immortalized cells. Type II latency is characterized by the expression of EBNA1, LMP1, LMP2, EBERs, and BARTs. In latency type I, only EBNA1, EBERs, and BARTs are expressed. In latency 0, EBV's protein expression is totally shut down, only expressing EBER and BART RNAs [20].

In individuals with competent immune systems, EBV rarely induces severe diseases but it can lead to severe consequences in people whose immune systems are compromised. HIV can severely destroy the human immune system and attenuate the immune suppression of EBV-induced abnormality in infected cells [107]. Incompetent immune surveillance and response will allow abnormal cell growth, proliferation, and tumorigenesis. It is frequently found that coinfections of EBV and HIV exist in malignancies associated with AIDS patients [90]. Frequency of multiple Epstein–Barr virus infections is changed in HIV-associated T-cell-immunocompromised individuals [155], which suggests that HIV infection may be able to modulate the status of EBV infection.

4.2 The Host Pathways Affected by EBV in Oncogenesis

EBV can transform B cells into lymphoblastoid cell lines (LCL) through the regulation of many pathways of cells including cell cycle, apoptosis, proliferation, chromatin and immune response repression [158]. Although EBV has the potential to induce B cell immortalization, not all of the B cells are transformed. A portion of EBV-infected cells are arrested in G1/S-phase [101]. Recently studies have found that the DNA damage response and metabolic stress induced by EBV can also be factors which suppresses B cell immortalization [101, 106]. It is also proposed that unknown host factors with the potential to suppress the immortalization process during EBV transformation remain to be discovered. Here, we reviewed several pathways and factors that affect transformation induced by EBV.

4.2.1 Resistance to Cell Apoptosis

EBV can prevent cell apoptosis through binding to the death receptors at the very early phase of infection. The best-characterized ligands and corresponding death receptors during apoptosis are FasL/FasR and TNF-α/TNFR1 pathways [40]. The death receptors recruit their adaptor proteins including TNF receptor-associated death domain (TRADD), Fas-associated death domain (FADD), and caspase-8. The interactions form a complex called death-inducing signaling complex (DISC) [11].

Following the formation of DISC, executioner caspases are cleaved by caspase 8 and the cell apoptosis program begins.

The expression of p53 in the presence of DNA damage is reduced and survivin can be induced by EBNA1 [70, 102]. Interestingly, EBNA2 can increase the protein levels of anti-apoptotic proteins such as Bfl-1, Bcl-xL, Bcl-2, and Mcl-1 [118]. The cellular proapoptotic BIK/NBK gene is also repressed by EBNA2 at transcription level to inhibit the proapoptotic program [22]. The Bcl-2 family member BIM1, which is known as a proapoptotic protein, is suppressed by the viral encoded essential nuclear antigens EBNA3A and EBNA3C [5]. P53-mediated activities can also be repressed by EBNA3C [20, 156] and P53 itself can be degraded through EBNA3C recruitment of the MDM2 E3-ubiquitin ligase [128]. EBNA3C also can inhibit cell apoptosis through regulation of E2F1, IRF4/8, Pim-1, Aurora kinase B [9, 10, 65, 127]. EBNA-LP contributes to the anti-apoptosis activities through interaction with an extensive number of host cell proteins which include PP2A, HAX-1, HSP70, HSP72, Rb, P53, P14ARF, and Fte1/S3a [11, 49, 60, 69–71, 96].

NF-κB activation is essential for the survival of EBV-transformed B lymphocytes [19] and can be activated by LMP1 recruited TNFR-associated factors (TRAFs) and TNFR-associated death domain protein (TRADD). LMP1 activated NF-κB can also induce A20 and BCL-2 to promote cell survival [81]. Down-regulation of BCL-2 in the absence of LMP1 and LMP1 induced upregulation of BCL-2 demonstrated that LMP1 can exert its anti-apoptotic function through BCL-2 [45, 93]. BCL-2 is also regulated by LMP2 which protects EBV protected cells from proapoptotic signals [117].

4.2.2 Cell Cycle and Proliferation

Proper cell cycle arrest is a safeguard to prevent premature cell division and uncontrolled proliferation. Cyclins, cyclin-dependent kinases (CDKs), and CDK inhibitors strictly regulate the cell cycle [46]. c-Myc activation, which is common in kinds of cancer, can increase the activities of Cyclin D and E2F and repress the expression of CDK inhibitors ($p27^{Kip1}$, $p21^{Cip1}$, and $p15^{Ink4b}$) [17, 44, 61]. Increased expression of c-Myc can also be induced by EBV proteins, such as EBNA2 [67], LMP2A [46] and EBNA3C [7]. pRb, the ubiquitin ligase SCFSKP2, cyclin D1, cyclin A, c-Myc, MDM2, p53, CHK2, E2F1, and E2F6 are all directly regulated by EBNA3C [78, 79, 128, 139]. These interactions are involved in G1-S and G2-M transitions. Recently, it is noticed that mitotic checkpoint proteins, such as BubR1 expression and stability are affected by EBNA3C and these disruptions of mitotic checkpoints also contribute to cell proliferation [57, 65, 108]. Inactivation of G1-S, G2-M, and mitotic spindle checkpoints can enhance the propagation of damaged DNA and provide EBV increased diversity of the genetic context of progeny cells, which also contributes to oncogenesis.

4.2.3 Promotion of Cell Metabolism

Disruption of cellular metabolism is a hallmark of cancer [111]. Viruses rely on host cells for the energy needed during their lifecycle. It is known that viruses are able to change the profile of host cell metabolism to facilitate assimilation of carbon into macromolecules for viral and host cell activities during infection [132]. Many cancers undergo hypoxia, which diminishes the use of ATP by downregulating Na-K-ATPase [148]. During EBV latent infection, LMP-1 activates the expression of hexokinase 2 (HK2). Expression of HK2 leads to the induction of glycolysis and this proves to be necessary for nasopharyngeal carcinoma cell survival [152]. B cell transformation mediated by EBV, glucose import and surface glucose transporter 1 (GLUT1) levels were increased in hyper-proliferated B cell subsets [101]. However, how cellular metabolism is regulated by EBV is largely unknown.

4.2.4 Evasion from Immune Surveillance

Innate immunity and adaptive immunity constitute the defense line against pathogen infections. Pattern recognition receptors, such as Toll-like receptors (TLRs) and cytoplasmic foreign DNA and RNA sensors (including IFI16, cGAS, and RIG-I-like receptors) are important components of innate immunity [72]. Once the virus establishes primary infection and resides inside cells, adaptive immunity is needed to detect and clear virus-infected cells. T cells which can recognize virus expressed peptides through surface HLA molecules do play important roles in the process of elimination [121].

Although the host immune system can effectively inhibit EBV activities, EBV survives in the human body by establishing long-term latent infection with limited gene expression [141]. To survive in host cells, EBV has developed multiple pathways to suppress, counter and evade the host immune surveillance [28]. There are several key latent proteins that aid in EBV's ability to evade immune detection. The EBNA1 antigen has a weak immunogenicity for MHC I presentation and interferes with the NF-κB, STAT1, and TGFβ pathways [16, 150]. EBNA2 inhibits IFNβ and ISG production through enhancement of STAT3 transcription [4, 105]. LMP1 can inhibit STAT2 activity and IFN production through preventing Tyk2 phosphorylation and induction of an inhibitory IRF7 splice variant [50, 159]. LMP-1 can also reduce TLR9 expression to block pattern recognition through NF-κB activation [43]. LMP2a can limit IFN-stimulated gene expression by interfering with signaling between IFNγ and IFNα and their receptors [136].

In addition to immune system evasion during latency, EBV can attenuate CD4+ and CD8+ T cell recognition of EBV positive cells through down-regulation of HLA I and HLA II during its replication. The process is mediated by BNLF2a, BILF1, and BGLF5 during lytic infection [103]. Viral interleukin-10 (vIL-10), a homolog of human IL-10, is produced by the EBV gene BCRF1 which stimulates B cell growth, inhibit antigen presentation, T cell growth and IFN-γ production [68]. EBV-infected lymphoma cells highly express inhibitory ligands for the

PD-1/CTLA-4 receptor, PD-L1 and PD-L2 to suppress EBV-specific T cell responses. PD-1/CTLA-4 blockade by drugs can effectively reduce the size of lymphomas induced by EBV [95].

4.2.5 Epigenetic Regulation Due to EBV Infection

Cancer cells and normal cells are largely different in epigenetic states, including chromatin remodeling, histone acetylation, histone methylation, and DNA methylation [116]. Hyper-methylated CpG islands at promoters, genome-wide hypomethylated DNA in the gene body, changes in histone modification expression and distribution are constantly associated with cancer cells [154]. These abnormal epigenetic profiles lead to aberrant gene expression and contribute to cancer development. DNA methyltransferase (DNMT), histone deacetylase (HDAC), and histone methyltransferase activities can also be regulated to modulate the hypomethylation patterns [124]. Tumor viruses can also manipulate the host epigenetic machinery and change host cellular genome through DNA methylations and histone modifications [116]. Emerging data from a number of studies have suggested that inappropriate epigenetic regulation underlies human tumor virus-mediated oncogenesis.

EBV influences host epigenetic profiles from different angles and viral-mediated epigenetic regulations on host chromatin is believed as an essential factor which contributes to oncogenesis. Hyper-methylated CpG islands are recruited at the promoters of tumor suppressor genes (TSGs) and result in global transcriptional repression of TSGs in EBV-infected resting B cells [126]. EBV latent proteins play essential roles in epigenetic deregulation during B cell lymphomagenesis. A chromatin remodeling complex, SWI/SNF, can be recruited by EBNA2, to create an open chromatin conformation and so induces c-Myc transcription [149]. EBNA3C can interact with HATs and HDACs including p300, CBP and HDAC1/2 [34, 76]. These interactions suggest a possible role in the regulation of histone acetylation and chromatin remodeling. Hypermethylation at promoters of tumor suppressor genes, such as p14ARF, p15INK4a, and p16INK4a are recruited by EBNA3A and EBNA 3C [2, 3]. MIZ1 and H3K27me3 histone modification can be recruited by EBNA3A to the promoter of p15INK4a to inhibit its expression [12]. EBNA3A and 3C can also mediate Polycomb repressive complex 2 (PRC2) binding and H3K27me3 modification to the promoter of tumor suppressor Bim/Bcl2L11 [5, 110]. LMP1 can regulate all three DNA methyltransferase (DNMT) enzymes, DNMT1, DNMT3A, and DNMT3B in either nasopharyngeal carcinoma (NPC) cell lines or germinal center B cell-derived malignancies and so influence DNA methylation of specific genes, such as RARB and CDH13 [86, 135, 145]. LMP2A is shown to increase expression of DNMT1 and methylation of PTEN gene. Although people begin to know about how EBV influences host transcription regulation, much is focused on the interactions between viral factors and the host epigenetic machinery. Furthermore, genome-wide epigenetic regulation mediated by EBV is largely unknown. Therefore, there is a great deal yet to be explored regarding epigenetic regulation related to EBV infection.

4.3 The Functions of EBV Proteins During Oncogenesis

Viral oncogenes expressed in latent or lytic infection are potent in their ability to drive B cells proliferation and immortalization in vitro. During EBV primary infection in B cells, EBV initially enters a transient, lytic phase where lytic genes are expressed resulting in progeny virus [59, 119]. Within the first 24 h post infection, EBV also expresses latent genes as indicated by the expression of EBNA-LP and EBNA2[2] and establishes latent infection after 4–6 days with expressions of six EBV nuclear antigens and two membrane proteins [8, 98, 129, 160]. Each of these proteins play an important role in the oncogenic process during EBV infection (Fig. 4.1 and Table 4.1).

EBNA1 is regarded as a key protein important for replication and mitotic segregation of the viral genome [48] and is the only nuclear protein expressed in latency type I. Cooperating with viral origin of plasmid replication (oriP), EBNA1 supports viral replication using the host cell replication machinery as well as segregation into daughter cells upon mitosis [36, 48, 88]. Several cellular proteins that interact with EBNA1 have been identified. These include the cellular origin recognition complex and other components of the prereplication complex, replication protein A, and the telomere repeat binding factor 2 (TRF2) [35]. EBNA1 can also tether the EBV episomes to the cellular chromosomes by interacting with

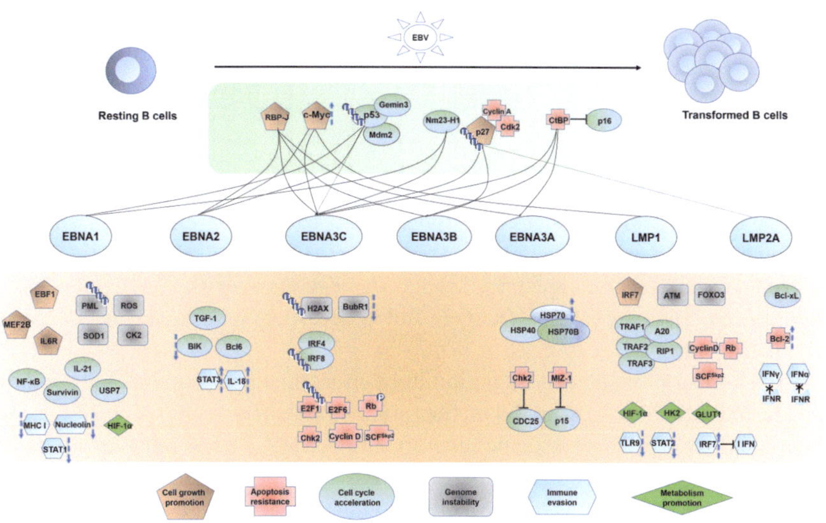

Fig. 4.1 EBV latent proteins induce hall marks of cancer through regulation of cellular proteins

Table 4.1 The role of viral proteins in regulating host cell activities

Viral proteins	Function pathways	Regulation of host proteins	Roles in cell transformation
EBNA1	Tethers viral genome la host cell chromosome, viral DNA replication, transcriptional activation, immune evasion	Cellular origin recognition complex, RPA, TRF2, RAG-1, RAG-2, Nm23-H1, AP-1, p53	Essential
EBNA2	Transcriptional regulation, cell proliferation and accelerating cell cycle	RBP-Jk, c-Myc, CDK2, Bcl6, SKIP	Indispensable
EBNA3A	Transcriptional regulation EBNA2 antagonist and coactivator	RBP-Jk, CtBP, Chk2, WDR48, WDR20, and USP46/UP12, MIZ-1, 20S proteasome, Chaperones, XAP-2, TCP-1, AhR, UK/UPRT	Indispensable
EBNA3B	Transcriptional regulation	RBP-Jk, Cyclin A, 20S proteasome, WDR48, WDR20, and USP46/UP12	
EBNA3C	Transcriptional regulation chromatin remodeling histone modification E3 ubiquitin degradation	RBP-Jk, Nm23-H1, SCF_{SKP2} ubiquitin ligase, Chk2, p300, CtBP, DDX20, HDAC1/2, mSin3A, NCoR, Cyclin A, D1, E, GSK3b, p53, Mdm2, ING4/5, E2F1, E2F6, Rb, c-myc, sumo1/3, MRPS18-2, Aurora kinase B, IRF4/8, H2AX, USP46/12, p73	Indispensable
LMP-1	Transcriptional modulation regulates cell cycle checkpoint cell survival cytotoxic T Lymphocyte modulation	NF-κB, JNK, ERK, and P38 MAPK, PI3-K/AKT, IRF7, p16, p27, BCL2	Indispensable
LMP-2	Maintenance of latency Disrupts B cell of latency receptor signaling	Bcl-xL, Ras/PI3K/Akt pathway	Essential

cellular protein EBP2 [151]. EBNA1 has roles in cell survival through regulating cellular gene expression, including upregulation of survivin expression and inhibition of the protein tyrosine phosphatase receptor kappa (PTPRK) [94]. EBNA1 also contributes to tumorigenesis through dysregulation of the host cell genome stability by inducing RAG1/RAG2 and increasing reactive oxygen species (ROS) [56, 146].

EBNA1 can promote cell migration through interacting with Nm23-H1 in lymphoblastoid cell lines and inhibits its ability to suppress cell migration [104]. Furthermore, loss of EBNA1-specific memory CD4+ and CD8+ T cells were found in HIV-infected patients progressing to AIDS-related non-Hodgkin lymphoma [115]. This suggests that EBNA1 may be a critical viral encoded antigen that increases the chance of development of an EBV positive lymphoma in HIV-infected patients.

EBNA2 has essential roles in the process of B cell immortalization and mainly promotes cell growth through induction of cell proliferation and accelerating cell cycle. EBNA2 functions as a transcriptional factor with a transactivation domain and interacts with DNA through other adaptors [74]. EBNA2 also downregulates c-Myc expression and up-regulates Cyclin D, E [67]. EBNA2-mediated transcriptional modulation is mainly exerted through association with other cellular DNA-binding proteins including RBP-Jκ and PU.1 [91, 112, 161]. EBNA2 contributes to B cell immortalization through constitutive activation of the Notch signaling pathway [80]. RBP-Jκ is a ubiquitous DNA-binding protein which recognizes the sequence CGTGGGAA [100]. The intracellular region of notch has been shown to possess transactivation ability when overexpressed in various cell lines. Binding of notch to the nuclear protein RBP-Jκ is dependent on two regions: the RAM domain located immediately C-terminal to the transmembrane region and the CDC 10/ankyrin repeats [130]. EBNA2 can bind to RBP-Jκ at the same regions and can replace the binding of intracellular region of notch. Therefore, EBNA2 acts as a constitutive component of notch-I signaling complex and results in activation of the notch-I signaling pathway [8]. The ability of EBNA2 to drive B cell immortalization relies on its ability to antagonize the transcriptional repression function of RBP-Jκ. Interestingly, EBNA2 transactivates the HIV LTR, and this transactivation is dependent on the NF-κB sites in the HIV LTR [133]. The exact mechanisms of this transactivation are still not well demonstrated.

EBNA3 family proteins: EBNA3 family proteins are EBV latent proteins that include EBNA3A, 3B, and 3C. The three EBNA3 genes encoded by EBV are expressed from adjacent loci in the EBV genome and it is believed that these genes have evolved from a common ancestral gene to mediate slightly divergent functions [8]. The amino termini of all three genes have a conserved domain that binds to the transcriptional corepressor RBP-Jκ and enables these proteins to differentially regulate EBNA2-mediated transcription. The two members of EBNA3 family EBNA3A and 3C are vital for B cell transformation and lymphomagenesis [14]. Modified EBV containing stop codons demonstrated that EBV loses its ability to transform naive B cells without EBNA3A and 3C but not EBNA3B [14, 144]. EBNA3 proteins exert transcriptional regulation functions through interacting with host proteins. For host proteins known to interact with EBNA 3 family proteins, RBP-JK is the first host protein that was identified as a transcription repressor whose interaction is essential for lymphoblastoid cell growth [99, 122, 123]. EBNA3A can mediate viral resistance to BCL-2 antagonism and maintain B cell long-term growth [118]. EBNA3C extensively interacts with host proteins and regulates host cell activities including transcription regulation, chromatin remodeling, histone modification, and E3 ubiquitin degradation (Table 4.1). Interactions with histone deacetylases, [76] histone acetyltransferases p300, CtBP [138] and polycomb proteins [2, 109] demonstrate that EBNA3C are involved in chromatin structure regulation.

EBNA3C mediates dysregulation of E2F1, E2F6, CyclinA, CyclinD, CyclinE, Tp53, c-Myc, Rb, and p27 [7, 77, 78, 114, 125, 127], so that EBNA3C can accelerate cell cycle by overcoming the cell cycle checkpoints. EBNA3C can also contribute to genome instability by interaction with Aurora kinase B and H2AX [64, 65], which are hallmarks of cancer.

LMP Family proteins: The LMP family of proteins which include LMP1, LMP2A, and LMP2B are critical viral genes for B cell transformation [1]. LMP1 is an essential transmembrane protein with six transmembrane domains, a short intracellular N-terminus and a relatively long intracellular C-terminus [73]. The C-terminus of the protein is functionally homologous to a constitutively active CD40 receptor. It binds to members of the TRAF family and interacts with JAK3 and members of the STAT family [39, 51, 75]. All these interactions contribute to the uncontrolled proliferation of LMP1 expressing cells [8]. LMP1 can regulate EBV as well as host cell gene expression [97]. LMP1 mimics CD40 receptor signaling and is involved in regulating NF-κB, JNK, PI3 K/Akt, and MAPK pathways to promote cellular proliferation [30, 83, 120, 137]. LMP1 dysregulates cell cycle checkpoint through inhibition of p16 and p27 [42] and also promotes cell survival through the BCL2 pathway. Drugs inhibiting JNK1 can strongly decrease ΔNp73a, an antagonist of p53 which is induced by LMP1 [1].

LMP2A is a transmembrane protein with 12 transmembrane domains, a short intracellular C-terminal domain and a relatively long intracellular N-terminal domain [85, 131]. LMP2B is an amino-terminal truncated form of LMP2A, with transcription beginning just before the first transmembrane domain [131]. The N-terminus of LMP2A is homologous to the B cell receptor cytoplasmic domain and contains the same ITAM motifs [8]. LMP2A inhibits B cell signal transduction by mimicking an activated B cell receptor (BCR). LMP2A promotes cell survival through up-regulating Bcl-xL expression and activating the Ras/PI3K/Akt pathway [30].

4.4 HIV-Associated Lymphoma in HIV/AIDS Patients

Lack of competent immune surveillance will lead to severe diseases induced by EBV. In healthy individuals, the number of EBV-infected cells is restricted by EBV-specific CD8+ T lymphocytes. Once the host lacks an effective number of T cells or competent immune response, the growth and proliferation of infected B cells will be out of control and eventually lead to proliferative disorders. The incidence risk of some lymphomas (non-Hodgkin lymphoma and central nervous system lymphoma) in HIV/AIDS patients is several hundred folds higher than in the healthy individuals [54]. Burkitt's Lymphoma, Diffuse large B cell lymphoma, Extranodal marginal zone lymphoma of MALT type, Peripheral T cell lymphoma, and Classical Hodgkin lymphoma which occur in immunocompromised patients account for 80% of HIV-associated lymphomas [15]. Although it was reported that HIV can directly immortalize B cell lines in EBV positive individuals [84], it is more closely accepted that the main effect of HIV in lymphomagenesis is immunosuppression, compromised immune surveillance and disturbed immune regulation.

Table 4.2 Features of EBV-associated aids-associated B cell lymphoma

Lymphoma histology	% in HIV	EBV positive frequency (%)	Histology feature	Expressed viral factors
Burkitt lymphoma	55	55	CD20, CD10, BCL6, BCL2, and Ki67 positive	EBNA-1
Diffuse Large B cell Lymphoma (DLBCL)	30	30–90	Centroblastic variant: CD10+, BCL6+, CD138−, and MUM1− Immunoblast variant: CD20 −/+, BCL6−, CD138+, MUM1/IRF4+, CD45−, PAX5−	LMP-1
Primary CNS Lymphoma (PCNSL)	<5	100	CD45þ, Pan-B cell markers þ, CD138−. Variable positivity for CD10 and BCL6, IRF4/MUM1þ. IGH monoclonal	LMP1
Hodgkin Lymphoma (HL)	–	100	BCL-6−, syn-1+, CD15+, CD30+, CD45−	EBER, LMP1, LMP2
Plasmablastic Lymphoma (PBL)	<5	70–80	CD45−, CD20−, Pax5−, CD79a−/þ, CD138þ, CD38þ, MUM1/IRF4þ, BLIMP1þ, XBP1þ, clgG, IGH monoclonal	EBER
Primary Effusion Lymphoma (PEL)	<5	90–100	CD45þ, Pan-B cell markers −, CD30þ, CD138þ, clgM−, IGH monoclonal	EBER
Post-Transplant Lymphoproliferative Disease (PTLD)	–	90–100	CD20+, CD79a+, CD3−, CD5−	LMP1, LMP2

Besides the deficient T cell levels leading to occurrence of lymphoma, we now realize that the overactivated T cell response and exhausted immune responses are also related to initiation of B cell lymphomas [55]. This may be because of the dysfunction of T cells caused by the chronic and continuous activation induced by HIV replication. Here, we reviewed several EBV-associated diseases occurring in HIV-infected patients (Table 4.2).

Burkitt Lymphoma (**BL**): Among the lymphomas, Burkitt's Lymphoma is a common symptom in HIV/AIDS patients with 30–40% of the BL-tumors being EBV positive [66]. BL occurs during HIV infection even with a normal amount of circulating CD4+ T cell. Burkitt's Lymphomas are featured with expression of CD20, CD10, BCL6, BCL2, and Ki67 molecules [26]. Burkitt's Lymphomas are highly proliferative and there is a translocation placing the c-Myc gene adjacent to the region of heavy- or light-chain immunoglobulin loci [134]. Besides the dysregulation of the c-Myc gene, it is common to find point mutations in the gene body of tumor suppressor gene TP53 [87].

There are several EBV proteins and RNA detected in Burkitt's lymphoma. EBNA-1 is expressed in Burkitt's lymphoma to tether the viral genome to host chromosome and also contributes to the genetic regulations during Burkitt's lymphoma development [92]. Since EBNA-1 does not induce a potent cytotoxic T cell response, it provides the EBV positive cells a survival advantage. In addition, cell cycle is disturbed in Burkitt's lymphoma because of the tumor suppressor gene RBL2 inactivation.

Diffuse Large B cell Lymphoma (DLBCL): DLBCL is the most common lymphoma in HIV-infected patients [26]. There are two subtypes, centroblastic DLBCL, and immunoblastic DLBCL, featured with different morphology and phenotype [153]. The probability of these two subtypes occur in HIV/AIDS patients is almost equivalent. Within 30–40% EBV positive cases, centroblastic DLBCL occurs accompanied with mild immunosuppression. Centroblastic DLBCL is CD10 +, BCL6+, CD138−, and MUM1− showing a germinal center B cell phenotype [6, 32]. Immunoblastic DLBCL cells express B cell markers including CD19, CD20, CD79a, Pax5 and lacks germinal center phenotype [90]. Immunoblastic DLBCL usually occurs under the condition of marked immunosuppression and about 90% of these cases are EBV positive [28].

LMP-1 plays essential roles in lymphoma pathogenesis and is usually expressed in DLBCL [28]. LMP1 can strongly downregulate BCL-6 in DLBCL and promotes cell survival [23, 113]. NF-kB pathway activation induced by loss of the anti-apoptotic protein A20 contributes to pathogenesis of DLBCL [33]. However, in the presence of EBV infection, LMP1 can induce the expression of A20 and the upregulation of A20 prevents cell apoptosis, and thus promote tumorigenesis [52]. The relationship between A20 expression and pathogenesis of DLBCL reflects the complex regulatory processes which occur during lymphomagenesis under conditions of HIV and EBV coinfection. In addition, the dysregulation and hypermutations of c-Myc, TP53, Pim1, Pax5, and RhoH/TTF genes occur quite frequently in HIV-related DLBCL cases [90].

Primary CNS Lymphoma (PCNSL) occurs in the brain, leptomeninges, eyes or spinal cord, and is also categorized to be one kind of DLBCL [89]. The frequency of PCNSL in normal population is less than 1%, but it is up to 20% of the total lymphomas occurring in HIV-infected patients who have rather low CD4 T cell counts, over 1000 times greater than in the non-HIV population [47].

EBV has been found in PCNSL cases at 100% frequency in HIV-infected individuals and all EBV latent proteins are expressed in PCNSL [15]. Interestingly, studies have reported that EBV detection combined with CNS lesions can be used as a kind of diagnosis of PCNSL [62].

Hodgkin Lymphoma (HL): HL is one of the most common lymphoma in HIV positive patients, with over 10 folds higher risk than the general population [53]. The occurrence of HIV-associated HL often accompany a higher ratio of CD8+ T cells over CD4+ lymphocytes compared with HIV negative HL. Hodgkin Reed–Sternberg (HRS) cells in HIV-associated HL represent the typical cells in HL and they are featured with BCL-6−, syn-1+, CD15+, CD30+, CD45− phenotype and thus reflect post-GC B cells [25].

There are high titers of EBV in patients with Hodgkin lymphoma, indicating that HIV and EBV may cooperate closely with each other to promote development of HL [24]. Both EBER in situ hybridization and LMP1/2 are positive in HIV-associated HL [25]. LMP1 and LMP2 play important roles in HL pathogenesis. LMP1 up-regulates the expression levels of the multiple function polycomb protein BMI-1, which plays a role in inhibiting tumor suppressors [38]. LMP1 also promotes B cell proliferation through mimicking CD40 and activating NF-κB, PI3K, and JAK/STAT pathways [82, 147]. LMP2 promotes immature BCR negative B cells survival via mimicking B cell receptor (BCR) signaling and migration from bone marrow/colonize peripheral lymphoid organs [21].

Plasmablastic Lymphoma (PBL): PBL is related to HIV-associated lymphoma, post-transplant immunodeficiency and immunosenescence [27]. PBL constitutes about 3% of lymphomas in HIV positive patients and 70–80% of these PBL are EBV positive.

PBL cells are positive for plasma cell markers (CD79a, IRF4/MUM1, BLIMP1, CD38, and CD138) and express low level or have no B cell specific markers, for example, CD19, CD20, and Pax5 [27]. IRF4/MUM1 and BLIMP1 are more frequently detected than CD38 and CD138 in PBL [90]. CD10 expression may also accompany c-Myc translocation in some cases. In more than 30% of cases, PBL can express CD10 while 78% of PBL cases are c-Myc translocated or amplified and nearly all the cases have overexpressed c-Myc protein levels. Moreover, c-Myc overexpression is believed to facilitate PBL cell proliferation and survival. The overexpression of c-Myc is also a result of the mutant PRDM1/Blimp1α protein, which lacks the functional regulatory domains of the c-Myc gene.

EBV factors may also be involved in the pathogenesis of plasmablastic lymphoma. The frequency of c-Myc translocation is found in about 70% of EBV positive PBL cells versus 40% in EBV negative cases. It is reported that EBER can be 100% percent positive in PBL cases with HIV infection while they are generally negative for EBNA2 and LMP-1 [37].

Primary Effusion Lymphoma (PEL): Almost only found in HIV-infected patients, PEL constitutes less than 5% of all HIV-associated non-Hodgkin lymphoma. Although it is believed that KSHV is the determining factor driving pathogenesis of PEL, EBV coinfection exists in most cases of PEL [102]. PEL cells lack B cell or T cell lineage phenotype, negative for CD19, CD20, CD79a, and immunoglobulins [31]. PEL cells frequently express activation or plasmacytic differentiation-related antigens, including CD30, HLA-DR, EMA, CD38, and CD138 [13]. The immunoglobulin gene and the non-coding regions in BCL6 gene are frequently mutated. Genes involved in inflammation, cell adhesion, and invasion are also highly expressed in PEL cells [63] and KSHV factors play essential roles in the pathogenesis of PEL. LANA-1, v-cyclin, vFLIP, LANA-2, and vIL6 can facilitate cell transformation through promoting proliferative, anti-apoptotic, pro-inflammatory, and angiogenic effects [29]. The EBV viral RNA factor EBER can be detected in most cases of PEL. It demonstrates that EBV is an important coinfecting factor, but it is not clear about the roles of EBV in pathogenesis of PEL.

Post-Transplant Lymphoproliferative Disease (PTLD): PTLD can occur in patients with primary immunosuppression, drug-mediated immunosuppression following transplant to prevent the rejection and HIV infection induced immunodeficiency. It is different from HIV-associated lymphoma because PTLD has more limited disease distribution, normal tumor suppressor gene expression and is negative for oncogenes.

EBV infection is tightly linked to PTLDs and nearly 100% positive in PTLD-associated Hodgkin Lymphoma cases. The deficiency of killer T cells in PTLD patients also contributes to the activation of EBV. Widely expressed EBV latent proteins promote cell survival and resistance to programmed cell death [143]. LMP1 and LMP2 expressed in PTLD cells can induce neoplastic cell proliferation through dysregulation of BCR, CD40 and NF-kB regulatory pathways [58].

4.5 Conclusions

Numerous studies have revealed multiple mechanisms involved in EBV and HIV/AIDS-associated oncogenesis in the last couple of decades. Strategies utilized in the discovery of viruses provide clinical investigators clues as to the development of therapies and methodologies to decrease the risk of virus-associated diseases. New therapies to combat HIV and EBV infections are emerging. Highly active antiretroviral therapy (HAART) has highly reduced the mortality in the HIV-infected patients. HDAC inhibitors, which can induce virus reactivation from latency infected cells, are used as a "shock and kill" strategy and have been proved an efficient method to decrease HIV and EBV particles in vivo. Crispr-cas9 is also used to effectively edit target genes and so this could be used to destroy HIV or EBV which contribute to lymphoma pathogenesis. In the future, combined therapeutic approaches, including chemotherapy and biotherapy may be adopted to clear virus from infected patients and repress virus-induced diseases.

References

1. Accardi R, Fathallah I, Gruffat H, Mariggio G, Le Calvez-Kelm F, Voegele C, Bartosch B, Hernandez-Vargas H, McKay J, Sylla BS, Manet E, Tommasino M (2013) Epstein-Barr virus transforming protein LMP-1 alters B cells gene expression by promoting accumulation of the oncoprotein ΔNp73α. PLoS Pathog 9:e1003186
2. Allday MJ (2013) EBV finds a polycomb-mediated, epigenetic solution to the problem of oncogenic stress responses triggered by infection. Front Genet 4:212
3. Allday MJ, Bazot Q, White RE (2015) The EBNA3 family: two oncoproteins and a tumour suppressor that are central to the biology of EBV in B cells. In: Epstein Barr virus, Vol 2: one Herpes virus: many diseases, 391. pp 61–117

4. Aman P, Vongabain A (1990) An Epstein-Barr virus immortalization associated gene segment interferes specifically with the IFN-induced anti-proliferative response in human B-lymphoid cell-lines. EMBO J 9:147–152
5. Anderton E, Yee J, Smith P, Crook T, White RE, Allday MJ (2008) Two Epstein-Barr virus (EBV) oncoproteins cooperate to repress expression of the proapoptotic tumour-suppressor Bim: clues to the pathogenesis of Burkitt's lymphoma. Oncogene 27:421–433
6. Bai M, Skyrlas A, Agnantis NJ, Kamina S, Tsanou E, Grepi C, Galani V, Kanavaros P (2004) Diffuse large B-cell lymphomas with germinal center B-cell-like differentiation immunophenotypic profile are associated with high apoptotic index, high expression of the proapoptotic proteins bax, bak and bid and low expression of the antiapoptotic protein Bcl-Xl. Mod Pathol 17:847–856
7. Bajaj BG, Murakami M, Cai Q, Verma SC, Lan K, Robertson ES (2008) Epstein-Barr virus nuclear antigen 3C interacts with and enhances the stability of the c-Myc oncoprotein. J Virol 82:4082–4090
8. Bajaj BG, Murakami M, Robertson ES (2007) Molecular biology of EBV in relationship to aids-associated oncogenesis. Cancer Treat Res 133:141–162
9. Banerjee S, Lu J, Cai QL, Saha A, Jha HC, Dzeng RK, Robertson ES (2013) The EBV latent antigen 3C inhibits apoptosis through targeted regulation of interferon regulatory factors 4 and 8. PLoS Pathog 9
10. Banerjee S, Lu J, Cai QL, Sun ZG, Jha HC, Robertson ES (2014) EBNA3C augments Pim-1 mediated phosphorylation and degradation of P21 to promote B-cell proliferation. PLoS Pathog 10
11. Banerjee S, Uppal T, Strahan R, Dabral P, Verma SC (2016) The modulation of apoptotic pathways by gammaherpesviruses. Front Microbiol 7:585
12. Bazot Q, Deschamps T, Tafforeau L, Siouda M, Leblanc P, Harth-Hertle ML, Rabourdin-Combe C, Lotteau V, Kempkes B, Tommasino M, Gruffat H, Manet E (2014) Epstein-Barr virus nuclear antigen 3A protein regulates CDKN2B transcription via interaction with MIZ-1. Nucleic Acids Res 42:9700–9716
13. Bhatt S, Ashlock BM, Natkunam Y, Sujoy V, Chapman JR, Ramos JC, Mesri EA, Lossos IS (2013) CD30 targeting with brentuximab vedotin: a novel therapeutic approach to primary effusion lymphoma. Blood 122:1233–1242
14. Bhattacharjee S, Ghosh Roy S, Bose P, Saha A (2016) Role of EBNA-3 family proteins in EBV associated B-cell lymphomagenesis. Front Microbiol 7:457
15. Bibas M, Antinori A (2009) EBV and HIV-related lymphoma. Mediterr J Hematol Infect Dis 1:e2009032
16. Blake N, Lee S, Redchenko I, Thomas W, Steven N, Leese A, Steigerwald-Mullen P, Kurilla MG, Frappier L, Rickinson A (1997) Human Cd8+ T cell responses to EBV EBNA1: HLA class I presentation of the (Gly-Ala)-containing protein requires exogenous processing. Immunity 7:791–802
17. Bouchard C, Thieke K, Maier A, Saffrich R, Hanley-Hyde J, Ansorge W, Reed S, Sicinski P, Bartek J, Eilers M (1999) Direct induction of cyclin D2 by Myc contributes to cell cycle progression and sequestration of P27. EMBO J 18:5321–5333
18. Burkitt DP (1969) Etiology of Burkitt's lymphoma–an alternative hypothesis to a vectored virus. J Natl Cancer Inst 42:19–28
19. Cahir-McFarland ED, Davidson DM, Schauer SL, Duong J, Kieff E (2000) NF-κB inhibition causes spontaneous apoptosis in Epstein-Barr virus-transformed lymphoblastoid cells. Proc Natl Acad Sci U S A 97:6055–6060
20. Cai Q, Guo Y, Xiao B, Banerjee S, Saha A, Lu J, Glisovic T, Robertson ES (2011) Epstein-Barr virus nuclear antigen 3C stabilizes Gemin3 to block P53-mediated apoptosis. PLoS Pathog 7:e1002418
21. Caldwell RG, Wilson JB, Anderson SJ, Longnecker R (1998) Epstein-Barr virus LMP2A drives B cell development and survival in the absence of normal B cell receptor signals. Immunity 9:405–411

22. Campion EM, Hakimjavadi R, Loughran ST, Phelan S, Smith SM, D'Souza BN, Tierney RJ, Bell AI, Cahill PA, Walls D (2014) Repression of the proapoptotic cellular BIK/NBK gene by Epstein-Barr virus antagonizes transforming growth factor β1-induced B-cell apoptosis. J Virol 88:5001–5013
23. Carbone A, Gaidano G, Gloghini A, Pastore C, Saglio G, Tirelli U, Dalla-Favera R, Falini B (1997) BCL-6 protein expression in aids-related Non-Hodgkin's lymphomas: inverse relationship with Epstein-Barr virus-encoded latent membrane protein-1 expression. Am J Pathol 150:155–165
24. Carbone A, Gloghini A, Caruso A, De Paoli P, Dolcetti R (2017) The impact of EBV and HIV infection on the microenvironmental Niche underlying Hodgkin lymphoma pathogenesis. Int J Cancer 140:1233–1245
25. Carbone A, Gloghini A, Larocca LM, Antinori A, Falini B, Tirelli U, Dalla-Favera R, Gaidano G (1999) Human immunodeficiency virus-associated Hodgkin's disease derives from post-germinal center B cells. Blood 93:2319–2326
26. Carbone A, Vaccher E, Gloghini A, Pantanowitz L, Abayomi A, de Paoli P, Franceschi S (2014) Diagnosis and management of lymphomas and other cancers in HIV-infected patients. Nat Rev Clin Oncol 11:223–238
27. Castillo JJ, Bibas M, Miranda RN (2015) The biology and treatment of plasmablastic lymphoma. Blood 125:2323–2330
28. Cesarman E (2014) Gammaherpesviruses and lymphoproliferative disorders. Annu Rev Pathol 9:349–372
29. Cesarman E, Chang Y, Moore PS, Said JW, Knowles DM (1995) Kaposi's sarcoma-associated herpesvirus-like DNA sequences in aids-related body-cavity-based lymphomas. N Engl J Med 332:1186–1191
30. Chen J (2012) Roles of the PI3K/Akt pathway in Epstein-Barr virus-induced cancers and therapeutic implications. World J Virol 1:154–161
31. Chen YB, Rahemtullah A, Hochberg E (2007) Primary effusion lymphoma. Oncologist 12:569–576
32. Colomo L, Lopez-Guillermo A, Perales M, Rives S, Martinez A, Bosch F, Colomer D, Falini B, Montserrat E, Campo E (2003) Clinical impact of the differentiation profile assessed by immunophenotyping in patients with diffuse large B-cell lymphoma. Blood 101:78–84
33. Compagno M, Lim WK, Grunn A, Nandula SV, Brahmachary M, Shen Q, Bertoni F, Ponzoni M, Scandurra M, Califano A, Bhagat G, Chadburn A, Dalla-Favera R, Pasqualucci L (2009) Mutations of multiple genes cause deregulation of NF-κB in diffuse large B-cell lymphoma. Nature 459:717–721
34. Cotter MA, Robertson ES (2000) Modulation of histone acetyltransferase activity through interaction of Epstein-Barr nuclear antigen 3C with prothymosin alpha. Mol Cell Biol 20:5722–5735
35. Deng Z, Lezina L, Chen CJ, Shtivelband S, So W, Lieberman PM (2002) Telomeric proteins regulate episomal maintenance of Epstein-Barr virus origin of plasmid replication. Mol Cell 9:493–503
36. Dheekollu J, Lieberman PM (2011) The replisome pausing factor timeless is required for episomal maintenance of latent Epstein-Barr virus. J Virol 85:5853–5863
37. Dong HY, Scadden DT, de Leval L, Tang Z, Isaacson PG, Harris NL (2005) Plasmablastic lymphoma in HIV-positive patients: an aggressive Epstein-Barr virus-associated extramedullary plasmacytic neoplasm. Am J Surg Pathol 29:1633–1641
38. Dutton A, Woodman CB, Chukwuma MB, Last JI, Wei W, Vockerodt M, Baumforth KR, Flavell JR, Rowe M, Taylor AM, Young LS, Murray PG (2007) Bmi-1 Is induced by the Epstein-Barr virus oncogene LMP1 and regulates the expression of viral target genes in Hodgkin lymphoma cells. Blood 109:2597–2603

39. Eliopoulos AG, Waites ER, Blake SM, Davies C, Murray P, Young LS (2003) TRAF1 is a critical regulator of JNK signaling by the TRAF-binding domain of the Epstein-Barr virus-encoded latent infection membrane protein 1 but not CD40. J Virol 77:1316–1328
40. Elmore S (2007) Apoptosis: a review of programmed cell death. Toxicol Pathol 35:495–516
41. Epstein MA, Achong BG, Barr YM (1964) Virus particles in cultured lymphoblasts from Burkitt's lymphoma. Lancet 1:702–703
42. Everly DN Jr, Mainou BA, Raab-Traub N (2004) Induction of Id1 and Id3 by latent membrane protein 1 of Epstein-Barr virus and regulation of P27/Kip and cyclin-dependent kinase 2 in rodent fibroblast transformation. J Virol 78:13470–13478
43. Fathallah I, Parroche P, Gruffat H, Zannetti C, Johansson H, Yue JP, Manet E, Tommasino M, Sylla BS, Hasan UA (2010) EBV latent membrane protein 1 is a negative regulator of TLR9. J Immunol 185:6439–6447
44. Fernandez PC, Frank SR, Wang L, Schroeder M, Liu S, Greene J, Cocito A, Amati B (2003) Genomic targets of the human c-Myc protein. Genes Dev 17:1115–1129
45. Finke J, Fritzen R, Ternes P, Trivedi P, Bross KJ, Lange W, Mertelsmann R, Dolken G (1992) Expression of Bcl-2 in Burkitt's lymphoma cell lines: induction by latent Epstein-Barr virus genes. Blood 80:459–469
46. Fish K, Chen J, Longnecker R (2014) Epstein-Barr virus latent membrane protein 2A enhances Myc-driven cell cycle progression in a mouse model of B lymphoma. Blood 123:530–540
47. Flinn IW, Ambinder RF (1996) Aids primary central nervous system lymphoma. Curr Opin Oncol 8:373–376
48. Frappier L (2012) EBNA1 and host factors in Epstein-Barr virus latent DNA replication. Curr Opin Virol 2:733–739
49. Garibal J, Hollville E, Bell AI, Kelly GL, Renouf B, Kawaguchi Y, Rickinson AB, Wiels J (2007) Truncated form of the Epstein-Barr virus protein EBNA-LP protects against caspase-dependent apoptosis by inhibiting protein phosphatase 2A. J Virol 81:7598–7607
50. Geiger TR, Martin JM (2006) The Epstein-Barr virus-encoded LMP-1 oncoprotein negatively affects Tyk2 phosphorylation and interferon signaling in human B cells. J Virol 80:11638–11650
51. Gires O, Kohlhuber F, Kilger E, Baumann M, Kieser A, Kaiser C, Zeidler R, Scheffer B, Ueffing M, Hammerschmidt W (1999) Latent membrane protein 1 of Epstein-Barr virus interacts with JAK3 and activates stat proteins. EMBO J 18:3064–3073
52. Giulino L, Mathew S, Ballon G, Chadburn A, Barouk S, Antonicelli G, Leoncini L, Liu YF, Gogineni S, Tam W, Cesarman E (2011) A20 (TNFAIP3) genetic alterations in EBV-associated aids-related lymphoma. Blood 117:4852–4854
53. Goedert JJ, Cote TR, Virgo P, Scoppa SM, Kingma DW, Gail MH, Jaffe ES, Biggar RJ (1998) Spectrum of aids-associated malignant disorders. Lancet 351:1833–1839
54. Grogg KL, Miller RF, Dogan A (2007) Hiv infection and lymphoma. J Clin Pathol 60: 1365–1372
55. Grossman Z, Meier-Schellersheim M, Sousa AE, Victorino RM, Paul WE (2002) CD4+ T-cell depletion in HIV infection: are we closer to understanding the cause? Nat Med 8:319–323
56. Gruhne B, Sompallae R, Marescotti D, Kamranvar SA, Gastaldello S, Masucci MG (2009) The Epstein-Barr virus nuclear antigen-1 promotes genomic instability via induction of reactive oxygen species. Proc Natl Acad Sci U S A 106:2313–2318
57. Gruhne B, Sompallae R, Masucci MG (2009) Three Epstein-Barr virus latency proteins independently promote genomic instability by inducing DNA damage, inhibiting DNA repair and inactivating cell cycle checkpoints. Oncogene 28:3997–4008
58. Guasparri I, Bubman D, Cesarman E (2008) EBV LMP2A affects LMP1-mediated NF-κB signaling and survival of lymphoma cells by regulating TRAF2 expression. Blood 111:3813–3820

59. Halder S, Murakami M, Verma SC, Kumar P, Yi F, Robertson ES (2009) Early events associated with infection of Epstein-Barr virus infection of primary B-cells. PLoS ONE 4: e7214
60. Han I, Harada S, Weaver D, Xue Y, Lane W, Orstavik S, Skalhegg B, Kieff E (2001) EBNA-LP associates with cellular proteins including DNA-PK and HA95. J Virol 75:2475–2481
61. Hipfner DR, Cohen SM (2004) Connecting proliferation and apoptosis in development and disease. Nat Rev Mol Cell Biol 5:805–815
62. Ivers LC, Kim AY, Sax PE (2004) Predictive value of polymerase chain reaction of cerebrospinal fluid for detection of Epstein-Barr virus to establish the diagnosis of HIV-related primary central nervous system lymphoma. Clin Infect Dis 38:1629–1632
63. Jenner RG, Maillard K, Cattini N, Weiss RA, Boshoff C, Wooster R, Kellam P (2003) Kaposi's Sarcoma-associated herpesvirus-infected primary effusion lymphoma has a plasma cell gene expression profile. Proc Natl Acad Sci U S A 100:10399–10404
64. Jha HC, Mahadesh Prasad AJ, Saha A, Banerjee S, Lu J, Robertson ES (2014) Epstein-Barr virus essential antigen EBNA3C attenuates H2AX expression. J Virol 88:3776–3788
65. Jha HC, Lu J, Saha A, Cai Q, Banerjee S, Prasad MA, Robertson ES (2013) EBNA3C-mediated regulation of aurora kinase B contributes to Epstein-Barr virus-induced B-cell proliferation through modulation of the activities of the retinoblastoma protein and apoptotic caspases. J Virol 87:12121–12138
66. Jha HC, Pei Y, Robertson ES (2016) Epstein-Barr virus: diseases linked to infection and transformation. Front Microbiol 7:1602
67. Kaiser C, Laux G, Eick D, Jochner N, Bornkamm GW, Kempkes B (1999) The proto-oncogene c-Myc is a direct target gene of Epstein-Barr virus nuclear antigen 2. J Virol 73:4481–4484
68. Kanegane H, Wakiguchi H, Kanegane C, Kurashige T, Tosato G (1997) Viral interleukin-10 in chronic active Epstein-Barr virus infection. J Infect Dis 176:254–257
69. Kashuba E, Mattsson K, Pokrovskaja K, Kiss C, Protopopova M, Ehlin-Henriksson B, Klein G, Szekely L (2003) EBV-encoded EBNA-5 associates with P14ARF in extranucleolar inclusions and prolongs the survival of P14ARF-expressing cells. Int J Cancer 105:644–653
70. Kashuba E, Yurchenko M, Szirak K, Stahl J, Klein G, Szekely L (2005) Epstein-Barr virus-encoded EBNA-5 binds to Epstein-Barr virus-induced Fte1/S3a protein. Exp Cell Res 303:47–55
71. Kawaguchi Y, Nakajima K, Igarashi M, Morita T, Tanaka M, Suzuki M, Yokoyama A, Matsuda G, Kato K, Kanamori M, Hirai K (2000) Interaction of Epstein-Barr virus nuclear antigen leader protein (EBNA-LP) with HS1-associated protein X-1: implication of cytoplasmic function of EBNA-LP. J Virol 74:10104–10111
72. Kawai T, Akira S (2009) The roles of TLRs, RLRs and NLRs in pathogen recognition. Int Immunol 21:317–337
73. Kaye KM, Izumi KM, Mosialos G, Kieff E (1995) The Epstein-Barr-virus LMP1 cytoplasmic carboxy-terminus is essential for B-lymphocyte transformation—fibroblast cocultivation complements a critical function within the terminal-155 residues. J Virol 69:675–683
74. Kempkes B, Ling PD (2015) EBNA2 and its coactivator EBNA-LP. Curr Top Microbiol Immunol 391:35–59
75. Kieser A, Kaiser C, Hammerschmidt W (1999) LMP1 signal transduction differs substantially from TNF receptor 1 signaling in the molecular functions of TRADD and TRAF2. EMBO J 18:2511–2521
76. Knight JS, Lan K, Subramanian C, Robertson ES (2003) Epstein-Barr virus nuclear antigen 3C recruits histone deacetylase activity and associates with the corepressors mSin3a and NCoR in human B-cell lines. J Virol 77:4261–4272

77. Knight JS, Sharma N, Kalman DE, Robertson ES (2004) A cyclin-binding motif within the amino-terminal homology domain of EBNA3C binds cyclin a and modulates cyclin a-dependent kinase activity in Epstein-Barr virus-infected cells. J Virol 78:12857–12867
78. Knight JS, Sharma N, Robertson ES (2005) Epstein-Barr virus latent antigen 3C can mediate the degradation of the retinoblastoma protein through an SCF cellular ubiquitin ligase. Proc Natl Acad Sci U S A 102:18562–18566
79. Knight JS, Sharma N, Robertson ES (2005) SCFSkp2 complex targeted by Epstein-Barr virus essential nuclear antigen. Mol Cell Biol 25:1749–1763
80. Kohlhof H, Hampel F, Hoffmann R, Burtscher H, Weidle UH, Holzel M, Eick D, Zimber-Strobl U, Strobl LJ (2009) Notch1, Notch2, and Epstein-Barr virus-encoded nuclear antigen 2 signaling differentially affects proliferation and survival of Epstein-Barr virus-infected B cells. Blood 113:5506–5515
81. Kulwichit W, Edwards RH, Davenport EM, Baskar JF, Godfrey V, Raab-Traub N (1998) Expression of the Epstein-Barr virus latent membrane protein 1 induces B cell lymphoma in transgenic mice. Proc Natl Acad Sci U S A 95:11963–11968
82. Lam N, Sugden B (2003) CD40 and its viral mimic, LMP1: similar means to different ends. Cell Signal 15:9–16
83. Lambert SL, Martinez OM (2007) Latent membrane protein 1 of EBV activates phosphatidylinositol 3-kinase to induce production of IL-10. J Immunol 179:8225–8234
84. Laurence J, Astrin SM (1991) Human immunodeficiency virus induction of malignant transformation in human B lymphocytes. Proc Natl Acad Sci U S A 88:7635–7639
85. Laux G, Perricaudet M, Farrell PJ (1988) A spliced Epstein-Barr virus gene expressed in immortalized lymphocytes is created by circularization of the linear viral genome. EMBO J 7:769–774
86. Leonard S, Wei WB, Anderton J, Vockerodt M, Rowe M, Murray PG, Woodman CB (2011) Epigenetic and transcriptional changes which follow Epstein-Barr virus infection of germinal center B cells and their relevance to the pathogenesis of Hodgkin's lymphoma. J Virol 85:9568–9577
87. Leventaki V, Rodic V, Tripp SR, Bayerl MG, Perkins SL, Barnette P, Schiffman JD, Miles RR (2012) TP53 pathway analysis in paediatric Burkitt lymphoma reveals increased MDM4 expression as the only TP53 pathway abnormality detected in a subset of cases. Br J Haematol 158:763–771
88. Lieberman PM, Hu J, Renne R (2007) Maintenance and replication during latency. In Arvin A, Campadelli-Fiume G, Mocarski E, Moore PS, Roizman B, Whitley R, Yamanishi K (eds) Human herpesviruses: biology, therapy, and immunoprophylaxis. Cambridge University Press, Cambridge
89. Lim T, Kim SJ, Kim K, Lee JI, Lim DH, Lee DJ, Baek KK, Lee HY, Han B, Uhm JE, Ko YH, Kim WS (2011) Primary CNS lymphoma other than DLBCL: a descriptive analysis of clinical features and treatment outcomes. Ann Hematol 90:1391–1398
90. Linke-Serinsoz E, Fend F, Quintanilla-Martinez L (2017) Human immunodeficiency virus (HIV) and Epstein-Barr Virus (EBV) related lymphomas, pathology view point. Semin Diagn Pathol 34(4):352–363
91. Lu F, Chen HS, Kossenkov AV, DeWispeleare K, Won KJ, Lieberman PM (2016) EBNA2 drives formation of new chromosome binding sites and target genes for B-cell master regulatory transcription factors RBP-Jk and EBF1. PLoS Pathog 12:e1005339
92. Lu F, Wikramasinghe P, Norseen J, Tsai K, Wang P, Showe L, Davuluri RV, Lieberman PM (2010) Genome-wide analysis of host-chromosome binding sites for Epstein-Barr virus nuclear antigen 1 (EBNA1). Virol J 7:262
93. Lu JJY, Chen JY, Hsu TY, Yu WCY, Su IJ, Yang CS (1997) Cooperative interaction between Bcl-2 and Epstein-Barr virus latent membrane protein 1 in the growth transformation of human epithelial cells. J Gen Virol 78:2975–2985

94. Lu J, Murakami M, Verma SC, Cai Q, Haldar S, Kaul R, Wasik MA, Middeldorp J, Robertson ES (2011) Epstein-Barr virus nuclear antigen 1 (EBNA1) confers resistance to apoptosis in EBV-positive B-lymphoma cells through up-regulation of survivin. Virology 410:64–75
95. Ma SD, Xu X, Jones R, Delecluse HJ, Zumwalde NA, Sharma A, Gumperz JE, Kenney SC (2016) PD-1/CTLA-4 blockade inhibits Epstein-Barr virus-induced lymphoma growth in a cord blood humanized-mouse model. PLoS Pathog 12:e1005642
96. Mannick JB, Tong X, Hemnes A, Kieff E (1995) The Epstein-Barr virus nuclear antigen leader protein associates with Hsp72/Hsc73. J Virol 69:8169–8172
97. Martin KA, Lupey LN, Tempera I (2016) Epstein-Barr virus oncoprotein LMP1 mediates epigenetic changes in host gene expression through PARP1. J Virol 90:8520–8530
98. Maruo S, Wu Y, Ishikawa S, Kanda T, Iwakiri D, Takada K (2006) Epstein-Barr virus nuclear protein EBNA3C is required for cell cycle progression and growth maintenance of lymphoblastoid cells. Proc Natl Acad Sci U S A 103:19500–19505
99. Maruo S, Wu Y, Ito T, Kanda T, Kieff ED, Takada K (2009) Epstein-Barr virus nuclear protein EBNA3C residues critical for maintaining lymphoblastoid cell growth. Proc Natl Acad Sci U S A 106:4419–4424
100. Masy E, Adriaenssens E, Montpellier C, Crepieux P, Mougel A, Quatannens B, Goormachtigh G, Faumont N, Meggetto F, Auriault C, Groux H, Coll J (2002) Human monocytic cell lines transformed in vitro by Epstein-Barr virus display a type II latency and LMP-1-dependent proliferation. J Virol 76:6460–6472
101. McFadden K, Hafez AY, Kishton R, Messinger JE, Nikitin PA, Rathmell JC, Luftig MA (2016) Metabolic stress is a barrier to Epstein-Barr virus-mediated B-cell immortalization. Proc Natl Acad Sci U S A 113:E782–E790
102. McHugh D, Caduff N, Barros MHM, Ramer PC, Raykova A, Murer A, Landtwing V, Quast I, Styles CT, Spohn M, Fowotade A, Delecluse HJ, Papoudou-Bai A, Lee YM, Kim JM, Middeldorp J, Schulz TF, Cesarman E, Zbinden A, Capaul R, White RE, Allday MJ, Niedobitek G, Blackbourn DJ, Grundhoff A, Munz C (2017) Persistent KSHV infection increases EBV-associated tumor formation in vivo via enhanced EBV lytic gene expression. Cell Host Microbe 22:61–73
103. Merlo A, Turrini R, Dolcetti R, Martorelli D, Muraro E, Comoli P, Rosato A (2010) The interplay between Epstein-Barr virus and the immune system: a rationale for adoptive cell therapy of EBV-related disorders. Haematologica 95:1769–1777
104. Murakami M, Lan K, Subramanian C, Robertson ES (2005) Epstein-Barr virus nuclear antigen 1 interacts with Nm23-H1 in lymphoblastoid cell lines and inhibits its ability to suppress cell migration. J Virol 79:1559–1568
105. Murornoto R, Ikeda O, Okabe K, Togi S, Kamitani S, Fujimuro M, Harada S, Oritani K, Matsud T (2009) Epstein-Barr virus-derived EBNA2 regulates STAT3 activation. Biochem Biophys Res Commun 378:439–443
106. Nikitin PA, Yan CM, Forte E, Bocedi A, Tourigny JP, White RE, Allday MJ, Patel A, Dave SS, Kim W, Hu K, Guo J, Tainter D, Rusyn E, Luftig MA (2010) An ATM/Chk2-Mediated DNA damage-responsive signaling pathway suppresses Epstein-Barr virus transformation of primary human B cells. Cell Host Microbe 8:510–522
107. Olszewski AJ, Fallah J, Castillo JJ (2016) Human immunodeficiency virus-associated lymphomas in the antiretroviral therapy era: analysis of the national cancer data base. Cancer 122:2689–2697
108. Parker GA, Touitou R, Allday MJ (2000) Epstein-Barr virus EBNA3C can disrupt multiple cell cycle checkpoints and induce nuclear division divorced from cytokinesis. Oncogene 19:700–709
109. Paschos K, Parker GA, Watanatanasup E, White RE, Allday MJ (2012) BIM promoter directly targeted by EBNA3C in polycomb-mediated repression by EBV. Nucleic Acids Res 40:7233–7246

110. Paschos K, Smith P, Anderton E, Middeldorp JM, White RE, Allday MJ (2009) Epstein-Barr virus latency in B cells leads to epigenetic repression and CpG methylation of the tumour suppressor gene Bim. PLoS Pathog 5(6):e1000492
111. Pavlova NN, Thompson CB (2016) The emerging hallmarks of cancer metabolism. Cell Metab 23:27–47
112. Pegman PM, Smith SM, D'Souza BN, Loughran ST, Maier S, Kempkes B, Cahill PA, Simmons MJ, Gelinas C, Walls D (2006) Epstein-Barr virus nuclear antigen 2 trans-activates the cellular antiapoptotic Bfl-1 gene by a CBF1/RBPJk-dependent pathway. J Virol 80:8133–8144
113. Pei Y, Banerjee S, Jha HC, Sun Z, Robertson ES (2017) An essential EBV latent antigen 3C binds Bcl6 for targeted degradation and cell proliferation. PLoS Pathog 13:e1006500
114. Pei YG, Banerjee S, Sun ZG, Jha HC, Saha A, Robertson ES (2016) EBV nuclear antigen 3C mediates regulation of E2F6 to Inhibit E2F1 transcription and promote cell proliferation. PLoS Pathog 12(8):e1005844
115. Piriou E, van Dort K, Nanlohy NM, van Oers MH, Miedema F, van Baarle D (2005) Loss of EBNA1-specific memory Cd4+ and Cd8+ T cells in HIV-infected patients progressing to aids-related non-hodgkin lymphoma. Blood 106:3166–3174
116. Poreba E, Broniarczyk JK, Gozdzicka-Jozefiak A (2011) Epigenetic mechanisms in virus-induced tumorigenesis. Clin Epigenetics 2:233–247
117. Portis T, Longnecker R (2004) Epstein-Barr Virus (EBV) LMP2A mediates B-lymphocyte survival through constitutive activation of the Ras/PI3K/Akt pathway. Oncogene 23:8619–8628
118. Price AM, Dai J, Bazot Q, Patel L, Nikitin PA, Djavadian R, Winter PS, Salinas CA, Barry AP, Wood KC, Johannsen EC, Letai A, Allday MJ, Luftig MA (2017) Epstein-Barr virus ensures B cell survival by uniquely modulating apoptosis at early and late times after infection. Elife 6:e22509
119. Price AM, Luftig MA (2014) Dynamic Epstein-Barr virus gene expression on the path to B-Cell transformation. Adv Virus Res 88(88):279–313
120. Price AM, Tourigny JP, Forte E, Salinas RE, Dave SS, Luftig MA (2012) Analysis of Epstein-Barr virus-regulated host gene expression changes through primary B-cell outgrowth reveals delayed kinetics of latent membrane protein 1-mediated NF-κB activation. J Virol 86:11096–11106
121. Ressing ME, van Gent M, Gram AM, Hooykaas MJ, Piersma SJ, Wiertz EJ (2015) Immune evasion by Epstein-Barr virus. Curr Top Microbiol Immunol 391:355–381
122. Robertson ES, Grossman S, Johannsen E, Miller C, Lin J, Tomkinson B, Kieff E (1995) Epstein-Barr virus nuclear protein 3C modulates transcription through interaction with the sequence-specific DNA-binding protein J kappa. J Virol 69:3108–3116
123. Robertson ES, Lin J, Kieff E (1996) The amino-terminal domains of Epstein-Barr virus nuclear proteins 3A, 3B, and 3C interact with RBPJ (Kappa). J Virol 70:3068–3074
124. Roy SG, Robertson ES, Saha A (2016) Epigenetic impact on EBV associated B-cell lymphomagenesis. Biomolecules 6(4):46
125. Saha A, Halder S, Upadhyay SK, Lu J, Kumar P, Murakami M, Cai QL, Robertson ES (2011) Epstein-Barr virus nuclear antigen 3C facilitates G1-S transition by stabilizing and enhancing the function of cyclin D1. PLoS Pathog 7(2):e1001275
126. Saha A, Jha HC, Upadhyay SK, Robertson ES (2015) Epigenetic silencing of tumor suppressor genes during in vitro Epstein-Barr virus infection. Proc Natl Acad Sci U S A 112:E5199–E5207
127. Saha A, Lu J, Morizur L, Upadhyay SK, Prasad AJM, Robertson ES (2012) E2F1 mediated apoptosis induced by the DNA damage response is blocked by EBV nuclear antigen 3C in lymphoblastoid cells. PLoS Pathog 8(3):e1002573
128. Saha A, Murakami M, Kumar P, Bajaj B, Sims K, Robertson ES (2009) Epstein-Barr virus nuclear antigen 3C augments Mdm2-mediated P53 ubiquitination and degradation by deubiquitinating Mdm2. J Virol 83:4652–4669

129. Saha A, Robertson ES (2011) Epstein-Barr virus-associated B-cell lymphomas: pathogenesis and clinical outcomes. Clin Cancer Res 17:3056–3063
130. Sakai T, Taniguchi Y, Tamura K, Minoguchi S, Fukuhara T, Strobl LJ, Zimber-Strobl U, Bornkamm GW, Honjo T (1998) Functional replacement of the intracellular region of the Notch1 receptor by Epstein-Barr virus nuclear antigen 2. J Virol 72:6034–6039
131. Sample J, Liebowitz D, Kieff E (1989) Two related Epstein-Barr virus membrane proteins are encoded by separate genes. J Virol 63:933–937
132. Sanchez EL, Lagunoff M (2015) Viral activation of cellular metabolism. Virology 479–480:609–618
133. Scala G, Quinto I, Ruocco MR, Mallardo M, Ambrosino C, Squitieri B, Tassone P, Venuta S (1993) Epstein-Barr virus nuclear antigen 2 transactivates the long terminal repeat of human immunodeficiency virus type 1. J Virol 67:2853–2861
134. Schmitz R, Ceribelli M, Pittaluga S, Wright G, Staudt LM (2014) Oncogenic mechanisms in Burkitt lymphoma. Cold Spring Harb Perspect Med 4(2):a014282
135. Seo SY, Kim EO, Jang KL (2008) Epstein-Barr virus latent membrane protein 1 suppresses the growth-inhibitory effect of retinoic acid by inhibiting retinoic acid receptor-β2 expression via DNA methylation. Cancer Lett 270:66–76
136. Shah KM, Stewart SE, Wei W, Woodman CBJ, O'Neil JD, Dawson CW, Young LS (2009) The EBV-encoded latent membrane proteins, LMP2A and LMP2B, limit the actions of interferon by targeting interferon receptors for degradation. Oncogene 28:3903–3914
137. Shair KH, Bendt KM, Edwards RH, Bedford EC, Nielsen JN, Raab-Traub N (2007) EBV latent membrane protein 1 activates Akt, NFκB, and Stat3 in B cell lymphomas. PLoS Pathog 3:e166
138. Skalska L, White RE, Franz M, Ruhmann M, Allday MJ (2010) Epigenetic repression of p16 (INK4A) by latent Epstein-Barr virus requires the interaction of EBNA3A and EBNA3C with CtBP. PLoS Pathog 6:e1000951
139. Skalska L, White RE, Parker GA, Turro E, Sinclair AJ, Paschos K, Allday MJ (2013) Induction of p16(INK4A) is the major barrier to proliferation when Epstein-Barr Virus (EBV) transforms primary B cells into lymphoblastoid cell lines. PLoS Pathog 9:e1003187
140. Tempera I, Klichinsky M, Lieberman PM (2011) EBV latency types adopt alternative chromatin conformations. PLoS Pathog 7:e1002180
141. Thorley-Lawson DA (2001) Epstein-Barr virus: exploiting the immune system. Nat Rev Immunol 1:75–82
142. Thorley-Lawson DA, Gross A (2004) Persistence of the Epstein-Barr virus and the origins of associated lymphomas. N Engl J Med 350:1328–1337
143. Timms JM, Bell A, Flavell JR, Murray PG, Rickinson AB, Traverse-Glehen A, Berger F, Delecluse HJ (2003) Target cells of Epstein-Barr-Virus (EBV)-positive post-transplant lymphoproliferative disease: similarities to EBV-positive Hodgkin's lymphoma. Lancet 361:217–223
144. Tomkinson B, Robertson E, Kieff E (1993) Epstein-Barr virus nuclear proteins EBNA-3A and EBNA-3C are essential for B-lymphocyte growth transformation. J Virol 67:2014–2025
145. Tsai CN, Tsai CL, Tse KP, Chang HY, Chang YS (2002) The Epstein-Barr Virus oncogene product, latent membrane protein 1, induces the downregulation of E-cadherin gene expression via activation of DNA methyltransferases. Proc Natl Acad Sci U S A 99:10084–10089
146. Tsimbouri P, Drotar ME, Coy JL, Wilson JB (2002) Bcl-Xl and rag genes are induced and the response to Il-2 enhanced in EμEBNA-1 transgenic mouse lymphocytes. Oncogene 21:5182–5187
147. Vaysberg M, Lambert SL, Krams SM, Martinez OM (2009) Activation of the JAK/STAT pathway in Epstein Barr Virus+-associated posttransplant lymphoproliferative disease: role of interferon-gamma. Am J Transplant 9:2292–2302
148. Wheaton WW, Chandel NS (2011) Hypoxia. 2. Hypoxia regulates cellular metabolism. Am J Physiol Cell Physiol 300:C385–C393

149. Wood CD, Veenstra H, Khasnis S, Gunnell A, Webb HM, Shannon-Lowe C, Andrews S, Osborne CS, West MJ (2016) Myc activation and BCL2L11 silencing by a tumour virus through the large-scale reconfiguration of enhancer-promoter hubs. Elife 5:e18270
150. Wood VHJ, O'Neil JD, Wei W, Stewart SE, Dawson CW, Young LS (2007) Epstein-Barr virus-encoded EBNA1 regulates cellular gene transcription and modulates the STAT1 and TGFβ signaling pathways. Oncogene 26:4135–4147
151. Wu H, Ceccarelli DF, Frappier L (2000) The DNA segregation mechanism of Epstein-Barr virus nuclear antigen 1. EMBO Rep 1:140–144
152. Xiao L, Hu ZY, Dong X, Tan Z, Li W, Tang M, Chen L, Yang L, Tao Y, Jiang Y, Li J, Yi B, Li B, Fan S, You S, Deng X, Hu F, Feng L, Bode AM, Dong Z, Sun LQ, Cao Y (2014) Targeting Epstein-Barr virus oncoprotein LMP1-mediated glycolysis sensitizes nasopharyngeal carcinoma to radiation therapy. Oncogene 33:4568–4578
153. Xie Y, Pittaluga S, Jaffe ES (2015) The histological classification of diffuse large B-cell lymphomas. Semin Hematol 52:57–66
154. Yang XJ, Han H, De Carvalho DD, Lay FD, Jones PA, Liang GN (2014) Gene body methylation can alter gene expression and is a therapeutic target in cancer. Cancer Cell 26:577–590
155. Yao QY, Tierney RJ, Croom-Carter D, Dukers D, Cooper GM, Ellis CJ, Rowe M, Rickinson AB (1996) Frequency of multiple Epstein-Barr virus infections in T-cell-immunocompromised individuals. J Virol 70:4884–4894
156. Yi F, Saha A, Murakami M, Kumar P, Knight JS, Cai Q, Choudhuri T, Robertson ES (2009) Epstein-Barr virus nuclear antigen 3C targets p53 and modulates its transcriptional and apoptotic activities. Virology 388:236–247
157. Young LS, Dawson CW, Eliopoulos AG (2000) The expression and function of Epstein-Barr virus encoded latent genes. Mol Pathol 53:238–247
158. Young LS, Yap LF, Murray PG (2016) Epstein-Barr virus: more than 50 years old and still providing surprises. Nat Rev Cancer 16:789–802
159. Zhang L, Pagano JS (2000) Interferon regulatory factor 7 is induced by Epstein-Barr virus latent membrane protein 1. J Virol 74:1061–1068
160. Zhao B, Maruo S, Cooper A, Chase MR, Johannsen E, Kieff E, Cahir-McFarland E (2006) RNAs induced by Epstein-Barr virus nuclear antigen 2 in lymphoblastoid cell lines. Proc Natl Acad Sci U S A 103: 1900–1905
161. Zimber-Strobl U, Strobl LJ (2001) EBNA2 and Notch signalling in Epstein-Barr Virus mediated immortalization of B lymphocytes. Semin Cancer Biol 11:423–434

Human Papillomavirus Infection and Cervical Cancer in HIV+ Women

5

Ping Du

Contents

P. Du (✉)
Department of Medicine, Department of Public Health Sciences,
Penn State Hershey College of Medicine, 90 Hope Drive,
Suite 2200, A210, Hershey, PA, USA
e-mail: pud15@psu.edu; pingdu@phs.psu.edu

C. Meyers (ed.), *HIV/AIDS-Associated Viral Oncogenesis*, Cancer Treatment and Research 177, https://doi.org/10.1007/978-3-030-03502-0_5

Abstract

Human papillomavirus (HPV) is the first identified necessary cause of human cancers and is associated with nearly 100% of all cervical cancers. Compared to the general female populations, HIV+ women have higher prevalence and incidence of cervical HPV infections, higher risks of persistent HPV infections and subsequent cervical intraepithelial lesions, and a higher incidence of cervical cancer. Although the wide use of combined antiretroviral therapy (cART) has improved the immune function and the longevity of HIV+ women, the incidence of cervical cancer in HIV+ women has not declined. For HIV+ women who follow routine cervical cancer screenings, their incidence of cervical cancer is comparable to that in HIV-negative women. Thus, adherence to the recommended cervical cancer screening is still critical for HIV+ women to prevent cervical cancer. Prophylactic HPV vaccines may also benefit HIV+ women, but prospective studies are needed to determine the effectiveness of HPV vaccination on reducing cervical cancer incidence in HIV+ women.

5.1 Biology of Human Papillomavirus and HPV-Associated Cervical Cancer

Human papillomavirus (HPV) is a small, non-enveloped DNA virus that can infect squamous epithelium of skin and mucous membranes. HPV consists of 8000 base-pair long circular DNA. The viral genome codes six early proteins (E1, E2, and E4–E7) and two late proteins (L1 and L2). The early proteins E1 and E2 are essential for viral replication within the infected cells, and E6 and E7 play critical roles in HPV-related carcinogenesis [1–5]. The late proteins L1 and L2 are major structural proteins that form the viral capsid. L1 also contains type-specific neutralization epitopes that can induce host humoral immune responses against HPV infection. There are more than 150 types of HPV, classified based on differences in L1 genome sequence, and over 40 HPV types infect the human anogenital tract [6]. HPV life cycle includes the following major steps (1) infection of stem cells at the basal layer of the epithelium through microabrasion, (2) maintenance of infection with low viral replication activities at the basal layer, (3) increased viral replication through E6 and E7 viral proteins when basal cells are pushed to the superficial layers, (4) the interaction of E6 and E7 with host cellular proteins (the binding of E6 to p53 and/or E7 to pRB) that induces cell proliferation, reduces DNA damage repair, and inhibits apoptosis of the infected cells, (5) viral genomes integration into the host cells and persistent activities of E6 and E7 that lead to the loss of cell-growth control, genomic instability, and eventually malignant transformation of host cells [3, 4].

HPV is the first identified necessary cause of human cancers by the International Agency for Research on Cancer (IARC) that is associated with nearly 5% of all cancers [7, 8]. As HPV DNA was detected in nearly all cervical cancer cases, a strong association between the presence of HPV DNA and cervical cancer has been

Table 5.1 Global burden of HPV-associated cancers and the prevalence of HPV DNA positivity in the tumors (3, 14)

Cancer site	Annual number of new cases	HPV DNA positivity	Common HPV types detected in the tumors
Cervix	530,000	99.7%	16,18, 45, 31, 33, 52, 58, 35
Anus	40,000	88–94%	16, 18, 31, 33
Vulva	34,000	60–90%	16, 18, 31, 33
Penis	26,000	60–90%	16, 18, 31, 33
Vagina	15,000	64–91%	16, 18, 31, 33
Oropharynx	96,000	35.6%	16, 18
Oral cavity	200,000	23.5%	16, 18
Larynx	160,000	24%	16, 18

reported from multiple case-control studies across different countries (overall adjusted odds ratio [aOR] = 90 for squamous carcinoma and aOR = 81 for adeno- and adenosquamous carcinoma) [3]. Bosch et al. summarized the causal relationship between HPV and cervical cancer based on basic science mechanisms and consistent findings from human research, and Moscicki et al. further provided updated findings [2, 9]. According to the association (aOR) between the presence of type-specific HPV DNA and cancer, 12 HPV types (16, 18, 31, 33, 35, 39, 45, 51, 52, 56, 58, 59) are classified as high-risk (or oncogenic) HPV types (HR-HPV) that are associated with over half a million new cancer cases worldwide, including cervical, anogenital, and oropharyngeal cancers (Table 5.1); three types (5, 8, and 68) are considered as probable or possible HR-HPV that are associated with cervical cancer or skin cancers, 12 types (26, 30, 34, 53, 66, 67, 69, 70, 73, 82, 85, and 97) might have carcinogenic effects but their roles remain unclear, and 12 types (6, 11, 40, 42, 43, 44, 54, 61, 72, 81, and 89) are classified as low-risk HPV (LR-HPV) types that have not been found to be related to malignancies [1–3, 5, 8, 10–13].

While the majority of HPV infections are asymptomatic and self-limited, HPV infection can lead to a wide range of genital diseases, including genital warts, benign lesions, and invasive cancers. The natural history of cervical HPV infection has been well studied and comprehensive reviews are readily available [3, 5, 9, 15]. HPV-related cervical carcinogenesis begins with HR-HPV infection of the cervical epithelium. The majority of infected women (90%) can clear HPV infections within a few years; however, a small proportion of women with persistent HPV infections will develop cervical epithelial neoplasia (CIN). CIN includes three grades depending on the degree of histological abnormalities: CIN1 involves mild dysplasia or abnormal cell growth that is confined to one-thirds of the basal epithelium, and CIN2 or 3 represents moderate or severe dysplasia that spreads to two-thirds or more of the cervical epithelium. CIN3 sometimes is also referred as cervical carcinoma in site and is commonly used in human studies as the disease endpoint of cervical HPV infection. In cytology, CIN1 corresponds to the low-grade squamous intraepithelial lesion (LSIL) and CIN2/3 relates to the high-grade squamous intraepithelial lesion (HSIL). Most CIN1 lesions ($\sim$90%) regress but persistent

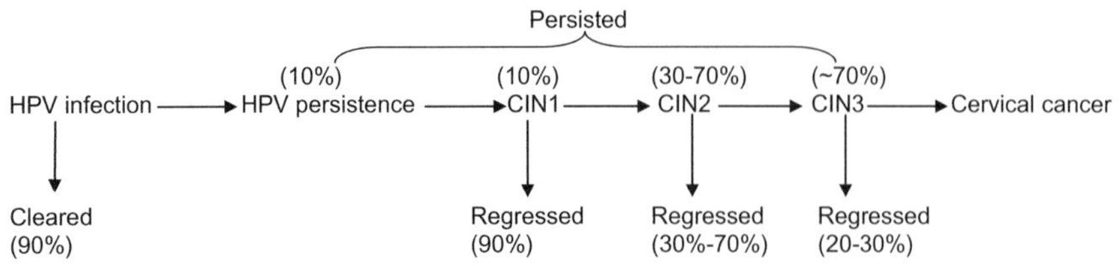

Fig. 5.1 The natural history of cervical HPV infection

CIN1 can lead to CIN2/3 lesions. CIN2/3 lesions are less likely to regress (30–70% for CIN2 and 20–30% for CIN3) compared with CIN1 and nearly 50% of persistent CIN3 progress to cervical cancer [9]. Because the time frame from CIN2/3 to cervical cancer is usually decades-long, treatment of CIN2/3 lesions is recommended for preventing cervical cancer [16] (Fig. 5.1).

In addition to viral factors, host risk factors for cervical cancer include parity, tobacco smoking, oral contraceptive (OC) use, and co-infection with other sexually transmitted diseases [3, 5]. The association between these risk factors and cervical cancer, measured by relative risk (RR) or odds ratio (OR), ranges from 1.1 to 3.4 in the European Prospective Investigation into Cancer and Nutrition (EPIC) study, which is one of the largest cohort studies in the world with more than 300,000 women recruited from 10 European countries and followed for almost 10 years [17–19]. However, without the presence of HPV, cervical cancer would not develop. Thus, these risk factors mainly serve as cervical cancer co-factors that may promote cancer development in conjunction with HPV infection. The possible mechanisms of these cancer co-factors include maintenance of the cervical transformation zone that facilitate HPV infection and persistence (high parity), tobacco-related carcinogens that directly result in genetic damages (smoking), enhancement of HPV oncogene expression (long-term OC use), inflammatory responses that cause genetic instability (chlamydia or HSV-2 infection), and immunosuppression (HIV infection) [3, 20]. Genetic research also suggests that APOBEC-mediated mutagenesis may be associated with cervical cancer development, but the mechanisms need to be further investigated [21].

Worldwide cervical cancer is the fourth most common cancer in women, accounting for almost 8% of all female cancer cases and 7.5% of all female cancer deaths [14, 22]. Recently the annual number of new cervical cancer cases has gradually increased, possibly due to population growth. In 2012 there were approximately 530,000 new cases and 266,000 cervical deaths from cervical cancer [22]. The majority of cervical cancer cases ($\sim 86\%$) and cervical cancer-related deaths occur in less developed regions, including Eastern, Southern, and Middle Africa, where the burden of HIV infection is also high [23]. In the U.S. about 34,000 HPV-associated cancers are diagnosed annually and the direct medical costs for preventing and treating HPV-related diseases are estimated to be $8 billion every year [24, 25].

5.2 Epidemiology of HPV Infection in the General Female Population

HPV infection is the most common sexually transmitted infection with over 300 million infected women globally [13, 26–28]. Risk factors for acquiring HPV infection include early sexual debut ($\leq$ 15 years of age), parity, multiple sexual partners, use of contraceptives, and smoking [29]. In the pre-HPV vaccine era, worldwide about 10% of women with normal cervical cytology has an HPV infection [13, 30]. HPV prevalence also varies by age, with the highest prevalence seen in women younger than 25 years of age. However, a second-peak of HPV prevalence is observed in women aged 45 years or older [30]. While the true cumulative incidence of cervical HPV infection in the general population is difficult to assess because HPV infection is not a reportable disease, it is estimated that up to 75% of sexually active women will get an HPV infection during their life time [26, 31]. Longitudinal studies assessing the incidence of cervical HPV infection among women have reported a 20% or higher one-year cumulative incidence, depending on the age range of the study populations and types of HPV infections examined [32–38].

In the U.S. nearly 80 million people are infected with HPV and about 14 million people acquire new HPV infections each year [28]. Based on the National Health and Nutrition Examination Survey (NHANES) data, the overall prevalence of genital HPV infection among U.S. females aged 14–59 years was 42.5% in 2003–2006 and the prevalence highly varied by age, with the highest prevalence (53.8%) among 20–24-year-old females and the lowest prevalence among 14–19-year-old females (32.9%) [39]. Minority women, including non-Hispanic blacks and Mexican Americans, had the higher prevalence of HPV infection (59.2 and 44.2%, respectively) than non-Hispanic white women (39.2%).

5.3 HPV Infection in HIV+ Women

Compared with HIV-negative women or the general female population, HIV + women have higher prevalence and incidence of cervical HPV infection, higher risks of persistent HPV infection and subsequent cervical intraepithelial lesions, and a higher incidence of cervical cancer [40–44]. Large national or international multisite longitudinal cohort studies have been conducted to better understand the long-term health outcomes, including HPV-associated diseases, in HIV+ people. Several cohort studies are noteworthy: The North American AIDS Cohort Collaboration on Research and Design (NA-ACCORD), which was established in 2006, includes >130,000 HIV+ people from 25 large contributing cohorts throughout the United States and Canada [45]. Four studies in the NA-ACCORD, The AIDS Link to the IntraVenous Experience (ALIVE), Multicenter AIDS Cohort Study (MACS, males only), Polaris HIV Seroconversion Study (Polaris), Veterans Aging Cohort Study (VACS), and Women's Interagency HIV Study (WIHS), have recruited both HIV+ and HIV-negative people. Another similar multisite prospective cohort study in the U.S. is The HIV Epidemiology Research Study (HERS), which also includes both HIV+ women and risk-matched HIV-negative women to evaluate HIV-related diseases [46]. In Europe, The SWISS HIV Cohort study (SHCS), The Project for Electronic Clinical-Epidemiologic Follow-up of HIV-1 Infection and AIDS (PISCIS cohort), The French Hospital Database on HIV (FHDH-ANRS CO4), and the Study on HIV, cervical Abnormalities and infections in women in Denmark (SHADE) are long-term cohort studies that include HPV-related health outcomes in people living with HIV [41, 47–49]. The Management of Abnormal Cytology in HIV-1 Infected Women (MACH-1) study is an international collaboration between six European hospital centers and one community center in Cape Town in South African [50]. There are also numerous single-center prospective cohorts to investigate the natural history of HPV infection in HIV+ people.

5.3.1 Prevalence of HPV Infection and HR-HPV Types in HIV + Women

High prevalence of cervical HPV infection in HIV+ women have been observed, but the prevalence varies across studies because of differences in study periods, geographic areas, or study populations and HPV detection methods [51–64]. An earlier meta-analysis published by Clifford et al. in 2006 reported that in 3230 HIV + women with normal cervical cytology, the overall prevalence of cervical HPV infection was 36.3% for any HPV type and was 11.9% for multiple HPV infection (infection with >=2 types) [65]. However, the prevalence sharply increased in HIV + women with abnormal cervical cytology. Geographic variation in HPV prevalence was also observed: the prevalence was highest in South/Central America (57.3%), followed by Africa (56.6%), and was lowest in North America (31.4%) [65]. Interestingly, there were noticeable differences in HPV-type distributions across regions: the prevalence of HPV31 and HPV35 were significantly higher in Africa, HPV39 was more common in Asia, and HPV16 and HPV68 were more prevalence in South/Central America. Currently most studies utilize sensitive assays, such as polymerase chain reaction (PCR) or hybrid capture II, to detect the presence of over 30 HPV types. In a recent meta-analysis, Park et al. examined five large, population-based studies, mainly from the U.S. and Western Europe, published between 2011 and 2013 to assess the prevalence of cervical HPV infection in HIV + women [66]. The results indicated that the summary prevalence of any type(s) of cervical HPV infection in HIV+ women (64%, 95% confidence interval: 25–95%) was 20% higher than the prevalence in U.S. females aged 14–59 years (43%). HIV + women were also more likely to have HR-HPV infection (summary prevalence of HR-HPV infection: 46%, 95% CI: 34–58%) than the U.S. females (29%). Various risk factors associated with cervical HPV infection in HIV+ women have been reported, but the most consistent risk factors for both any type HPV infection and HR-HPV infection are low CD4 count (<350), younger age (<30 years), non-white race, smoking, and high HIV RNA viral load [49, 52, 67–71].

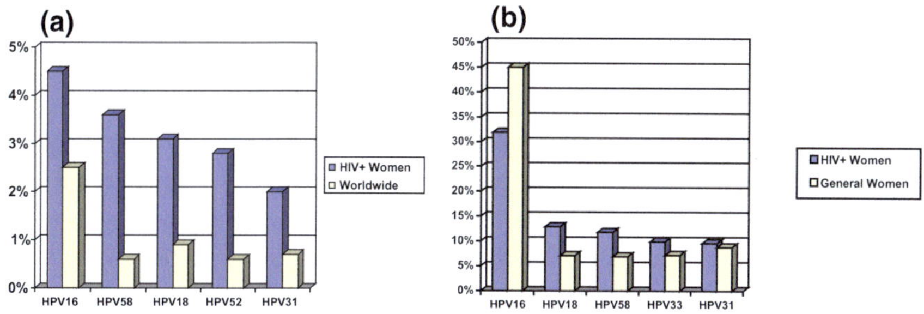

Fig. 5.2 **a** Comparison of prevalence of top five HPV types in HIV+ women with normal cervical cytology and in the general female population with normal cervical cytology worldwide. (30, 65) **b** Comparison of prevalence of top five HPV types in HIV + women with high-grade intraepithelial lesions (HSIL) and in the general female population with HSIL. (65)

While the overall type-specific distribution of cervical HPV infection in HIV + women with normal cervical cytology are similar to the distribution in the general population (Fig. 5.2a), [65, 72] HIV+ women are more likely to be infected with non-16 or non-18 HR-HPV (such as 51, 52, 53, 56, and 58), [73, 74] and the prevalence of infection with multiple HPV types is also higher in HIV+ women (14–78%) than in HIV-negative women (7–26%) [49, 52, 75–78]. However, HPV16 prevalence increases with the severity of cervical lesions: in HIV+ women with HSIL (Fig. 5.2b) or cervical cancer, HPV16 was the most predominant type (31.9%), followed by HPV18 (12.9%) and HPV58 (11.8%), suggesting that HPV16, 18, and 58 are likely to persist over time [65].

5.3.2 Incidence and Persistence of HPV Infection in HIV + Women

Prospective cohort studies demonstrate a high cumulative incidence (new detection) of cervical HPV infection in HIV+ women. Sun et al. initially reported the incidence rate of any cervical HPV infection was 11 per 100 person-visits in 220 HIV+ women [49]. Later Ahdieh et al. provided type-specific incidence among 862 HIV+ women with median 2.5 years of follow-up, indicating a higher incidence rate of LR-HPV infection (any type: 19.8 per 100 person-years [PY]) than the rate of HR-HPV infection (any type: 8 per 100 PY) [79]. The incidence rates for HPV16 and 18 were similar (2.0 per 100 PY). In the WIHS with up to 9 years of follow-up, the cumulative incidence of any HPV infection was 68% at one year in 2543 HIV+ women [80]. The incidence continued increasing over time and at 8 years 92% of HIV + women had experienced a new HPV infection. The trend of cumulative incidence of HR-HPV infection was similar: the incidence was 40% at one year, but rose to 67% at 8 years. HIV+ women, even those with CD4 count >500 cells/μL, had an approximately twofold risk of acquiring new HPV infection than HIV-negative women. [56, 68, 80, 81] Risk factors for incident HPV infection in HIV+ women include pre-existing HPV infection, younger age (<30 years), inconsistent condom use, multiple recent sexual partners, smoking, low CD4 count (<200 cells/μL), or detectable HIV RNA viral load. [56, 68, 79, 80, 82, 83].

As there is no "gold standard" definition of HPV persistence, a persistent HPV infection has been defined as continuous detection of the same type of HPV DNA for longer than a certain time period (6 months or 12 months) or repeated HPV DNA positivity at two or more clinic visits, which are at approximately 4–6-month intervals, during the study period [49, 59, 78, 79, 83–88]. The persistent rate of cervical HPV infection may not be comparable across studies, but a higher HPV persistence rate was observed in HIV+ women. In one of the WIHS cohort studies conducted by Sun et al. involving 220 HIV+ women and 231 HIV-negative women in New York City, HIV + women were significantly more likely to have persistent cervical HPV infections at follow-up visits than HIV-negative women (24.1 vs. 3.9%), and HPV16-associated types (16, 31, 33, 35, or 58) were the most common persistent types in both groups [49]. Similar findings were reported from a large, short-term (approximate 6-month follow-up) cohort study conducted in Nigeria that 24% of 321 HIV+ women had persistent cervical HPV infection, while only 10% of 309 HIV-negative women had persistent cervical HPV infection [78]. However, another short-term (6 months) natural history study using a convenience sample from the WIHS found a higher cervical HPV persistence rate in both HIV+ women (65%) and HIV-negative women (32%) [85]. Because of different definitions for persistent HPV infection, multiple risk factors for persistent cervical HPV infection in HIV+ women have been reported, but only low CD4 count (<200 cells/μL) or high HIV RNA viral load was consistently associated with HPV persistence [49, 79, 85].

A few studies have reported HPV DNA viral load, assessed by the real-time PCR assays and expressed as the number of HPV copies/μg of cellular DNA, is associated with HPV persistence in HIV+ women, but the sample size of HIV+ women with HPV DNA viral load data is generally small. In the Canadian Women's HIV Study, the persistence of HPV16 infection was significantly related to high HPV16 viral loads: in 20 HIV+ women with HPV16 viral loads $\geq 10^7$ copies/μg cellular DNA, the average duration of HPV16 infection was 21.3 months, while the duration was only 13.5 months in 32 HIV+ women with HPV16 viral load $<10^7$ copies/μg cellular DNA (p-value = 0.01). Among 15 HIV+ women who had persistent HPV16 infection and developed cervical lesions during the study period, HPV16 viral load also increased with the severity of the lesions [87].

5.4 Cervical Lesions and Cervical Cancer in HIV+ Women

5.4.1 Prevalence, Incidence and Progression of Cervical Lesions in HIV+ Women

Abnormal cervical cytologies, mostly atypical squamous cells with undetermined significance (AS-CUS), are commonly detected (up to 50%) in HIV+ women, but the prevalence of high-grade cervical lesion is low (<10%) [51, 52, 54–56, 61, 63, 89, 90]. Compared with HIV-negative women, HIV+ women have an approximately threefold risk of having an abnormal cervical cytology [91]. In the WIHS with 1661 HIV + women and 462 risk-matched HIV-negative women, 16.2% of HIV+ women had LSIL or worse (LSIL+), while in HIV-negative women the prevalence of LSIL+ was only 4% [92]. The presence of cervical lesion in HIV+ women is consistently associated with HR-HPV, low CD4 counts and high HIV viral loads, indicating the role of immunosuppression on the progression of HPV infection [91, 93, 94].

HIV+ women are also more likely to develop new cervical lesions or progress to high-grade lesions. In a recent systematic review based on the data from 15 longitudinal cohort studies involving 5882 HIV+ women with normal cytology at the baseline, the incidence rates ranged from 4.9 per 100 person-years (PY) in France to 21.1 per 100 PY in Thailand for any cervical lesion, and 0.4 per 100 PY in the U.S. to 8.8 per 100 PY in Italy for high-grade cervical lesions (HSIL or CIN2/3), but the incidence rates of any cervical lesion in HIV-negative women were only 1.1 per 100 PY to 4.7 per 100 PY [43]. The cumulative incidence of cervical lesion was difficult to estimate from these studies because of different follow-up periods (from <one year to 8 years), but up to 44% of HIV+ women developed cervical lesions within five years after the baseline visit. While most of the studies included in this systematic review recruited mainly HIV+ women younger than 45 years of age, progression from low-grade (LSIL or CIN1) to high-grade lesions (HSIL or CIN2/3) also occurred at rates ranging from 1.2 to 26.2 cases per 100 PY. Compared with risk-matched HIV-negative women, HIV+ women had an average of threefold risk (RR range: 1.5–10) of developing any cervical lesions. HIV+ women also had an increased risk of progression to high-grade cervical lesions, but because the progression rates in HIV-negative women were limited and highly varied across studies (from 0 to 5.7 per 100 PY), the RR of progression of cervical lesions between HIV+ women and HIV-negative women was not calculated to avoid any unstable estimation. There were no consistent findings on the effects of CD4 count or combined antiretroviral therapy (cART) on cervical lesion development, likely because CD4 count measurement was different across studies and the cART use was not well assessed. In seven studies with available information on CD4 count (either baseline CD4 count or nadir CD4 count), while there was a tendency that low CD4 count (<200 cells/μL) was associated with both incidence and progression of cervical lesions, only two studies reported significant, moderate associations [43]. The role of cART use was contradictory, as the association between cART and the incidence of cervical lesions ranged from 0.3 to 1.8 in five studies. But in the WHIS assessing the effects of cART adherence, HIV+ women who used cART as prescribed >=95% of the time had nearly 50% lower risk of having prevalent and incident cervical HPV infection and also had more rapid clearance of HR-HPV-positive cervical lesions compared with non-adherent HIV+ women [95]. Another nested case-control study from the SWISS HIV Cohort Study also showed a nearly 40% protective effect of long-term cART use (>2 years) against CIN2/3 [96]. These studies demonstrate the need to develop more thorough measurement of CD4 count and cART to better evaluate the natural history of HPV infection in HIV

+ women in the cART era. In addition to HIV-related factors, infection with HR-HPV types, especially HPV16 or 18, is also associated with two-four-fold risk of progression to high-grade cervical lesions [81, 82]. In several small studies with HPV DNA viral loads data, high HPV viral loads may also be related to the development of high-grade lesions [87, 97].

5.4.2 Incidence and Risk Factors of Cervical Cancer in HIV + Women

In early epidemic of HIV/AIDS, HIV+ women had high morbidity and mortality of cervical cancer, thus, in 1993 cervical cancer was included as one of AIDS-defining clinical conditions by the U.S. Centers For Disease Control (CDC) [98]. Prior to the cART era, nearly one in five of HIV-HPV co-infected women without evidence of cervical disease developed precancerous cervical lesions within three years of HIV diagnosis [40]. Worldwide HIV+ women have a greater risk of developing cervical cancer compared with the general population with a standardized incidence ratio (SIR) ranging from 2–40 [42]. However with the wide use of cART, the incidence of cervical cancer in HIV+ women has not declined [99–104]. Findings from a large international collaborative study even showed a possible increase of cervical cancer incidence between 1992–1996 (pre-cART) and 1997–1999 (post-cART) in HIV+ women (rate ratio = 1.87; 99% CI = 0.77–4.56; P-value = 0.07) [105]. The SIR of cervical cancer in HIV+ women remains high (3.3–5.0) and the number of cervical cancer cases may actually increase because of prolonged life span [106–111]. This excess risk of cervical cancer in HIV+ women in the cART era is likely due to the combinations of immunodeficiency, cumulative effect of genetic instability from long-term HR-HPV infection, and the high prevalence of cervical cancer co-factors (e.g., smoking) [41, 44, 104, 108–110, 112–123].

Due to the lack of preclinical models of HIV-HPV co-infection, research on HPV-related cancer risk in HIV+ people is generally based on large longitudinal cohort studies or through the linkage of HIV/AIDS case reports with the cancer registry data [40, 41, 44, 48, 108, 112, 116, 117, 124, 125]. Recent data from the WIHS showed that since the study population was established in 1994-5, only four cervical cancer cases had occurred in 1807 HIV+ women with a median 12.3 years of follow-up (incidence rate = 19.5 per 100,000 PY) [111]. Although no cervical cancer was observed in 488 HIV-negative women, because of the low number of cervical cancer cases, there was no significant difference in cervical cancer incidence rate between HIV+ and HIV-negative women (p-value = 0.53). However most women included in the WIHS had regular Papanicolaou (Pap) tests every 6 months, the results from the WIHS may not reflect the real-world experience of cervical cancer incidence in HIV+ women. Using the U.S. AIDS–Cancer Match Registry Study data, Frisch et al. reported a much higher incidence of cervical cancer during 1978–1996 in AIDS women (85.7 per 100,000 PY), and Chaturvedi et al. later found a similar incidence (90.4 per 100,000 PY) during 1996–2004 [108].

Although cervical HPV infection is highly prevalent, only a subset of HIV + women has persistent HPV infection and eventually develops cervical cancer [51, 52, 54, 56, 61, 63, 89]. Numerous studies have reported that severe immunodeficiency, represented by low CD4 count (<200 cells/μL) and AIDS diagnosis, are associated with the development of precancerous cervical lesions (such as CIN2/3) or cervical cancer [96, 104, 108, 112, 113, 118, 121, 125, 126]. In a large prospective cohort study (>13,000 women) using data from the NA-ACCORD, HIV+ women with baseline CD4 counts>200 cells/μL have a two-three-time the incidence of cervical cancer compared with HIV-women, but for HIV+ women with CD4 count <200 cells/μL, their cervical cancer incidence was nearly eight times the incidence in HIV-negative women [125]. While the associations between low CD4 count or cART use and cervical cancer risk have been extensively examined, the beneficial effects of high CD4 count and cART on cervical cancer remain unclear [42, 48, 108]. The inconsistent findings across studies could be due to the different measurement of CD4 count (nadir CD4 count, baseline CD4 count, or CD4 count prior to the cancer diagnosis) or the lack of detailed information on cART use (the timing and the length of cART use or the actual cART regimen). Because many HIV+ people are living healthier with the cART use, they are less likely to have CD4 count <200 cells/μL and may never progress to AIDS. Yet as there is no decrease in the incidence of cervical cancer in the cART era, it appears that the low CD4 count may not be the sole predictor of cervical cancer risk.

As a cervical cancer co-factor, tobacco-smoking is highly prevalent in HIV + people (54% versus 20–23% in US adults) and HIV+ smokers are less likely to quit smoking than the general population [66, 127]. Various studies from different countries have reported that smokers have a higher prevalence and incidence of cervical HPV infection, higher HPV DNA load, and almost two-fold risk of CIN2/3 or cervical cancer compared with non-smokers. [34, 128–131] Preclinical studies also suggest that tobacco-smoking not only enhances human papillomavirus synthesis, but also likely results in increased exposure of the cervical epithelium to potentially mutagenic metabolites of tobacco carcinogen [132, 133]. Thus, the increased risk of cervical cancer in HIV+ women could be due to the synergistic effects between tobacco carcinogens (such as benzo[*a*]pyrene) and HPV infection, regardless of host's immune status [20, 132].

5.4.3 The Interaction Between HPV and HIV

As HIV primarily targets the CD4 T-lymphocyte and HPV infects epithelial cells, a direct interaction between HIV and HPV is unlikely, but basic science discoveries suggest that HIV could indirectly affect HPV life cycle by upregulating the transcription of HPV early genes or enhancing the expression of viral oncoproteins E6 and E7 [134, 135]. The role of HIV in the natural history of HPV infection is considered through HIV-related immunodeficiency, increased susceptibility of HPV infection, or possibly reactivation of latent HPV infection [115, 136, 137]. The strong associations between low CD4 cell count, HPV persistence, and cervical cancer observed in human research support that immunosuppression induced by HIV infection reduces cell-mediated immunity and facilitates HPV pathogenesis [42, 137, 138]. Additionally high cervical HIV RNA viral load (> 545 copies/mL) was found to be associated with persistent infection with HR-HPV (OR = 2.84, p-value = 0.04) in HIV+ women, providing evidence that HIV infection may influence the natural history of HPV infection [139].

HPV infection also increases the risk of HIV infection in both men and women, likely due to the similar transmission route through sexual contacts, a possible local inflammatory response that recruits immune cells susceptible to HIV infection, or HPV-induced epithelial ruptures that promote HIV entry [138, 140–143]. Results from a systematic review and meta-analysis conducted by Lissouba et al. in 2013 indicated that individuals with any type(s) HPV infection had nearly two times the risk of acquiring HIV infection (summary OR = 1.96; 95% CI, 1.55–2.49) [143]. Additionally, the association was statistically significant between HR-HPV infection and HIV acquisition (summary OR, 1.92; 95% CI, 1.49–2.46), but was borderline with LR-HPV (summary OR, 1.53; 95% CI, 0.96–2.42). The observed strong association between HPV infection and HIV acquisition could have great public health implications for utilizing HPV vaccination as a tool for HIV prevention.

5.5 Prevention of HPV Infection and Cervical Cancer in HIV+ Women

5.5.1 The Role of CART

Current HIV treatment guidelines recommend that all HIV+ people be offered cART after HIV diagnosis to reconstitute the immune system and increase CD4 cell counts [144]. As HIV+ women with low CD4 counts have a greater risk of developing cervical cancer, initiation of cART at early stage of HIV infection may help prevent cervical cancer in HIV+ women. However, the effect of cART on the natural history of cervical HPV infection is controversial. Although the wide use of cART has dramatically improved the immune function and the longevity of HIV-infected (HIV+) people, there is no clear benefit of cART in reducing cervical cancer incidence in HIV+ women [103, 106, 145–148]. A recent study conducted by Rohner et al. in South Africa showed that without the implementation of cervical cancer screening program and access to treatment of cervical precancerous lesions, the incidence of cervical cancer remained high (>500/100,000 person-years) in HIV + women who initiated cART [149]. Thus, adherence to recommended cervical cancer screening is still critical for HIV+ women.

5.5.2 HPV Vaccination

Three prophylactic HPV vaccines (bivalent 16/18 Cervarix®, quadrivalent 16/18/6/11 Gardasil®, and nonavalent 16/18/31/33/45/52/58/6/11 Gardasil®9), which are recombinant vaccines based on the L1 virus-like particle technology, have been approved in many countries and have also been included in the national immunization program in over 50 countries [150]. Clinical trials data show a high efficacy of HPV vaccines in protection against cervical HPV infection with vaccine-covered types (>90%) and CIN1 or worse (>60%) [151–153]. The real-world observational studies also provide evidence of HPV vaccine effectiveness on reductions of vaccine types of HPV infection (~90% reduction) and cervical lesions (~45% reduction for low-grade lesions and ~85% reduction for high-grade lesions) in the general population [154–156].

Currently there are no data on the efficacy of HPV vaccination or the safety and immunogenicity of nonavalent vaccine in HIV+ people. The safety and immunogenicity of the bivalent and quadrivalent vaccines have been evaluated in HIV + pre-adolescent girls and boys (aged 7–12) and women up to age 45 years [157–161]. These studies have demonstrated that HPV vaccines are safe and immunogenic; the seroconversion rate is greater than 90% for HPV6, 11, and 16 and is also over 75% for HPV18, even in HIV+ women with CD4 count <200 cells/μl [162]. Therefore HPV vaccines are advocated as an effective cancer prevention strategy for HIV + people [162–166]. However, HPV vaccines are recommended only for adolescents and young adults up to age 26 years, regardless of HIV status, and the greatest vaccine effectiveness on reducing cervical abnormalities is observed in younger age groups [167]. It is not clear if HPV vaccination would benefit HIV+ women and result in a reduction in cervical cancer as many HIV+ women are beyond the recommended age range, may have already been infected with HPV, or are more likely to be infected with non-vaccine-covered HR-HPV types [162, 168]. Additionally, although the protective titers of HPV antibodies have not been established, the geometric mean titers (GMT) of HPV antibodies in HIV+ women was only half that of the general women aged 24–45 years and the GMT is even lower in HIV+ women with low CD4 count [162]. Long-term prospective studies are needed to determine the HPV vaccine effectiveness on reducing cervical lesions in HIV+ women.

5.5.3 Cervical Cancer Screening in HIV+ Women

Cervical cancer screening guidelines have been well developed for the general population by the U.S. Preventive Services Task Force (U.S.PSTF), the American Cancer Society (ACS), the American Society for Colposcopy and Cervical Pathology (ASCCP), and the American Society for Clinical Pathology (ASCP) [169, 170]. In 2015, ASCCP and the Society of Gynecologic Oncology (SGO) issued an interim guidance for the use of a human papillomavirus (HPV) test as the primary screening for cervical cancer, [171] and in 2017, the U.S.PSTF also recommended FDA-approved HR-HPV testing every 5 years for women aged 30–65 years as one of the primary cancer screening methods [172]. However, these guidelines do not address the increased risk of cervical cancer in HIV+ women, even those with normal Pap test results [173]. The U.S. CDC, the U.S. National Institutes of Health and the HIV Medicine Association of the Infectious Diseases

Society of America recommend a shorter time intervals for re-screenings for HIV + women [163]. Table 5.2 summarizes the differences between the U.S. PSTF and the U.S. CDC in cervical cancer screening guidelines. To evaluate which cervical cancer screening recommendation is appropriate HIV+ women, Robbins et al.

Table 5.2 Comparison of the U.S. PSTF and the U.S. CDC cervical cancer screening guidelines for general population and for HIV+ women (163, 172)

Population	U.S. PSTF for general population		U.S. CDC for HIV+ women	
	Pap testing	HPV testing	Pap testing	HPV testing
Women <21 years	Not recommended	Not recommended	Within 1 year of the sexual debut and no later than 21 years old	Not recommended
Women aged 21–29 years	Pap test alone every 3 years	Not recommended	At HIV diagnosis; Pap test every year; after 3 consecutive normal test, Pap test every 3 years	Not recommended
Women aged 30–65 years	If Pap test alone, every 3 yeas	If HR-HPV test alone, every 5 years	At HIV diagnosis; Pap test every year; after 3 consecutive normal test, Pap test every 3 years	Pap-HPV co-testing at HIV diagnosis or age 30: if co-test negative, screen every 3 years; If Pap test normal but HR-HPV+ , repeat co-testing in 1 year
Women >65 years	Not recommended if had adequate screenings or not at high risk	Not specified	Continue screening as recommended for women >=30 years	Continue screening as recommended for women >=30 years
Women who have had a hysterectomy	Not recommended for women with removal of the cervix and not having a history of CIN2+ or cervical cancer	Not specified	Not specified	Not specified
Women with abnormal Pap test or HR-HPV results	Not specified		Pap-HPV co-testing: if either one at one-year repeat test abnormal, refer to colposcopy; ASC-US: if HR-HPV+ , refer to colposcopy; if HR-HPV is unknown, rescreen in 6–12 months; LSIL or worse: refer to colposcopy	

compared the risks of precancerous lesions between US HIV+ women and the general population based on different screening intervals and management strategies [174]. The study results supported the CDC's recommendation for the shorter screening intervals as HIV+ women, especially those with CD4 counts <500 cells/μL, would have higher risks of developing CIN2 or 3 if they followed the same intervals that are recommended for the general population. Therefore, tailored cervical cancer screening strategies should be applied to HIV+ women. In addition, the U.S. CDC recommends more intensive clinical management of abnormal Pap test results or positive HPV co-testing results in HIV+ women as progression risk is high and recurrence of cervical lesions after treatment is also common [163, 175].

In the U.S. since the Pap test was implemented in the 1950s, the incidence of cervical cancer in the US has decreased by 50% in the past 40 years: in 1975 the incidence was 14.8 per 100,000 women; and it decreased to 7.4 per 100,000 women in 2014. A similar reduction is also observed in the mortality of cervical cancer (5.6 deaths per 100,000 in 1975 versus 2.3 per 100,000 in 2014) [176]. While the incidence of cervical cancer has largely decreased in the general population, the burden of cervical cancer is still much higher in HIV+ women [44, 149]. The NA-ACCORD study revealed that the majority (90%) of cervical cancer cases in HIV+ women were those who did not have a recent Pap test, did not follow-up for a colposcopy after an abnormal Pap test, or did not receive treatment after detection of precancerous lesions [125]. For HIV+ women adherent to the cervical cancer screening program and with normal cervical cytology, their incidences of CIN and cervical cancer were comparable to HIV-negative women [44]. Nonetheless, adherence to cervical cancer screening remains a challenge for HIV+ women, even in high-resource settings. Worldwide less than 50% of HIV+ women undertake recommended cervical cancer screening and they are almost 30% less likely to have regular Pap tests compared with the general population. [177, 178] To increase cervical cancer screening uptake and minimize the number of false negatives in HIV + women, HR-HPV-based screening algorithms should be applied [179].

References

1. Zur Hausen H (2000) Papillomaviruses causing cancer: evasion from host-cell control in early events in carcinogenesis. J Natl Cancer Inst 92(9):690–8
2. Bosch FX, Lorincz A, Munoz N, Meijer CJ, Shah KV (2002) The causal relation between human papillomavirus and cervical cancer. J Clin Pathol 55(4):244–265
3. Munoz N, Castellsague X, de Gonzalez AB, Gissmann L (2006) Chapter 1: HPV in the etiology of human cancer. Vaccine, 24 (3):1–10
4. Doorbar J, Quint W, Banks L, Bravo IG, Stoler M, Broker TR et al (2012) The biology and life-cycle of human papillomaviruses. Vaccine 30(Suppl 5):F55–F70
5. de Sanjose S, Brotons M, Pavon MA (2017) The natural history of human papillomavirus infection. Best Practice & Research Clinical Obstetrics & Gynaecology
6. Bernard HU, Burk RD, Chen Z, van Doorslaer K, Hausen H, de Villiers EM (2010) Classification of papillomaviruses (PVs) based on 189 PV types and proposal of taxonomic amendments. Virology 401(1):70–79

7. Humans IWGotEoCRt (1995) Human papillomaviruses. IARC Monographs on the evaluation of carcinogenic risks to humans. 64(1)
8. de Martel C, Ferlay J, Franceschi S, Vignat J, Bray F, Forman D et al (2012) Global burden of cancers attributable to infections in 2008: a review and synthetic analysis. Lancet Oncol. 13(6):607–615
9. Moscicki AB, Schiffman M, Burchell A, Albero G, Giuliano AR, Goodman MT et al (2012) Updating the natural history of human papillomavirus and anogenital cancers. Vaccine. 30 (Suppl 5):F24–F33
10. Munoz N, Bosch FX, de Sanjose S, Herrero R, Castellsague X, Shah KV et al (2003) Epidemiologic classification of human papillomavirus types associated with cervical cancer. N Engl J Med 348(6):518–527
11. Parkin DM (2006) The global health burden of infection-associated cancers in the year 2002. Int J Cancer 118(12):3030–3044
12. Parkin DM, Bray F (2006) Chapter 2: The burden of HPV-related cancers. Vaccine. 24 (3):11–25
13. Castellsagué SdS X, Aguado T, Louie KS, Bruni L, Muñoz J, Diaz M, Irwin K, Gacic M, Beauvais O, Albero G, Ferrer SB E, Bosch FX (2007) HPV and cervical cancer in the world. 2007 report. WHO/ICO Information Centre on HPV and Cervical Cancer (HPV Information Centre). World Health Organization
14. de Martel C, Plummer M, Vignat J, Franceschi S (2017) Worldwide burden of cancer attributable to HPV by site, country and HPV type. Int J Cancer 141(4):664–670
15. Chow LT, Broker TR, Steinberg BM (2010) The natural history of human papillomavirus infections of the mucosal epithelia. APMIS. 118(6–7):422–449
16. Wright TC Jr, Massad LS, Dunton CJ, Spitzer M, Wilkinson EJ, Solomon D (2007) 2006 consensus guidelines for the management of women with abnormal cervical screening tests. J Low Genit Tract Dis. 11(4):201–222
17. Castellsague X, Pawlita M, Roura E, Margall N, Waterboer T, Bosch FX et al (2014) Prospective seroepidemiologic study on the role of human papillomavirus and other infections in cervical carcinogenesis: evidence from the EPIC cohort. Int J Cancer 135 (2):440–452
18. Roura E, Castellsague X, Pawlita M, Travier N, Waterboer T, Margall N et al (2014) Smoking as a major risk factor for cervical cancer and pre-cancer: results from the EPIC cohort. Int J Cancer 135(2):453–466
19. Roura E, Travier N, Waterboer T, de Sanjose S, Bosch FX, Pawlita M et al (2016) The influence of hormonal factors on the risk of developing cervical cancer and pre-cancer: results from the EPIC Cohort. PLoS ONE 11(1):e0147029
20. Castellsague X, Bosch FX, Munoz N (2002) Environmental co-factors in HPV carcinogenesis. Virus Res 89(2):191–199
21. Roberts SA, Lawrence MS, Klimczak LJ, Grimm SA, Fargo D, Stojanov P et al (2013) An APOBEC cytidine deaminase mutagenesis pattern is widespread in human cancers. Nat Genet 45(9):970–976
22. Ferlay J, Soerjomataram I, Dikshit R, Eser S, Mathers C, Rebelo M et al (2015) Cancer incidence and mortality worldwide: sources, methods and major patterns in GLOBOCAN 2012. Int J Cancer 136(5):E359–E386
23. World Health Organization. Number of people (all ages) living with HIV estimates by country 2017. Available from: http://apps.who.int/gho/data/node.main.620?lang=en
24. Centers for Disease Control and Prevention (2012) Human papillomavirus-associated cancers—United States, 2004-2008. MMWR Morb Mortal Wkly Rep 61:258–261
25. Chesson HW, Ekwueme DU, Saraiya M, Watson M, Lowy DR, Markowitz LE (2012) Estimates of the annual direct medical costs of the prevention and treatment of disease associated with human papillomavirus in the United States. Vaccine 30(42):6016–6019
26. Cates W Jr (1999) Estimates of the incidence and prevalence of sexually transmitted diseases in the United States. American social health association panel. Sex Transm Dis 26(4):S2–S7

27. Dunne EF, Markowitz LE (2006) Genital human papillomavirus infection. Clin Infect Dis 43 (5):624–629
28. Satterwhite CL, Torrone E, Meites E, Dunne EF, Mahajan R, Ocfemia MC et al (2013) Sexually transmitted infections among US Women and Men: prevalence and incidence estimates, 2008. Sex Transm Dis 40(3):187–193
29. Vinodhini K, Shanmughapriya S, Das BC, Natarajaseenivasan K (2012) Prevalence and risk factors of HPV infection among women from various provinces of the world. Arch Gynecol Obstet 285(3):771–777
30. de Sanjose S, Diaz M, Castellsague X, Clifford G, Bruni L, Munoz N et al (2007) Worldwide prevalence and genotype distribution of cervical human papillomavirus DNA in women with normal cytology: a meta-analysis. Lancet Infect Dis. 7(7):453–459
31. Baseman JG, Koutsky LA (2005) The epidemiology of human papillomavirus infections. J Clin Virol 32(Suppl 1):S16–S24
32. Ho GY, Bierman R, Beardsley L, Chang CJ, Burk RD (1998) Natural history of cervicovaginal papillomavirus infection in young women. N Engl J Med 338(7):423–428
33. Moscicki AB, Hills N, Shiboski S, Powell K, Jay N, Hanson E et al (2001) Risks for incident human papillomavirus infection and low-grade squamous intraepithelial lesion development in young females. JAMA 285(23):2995–3002
34. Winer RL, Lee SK, Hughes JP, Adam DE, Kiviat NB, Koutsky LA (2003) Genital human papillomavirus infection: incidence and risk factors in a cohort of female university students. Am J Epidemiol 157(3):218–226
35. Rodriguez AC, Schiffman M, Herrero R, Hildesheim A, Bratti C, Sherman ME et al (2010) Longitudinal study of human papillomavirus persistence and cervical intraepithelial neoplasia grade 2/3: critical role of duration of infection. J Natl Cancer Inst 102(5):315–324
36. Winer RL, Hughes JP, Feng Q, Xi LF, Cherne S, O'Reilly S et al (2011) Early natural history of incident, type-specific human papillomavirus infections in newly sexually active young women. Cancer Epidemiol Biomarkers Prev 20(4):699–707
37. Rositch AF, Burke AE, Viscidi RP, Silver MI, Chang K, Gravitt PE (2012) Contributions of recent and past sexual partnerships on incident human papillomavirus detection: acquisition and reactivation in older women. Cancer Res 72(23):6183–6190
38. Winer RL, Hughes JP, Feng Q, Stern JE, Xi LF, Koutsky LA (2016) Incident Detection of high-risk human papillomavirus infections in a cohort of high-risk women aged 25–65 Years. J Infect Dis 214(5):665–675
39. Hariri S, Unger ER, Sternberg M, Dunne EF, Swan D, Patel S et al (2011) Prevalence of genital human papillomavirus among females in the United States, the national health and nutrition examination survey, 2003–2006. J Infect Dis 204(4):566–573
40. Ellerbrock TV, Chiasson MA, Bush TJ, Sun XW, Sawo D, Brudney K et al (2000) Incidence of cervical squamous intraepithelial lesions in HIV-infected women. JAMA 283(8):1031–1037
41. Clifford GM, Polesel J, Rickenbach M, Dal Maso L, Keiser O, Kofler A et al (2005) Cancer risk in the Swiss HIV cohort study: associations with immunodeficiency, smoking, and highly active antiretroviral therapy. J Natl Cancer Inst 97(6):425–432
42. De Vuyst H, Lillo F, Broutet N, Smith JS (2008) HIV, human papillomavirus, and cervical neoplasia and cancer in the era of highly active antiretroviral therapy. Eur. J. Cancer Prev. official J European Cancer Prevention Organisation (ECP). 17(6):545–554
43. Denslow SA, Rositch AF, Firnhaber C, Ting J, Smith JS (2014) Incidence and progression of cervical lesions in women with HIV: a systematic global review. Int J STD AIDS 25 (3):163–177
44. Thorsteinsson K, Ladelund S, Jensen-Fangel S, Katzenstein TL, Johansen IS, Pedersen G et al (2016) Incidence of cervical dysplasia and cervical cancer in women living with HIV in Denmark: comparison with the general population. HIV Med. 17(1):7–17

45. Gange SJ, Kitahata MM, Saag MS, Bangsberg DR, Bosch RJ, Brooks JT et al (2007) Cohort profile: the North American AIDS cohort collaboration on research and design (NA-ACCORD). Int J Epidemiol 36(2):294–301
46. Smith D, The HIV (1998) Epidemiology research study, HIV out-patient study, and the spectrum of disease studies. J Acquir Immune Defic Syndr Hum Retrovirol. 17(Suppl 1): S17–S19
47. Jaen A, Casabona J, Esteve A, Miro JM, Tural C, Ferrer E et al (2005) [Clinical-epidemiological characteristics and antiretroviral treatment trends in a cohort of HIV infected patients. The PISCIS Project]. Medicina clinica.124(14):525–31
48. Guiguet M, Boue F, Cadranel J, Lang JM, Rosenthal E, Costagliola D (2009) Effect of immunodeficiency, HIV viral load, and antiretroviral therapy on the risk of individual malignancies (FHDH-ANRS CO4): a prospective cohort study. Lancet Oncol. 10(12):1152–1159
49. Sun XW, Kuhn L, Ellerbrock TV, Chiasson MA, Bush TJ, Wright TC Jr (1997) Human papillomavirus infection in women infected with the human immunodeficiency virus. N Engl J Med 337(19):1343–1349
50. Kitchener H, Nelson L, Adams J, Mesher D, Sasieni P, Cubie H et al (2007) Colposcopy is not necessary to assess the risk to the cervix in HIV-positive women: an international cohort study of cervical pathology in HIV-1 positive women. Int J Cancer 121(11):2484–2491
51. Vernon SD, Holmes KK, Reeves WC (1995) Human papillomavirus infection and associated disease in persons infected with human immunodeficiency virus. Clin Infect Dis 21(Suppl 1):S121–S124
52. Palefsky JM, Minkoff H, Kalish LA, Levine A, Sacks HS, Garcia P et al (1999) Cervicovaginal human papillomavirus infection in human immunodeficiency virus-1 (HIV)-positive and high-risk HIV-negative women. J Natl Cancer Inst 91(3):226–236
53. Del Mistro A, Bonaldi L, Bertorelle R, Minucci D, Franzetti M, Cattelan A et al (2001) Genital human papillomavirus types in immunocompetent and immunodepressed women in Northeast Italy: prevalence and cytomorphological correlations. J Low Genit Tract Dis. 5 (1):12–20
54. de Sanjose S, Palefsky J (2002) Cervical and anal HPV infections in HIV positive women and men. Virus Res 89(2):201–211
55. Hessol NA, Seaberg EC, Preston-Martin S, Massad LS, Sacks HS, Silver S et al (2004) Cancer risk among participants in the women's interagency HIV study. J Acquir Immune Defic Syndr 36(4):978–985
56. Strickler HD, Burk RD, Fazzari M, Anastos K, Minkoff H, Massad LS et al (2005) Natural history and possible reactivation of human papillomavirus in human immunodeficiency virus-positive women. J Natl Cancer Inst 97(8):577–586
57. Luque AE, Jabeen M, Messing S, Lane CA, Demeter LM, Rose RC et al (2006) Prevalence of human papillomavirus genotypes and related abnormalities of cervical cytological results among HIV-1-infected women in Rochester. New York. J Infect Dis. 194(4):428–434
58. Jong E, Mulder JW, van Gorp EC, Wagenaar JK, Derksen J, Westerga J et al (2008) The prevalence of human papillomavirus (HPV) infection in paired urine and cervical smear samples of HIV-infected women. J Clin Virol 41(2):111–115
59. Fife KH, Wu JW, Squires KE, Watts DH, Andersen JW, Brown DR (2009) Prevalence and persistence of cervical human papillomavirus infection in HIV-positive women initiating highly active antiretroviral therapy. J Acquir Immune Defic Syndr 51(3):274–282
60. Grinsztejn B, Veloso VG, Levi JE, Velasque L, Luz PM, Friedman RK et al (2009) Factors associated with increased prevalence of human papillomavirus infection in a cohort of HIV-infected Brazilian women. Int J Infect Dis. 13(1):72–80
61. Kojic EM, Cu-Uvin S, Conley L, Bush T, Onyekwuluje J, Swan DC, et al (2010) Human papillomavirus infection and cytologic abnormalities of the anus and cervix among HIV-infected women in the study to understand the natural history of HIV/AIDS in the era of effective therapy (The SUN study). Sex Transm Dis

62. Kahn JA, Burk RD, Squires KE, Kapogiannis BG, Rudy B, Xu J, et al (2012) Prevalence and risk factors for HPV in HIV-positive young women receiving their first HPV vaccination. J Acquir Immune Defic Syndr
63. Hessol NA, Holly EA, Efird JT, Minkoff H, Weber KM, Darragh TM et al (2013) Concomitant anal and cervical human papillomavirusV infections and intraepithelial neoplasia in HIV-infected and uninfected women. AIDS 27(11):1743–1751
64. Ebrahim S, Mndende XK, Kharsany AB, Mbulawa ZZ, Naranbhai V, Frohlich J et al (2016) High burden of human papillomavirus (HPV) infection among young women in KwaZulu-Natal, South Africa. PLoS ONE 11(1):e0146603
65. Clifford GM, Goncalves MA, Franceschi S (2006) Human papillomavirus types among women infected with HIV: a meta-analysis. AIDS 20(18):2337–2344
66. Park LS, Hernandez-Ramirez RU, Silverberg MJ, Crothers K, Dubrow R (2016) Prevalence of non-HIV cancer risk factors in persons living with HIV/AIDS: a meta-analysis. Aids 30 (2):273–291
67. Hankins C, Coutlee F, Lapointe N, Simard P, Tran T, Samson J, et al (1999) Prevalence of risk factors associated with human papillomavirus infection in women living with HIV. Canadian Women's HIV Study Group. CMAJ: Canadian Medical Association journal = journal de l'Association medicale canadienne 160(2):185–91
68. Minkoff H, Feldman JG, Strickler HD, Watts DH, Bacon MC, Levine A et al (2004) Relationship between smoking and human papillomavirus infections in HIV-infected and -uninfected women. J Infect Dis 189(10):1821–1828
69. Sinayobye J, Sklar M, Hoover DR, Shi Q, Dusingize JC, Cohen M et al (2014) Prevalence and risk factors for high-risk human papillomavirus (hrHPV) infection among HIV-infected and uninfected Rwandan women: implications for hrHPV-based screening in Rwanda. Infect Agent Cancer. 9:40
70. Massad L, Keller M, Xie X, Minkoff H, Palefsky J, D'Souza G et al (2016) Multitype infections with human papillomavirus: impact of human immunodeficiency virus coinfection. Sex Transm Dis 43(10):637–641
71. Thorsteinsson K, Storgaard M, Katzenstein TL, Ladelund S, Ronsholt FF, Johansen IS et al (2016) Prevalence and distribution of cervical high-risk human papillomavirus and cytological abnormalities in women living with HIV in Denmark— the SHADE. BMC Cancer 16(1):866
72. Dunne EF, Unger ER, Sternberg M, McQuillan G, Swan DC, Patel SS et al (2007) Prevalence of HPV infection among females in the United States. JAMA 297(8):813–819
73. McKenzie ND, Kobetz EN, Hnatyszyn J, Twiggs LB, Lucci JA 3rd (2010) Women with HIV are more commonly infected with non-16 and -18 high-risk HPV types. Gynecol Oncol 116(3):572–577
74. Massad LS, Xie X, Burk RD, D'Souza G, Darragh TM, Minkoff H et al (2016) Association of cervical precancer with human papillomavirus types other than 16 among HIV co-infected women. Am J Obstet Gynecol 214(3):354.e1–354.e6
75. Levi JE, Kleter B, Quint WG, Fink MC, Canto CL, Matsubara R et al (2002) High prevalence of human papillomavirus (HPV) infections and high frequency of multiple HPV genotypes in human immunodeficiency virus-infected women in Brazil. J Clin Microbiol 40 (9):3341–3345
76. Levi JE, Fernandes S, Tateno AF, Motta E, Lima LP, Eluf-Neto J et al (2004) Presence of multiple human papillomavirus types in cervical samples from HIV-infected women. Gynecol Oncol 92(1):225–231
77. Chaturvedi AK, Myers L, Hammons AF, Clark RA, Dunlap K, Kissinger PJ et al (2005) Prevalence and clustering patterns of human papillomavirus genotypes in multiple infections. Cancer Epidemiol Biomark Prev 14(10):2439–2445
78. Adebamowo SN, Olawande O, Famooto A, Dareng EO, Offiong R, Adebamowo CA (2017) Persistent low-risk and high-risk human papillomavirus infections of the uterine cervix in HIV-negative and HIV-positive women. Front. publ. health. 5:178

79. Ahdieh L, Klein RS, Burk R, Cu-Uvin S, Schuman P, Duerr A et al (2001) Prevalence, incidence, and type-specific persistence of human papillomavirus in human immunodeficiency virus (HIV)-positive and HIV-negative women. J Infect Dis 184(6):682–690
80. Massad LS, Xie X, Burk R, Keller MJ, Minkoff H, D'Souza G et al (2014) Long-term cumulative detection of human papillomavirus among HIV seropositive women. Aids 28 (17):2601–2608
81. Branca M, Garbuglia AR, Benedetto A, Cappiello T, Leoncini L, Migliore G et al (2003) Factors predicting the persistence of genital human papillomavirus infections and PAP smear abnormality in HIV-positive and HIV-negative women during prospective follow-up. Int J STD AIDS 14(6):417–425
82. Whitham HK, Hawes SE, Chu H, Oakes JM, Lifson AR, Kiviat NB et al (2017) A comparison of the natural history of HPV infection and cervical abnormalities among HIV-positive and HIV-negative women in Senegal. Africa. Cancer Epidemiol Biomarkers Prev. 26(6):886–894
83. Travassos AG, Netto E, Xavier-Souza E, Nobrega I, Adami K, Timbo M et al (2017) Predictors of HPV incidence and clearance in a cohort of Brazilian HIV-infected women. PLoS ONE 12(10):e0185423
84. Nappi L, Carriero C, Bettocchi S, Herrero J, Vimercati A, Putignano G (2005) Cervical squamous intraepithelial lesions of low-grade in HIV-infected women: recurrence, persistence, and progression, in treated and untreated women. Eur J Obstet Gynecol Reprod Biol 121(2):226–232
85. D'Souza G, Fakhry C, Sugar EA, Seaberg EC, Weber K, Minkoff HL et al (2007) Six-month natural history of oral versus cervical human papillomavirus infection. Int J Cancer 121 (1):143–150
86. Gingelmaier A, Grubert T, Kaestner R, Mylonas I, Weissenbacher T, Bergauer F et al (2007) High recurrence rate of cervical dysplasia and persistence of HPV infection in HIV-1-infected women. Anticancer Res 27(4A):1795–1798
87. Fontaine J, Hankins C, Money D, Rachlis A, Pourreaux K, Ferenczy A et al (2008) Human papillomavirus type 16 (HPV-16) viral load and persistence of HPV-16 infection in women infected or at risk for HIV. J Clin Virol 43(3):307–312
88. Kjaer SK, Frederiksen K, Munk C, Iftner T (2010) Long-term absolute risk of cervical intraepithelial neoplasia grade 3 or worse following human papillomavirus infection: role of persistence. J Natl Cancer Inst 102(19):1478–1488
89. Piketty C, Kazatchkine MD (2005) Human papillomavirus-related cervical and anal disease in HIV-infected individuals in the era of highly active antiretroviral therapy. Curr HIV/AIDS Rep. 2(3):140–145
90. Denny L, Boa R, Williamson AL, Allan B, Hardie D, Stan R et al (2008) Human papillomavirus infection and cervical disease in human immunodeficiency virus-1-infected women. Obstet Gynecol 111(6):1380–1387
91. Massad LS, Riester KA, Anastos KM, Fruchter RG, Palefsky JM, Burk RD, et al (1999) Prevalence and predictors of squamous cell abnormalities in Papanicolaou smears from women infected with HIV-1. Women's Interagency HIV Study Group. J Acquir Immune Defic Syndr. 1999;21(1):33–41
92. Massad LS, Seaberg EC, Watts DH, Hessol NA, Melnick S, Bitterman P et al (2004) Low incidence of invasive cervical cancer among HIV-infected US women in a prevention program. Aids. 18(1):109–113
93. Yamada R, Sasagawa T, Kirumbi LW, Kingoro A, Karanja DK, Kiptoo M et al (2008) Human papillomavirus infection and cervical abnormalities in Nairobi, Kenya, an area with a high prevalence of human immunodeficiency virus infection. J Med Virol 80(5):847–855
94. Cardillo M, Hagan R, Abadi J, Abadi MA (2001) CD4 T-cell count, viral load, and squamous intraepithelial lesions in women infected with the human immunodeficiency virus. Cancer 93(2):111–114

95. Minkoff H, Zhong Y, Burk RD, Palefsky JM, Xue X, Watts DH et al (2010) Influence of adherent and effective antiretroviral therapy use on human papillomavirus infection and squamous intraepithelial lesions in human immunodeficiency virus-positive women. J Infect Dis 201(5):681–690
96. Clifford GM, Franceschi S, Keiser O, Schoni-Affolter F, Lise M, Dehler S et al (2016) Immunodeficiency and the risk of cervical intraepithelial neoplasia 2/3 and cervical cancer: a nested case-control study in the Swiss HIV cohort study. Int J Cancer 138(7):1732–1740
97. Rousseau MN, Costes V, Konate I, Nagot N, Foulongne V, Ouedraogo A et al (2007) Viral load and genomic integration of HPV 16 in cervical samples from HIV-1-infected and uninfected women in Burkina Faso. J Med Virol 79(6):766–770
98. Centers for Disease Control and Prevention (1992) 1993 revised classification system for HIV infection and expanded surveillance case definition for AIDS among adolescents and adults. MMWR Recomm Rep. 41(Rr-17):1–19
99. Schneider MF, Gange SJ, Williams CM, Anastos K, Greenblatt RM, Kingsley L et al (2005) Patterns of the hazard of death after AIDS through the evolution of antiretroviral therapy: 1984-2004. AIDS (London, England). 19(17):2009–2018
100. Lewden C, Chene G, Morlat P, Raffi F, Dupon M, Dellamonica P, et al (2007) HIV-infected adults with a CD4 cell count greater than 500 cells/mm3 on long-term combination antiretroviral therapy reach same mortality rates as the general population. J. Acquir. Immune Defic. Syndr. (1999) 46(1):72–7
101. Antiretroviral Therapy Cohort Collaboration (2008) Life expectancy of individuals on combination antiretroviral therapy in high-income countries: a collaborative analysis of 14 cohort studies. Lancet. 372(9635):293–9
102. Bhaskaran K, Hamouda O, Sannes M, Boufassa F, Johnson AM, Lambert PC et al (2008) Changes in the risk of death after HIV seroconversion compared with mortality in the general population. JAMA 300(1):51–59
103. Ray M, Logan R, Sterne JA, Hernandez-Diaz S, Robins JM, Sabin C et al (2010) The effect of combined antiretroviral therapy on the overall mortality of HIV-infected individuals. AIDS 24(1):123–137
104. Calabresi A, Ferraresi A, Festa A, Scarcella C, Donato F, Vassallo F et al (2013) Incidence of AIDS-defining cancers and virus-related and non-virus-related non-AIDS-defining cancers among HIV-infected patients compared with the general population in a large health district of Northern Italy, 1999-2009. HIV Med. 14(8):481–490
105. International Collaboration on HIV and Cancer (2000) Highly active antiretroviral therapy and incidence of cancer in human immunodeficiency virus-infected adults. J Natl Cancer Inst 92(22):1823–1830
106. Bonnet F, Chene G (2008) Evolving epidemiology of malignancies in HIV. Curr Opin Oncol 20(5):534–540
107. Bonnet F, Burty C, Lewden C, Costagliola D, May T, Bouteloup V et al (2009) Changes in cancer mortality among HIV-infected patients: the Mortalite 2005 Survey. Clin Infect Dis 48 (5):633–639
108. Chaturvedi AK, Madeleine MM, Biggar RJ, Engels EA (2009) Risk of human papillomavirus-associated cancers among persons with AIDS. J Natl Cancer Inst 101 (16):1120–1130
109. Simard EP, Pfeiffer RM, Engels EA (2010) Spectrum of cancer risk late after AIDS onset in the United States. Arch Intern Med 170(15):1337–1345
110. Shiels MS, Pfeiffer RM, Gail MH, Hall HI, Li J, Chaturvedi AK et al (2011) Cancer burden in the HIV-infected population in the United States. J Natl Cancer Inst 103(9):753–762
111. Massad LS, Hessol NA, Darragh TM, Minkoff H, Colie C, Wright RL et al (2017) Cervical cancer incidence after up to 20 years of observation among women with HIV. Int J Cancer 141(8):1561–1565

112. Frisch M, Biggar RJ, Goedert JJ (2000) Human papillomavirus-associated cancers in patients with human immunodeficiency virus infection and acquired immunodeficiency syndrome. J Natl Cancer Inst 92(18):1500–1510
113. Frisch M, Biggar RJ, Engels EA, Goedert JJ (2001) Association of cancer with AIDS-related immunosuppression in adults. JAMA 285(13):1736–1745
114. Engels EA, Pfeiffer RM, Goedert JJ, Virgo P, McNeel TS, Scoppa SM et al (2006) Trends in cancer risk among people with AIDS in the United States 1980–2002. AIDS 20(12):1645–1654
115. Palefsky J (2006) Biology of HPV in HIV infection. Adv Dent Res. 19(1):99–105
116. Engels EA, Biggar RJ, Hall HI, Cross H, Crutchfield A, Finch JL et al (2008) Cancer risk in people infected with human immunodeficiency virus in the United States. Int J Cancer 123 (1):187–194
117. Patel P, Hanson DL, Sullivan PS, Novak RM, Moorman AC, Tong TC et al (2008) Incidence of types of cancer among HIV-infected persons compared with the general population in the United States, 1992-2003. Ann Intern Med 148(10):728–736
118. Crum-Cianflone N, Hullsiek KH, Marconi V, Weintrob A, Ganesan A, Barthel RV et al (2009) Trends in the incidence of cancers among HIV-infected persons and the impact of antiretroviral therapy: a 20-year cohort study. AIDS 23(1):41–50
119. Shiels MS, Cole SR, Kirk GD, Poole C (2009) A meta-analysis of the incidence of non-AIDS cancers in HIV-infected individuals. J Acquir Immune Defic Syndr 52(5):611–622
120. Achenbach CJ, Cole SR, Kitahata MM, Casper C, Willig JH, Mugavero MJ et al (2011) Mortality after cancer diagnosis in HIV-infected individuals treated with antiretroviral therapy. AIDS 25(5):691–700
121. Silverberg MJ, Chao C, Leyden WA, Xu L, Horberg MA, Klein D, et al (2011) HIV infection, immunodeficiency, viral replication, and the risk of cancer. Cancer Epidemiol Biomark. Prev
122. Dubrow R, Silverberg MJ, Park LS, Crothers K, Justice AC (2012) HIV infection, aging, and immune function: implications for cancer risk and prevention. Curr Opin Oncol 24 (5):506–516
123. Rubinstein PG, Aboulafia DM, Zloza A (2014) Malignancies in HIV/AIDS: from epidemiology to therapeutic challenges. AIDS. 28(4):453–465
124. Cote TR, O'Brien TR, Ward JW, Wilson SE, Blattner WA (1995) AIDS and cancer registry linkage: measurement and enhancement of registry completeness. The National AIDS/Cancer match study group. Prev Med 24(4):375–7
125. Abraham AG, D'Souza G, Jing Y, Gange SJ, Sterling TR, Silverberg MJ et al (2013) Invasive cervical cancer risk among HIV-infected women: a North American multicohort collaboration prospective study. J Acquir Immune Defic Syndr 62(4):405–413
126. Harris TG, Burk RD, Palefsky JM, Massad LS, Bang JY, Anastos K et al (2005) Incidence of cervical squamous intraepithelial lesions associated with HIV serostatus, CD4 cell counts, and human papillomavirus test results. JAMA 293(12):1471–1476
127. Mdodo R, Frazier EL, Dube SR, Mattson CL, Sutton MY, Brooks JT et al (2015) Cigarette smoking prevalence among adults with HIV compared with the general adult population in the United States: cross-sectional surveys. Ann Intern Med 162(5):335–344. https://doi.org/10.7326/M14-0954
128. Oh JK, Ju YH, Franceschi S, Quint W, Shin HR (2008) Acquisition of new infection and clearance of type-specific human papillomavirus infections in female students in Busan, South Korea: a follow-up study. BMC Infect Dis 8:13
129. Xi LF, Koutsky LA, Castle PE, Edelstein ZR, Meyers C, Ho J et al (2009) Relationship between cigarette smoking and human papilloma virus types 16 and 18 DNA load. Cancer Epidemiol Biomark Prev 18(12):3490–3496

130. Mzarico E, Gomez-Roig MD, Guirado L, Lorente N, Gonzalez-Bosquet E (2015) Relationship between smoking, HPV infection, and risk of Cervical cancer. Eur J Gynaecol Oncol 36(6):677–680
131. Feng RM, Hu SY, Zhao FH, Zhang R, Zhang X, Wallach AI et al (2017) Role of active and passive smoking in high-risk human papillomavirus infection and cervical intraepithelial neoplasia grade 2 or worse. J. Gynecol. Oncol. 28(5):e47
132. Trushin N, Alam S, El-Bayoumy K, Krzeminski J, Amin SG, Gullett J et al (2012) Comparative metabolism of benzo[a]pyrene by human keratinocytes infected with high-risk human papillomavirus types 16 and 18 as episomal or integrated genomes. J. Carcinog. 11:1
133. Alam S, Conway MJ, Chen HS, Meyers C (2008) The cigarette smoke carcinogen benzo[a] pyrene enhances human papillomavirus synthesis. J Virol 82(2):1053–1058
134. Vernon SD, Hart CE, Reeves WC, Icenogle JP (1993) The HIV-1 tat protein enhances E2-dependent human papillomavirus 16 transcription. Virus Res 27(2):133–145
135. Arany I, Tyring SK (1998) Systemic immunosuppression by HIV infection influences HPV transcription and thus local immune responses in condyloma acuminatum. Int J STD AIDS 9 (5):268–271
136. Heard I, Palefsky JM, Kazatchkine MD (2004) The impact of HIV antiviral therapy on human papillomavirus (HPV) infections and HPV-related diseases. Antivir Ther. 9(1):13–22
137. Guidry JT, Scott RS (2017) The interaction between human papillomavirus and other viruses. Virus Res 231:139–147
138. Denny LA, Franceschi S, de Sanjose S, Heard I, Moscicki AB, Palefsky J (2012) Human papillomavirus, human immunodeficiency virus and immunosuppression. Vaccine 30(Suppl 5):F168–F174
139. Ghartey J, Kovacs A, Burk RD, Stewart Massad L, Minkoff H, Xie X et al (2014) Genital tract HIV RNA levels and their associations with human papillomavirus infection and risk of cervical precancer. J Acquir Immune Defic Syndr 66(3):316–323
140. de Jong MA, Geijtenbeek TB (2009) Human immunodeficiency virus-1 acquisition in genital mucosa: Langerhans cells as key-players. J Intern Med 265(1):18–28
141. Herfs M, Hubert P, Moutschen M, Delvenne P (2011) Mucosal junctions: open doors to HPV and HIV infections? Trends Microbiol 19(3):114–120
142. Houlihan CF, Larke NL, Watson-Jones D, Smith-McCune KK, Shiboski S, Gravitt PE et al (2012) Human papillomavirus infection and increased risk of HIV acquisition. A systematic review and meta-analysis. Aids 26(17):2211–2222
143. Lissouba P, Van de Perre P, Auvert B (2013) Association of genital human papillomavirus infection with HIV acquisition: a systematic review and meta-analysis. Sex Transm Infect. 89(5):350–356
144. Panel on Antiretroviral Guidelines for Adults and Adolescents (2016) Guidelines for the use of antiretroviral agents in HIV-1-infected adults and adolescents
145. Palefsky JM (2003) Cervical human papillomavirus infection and cervical intraepithelial neoplasia in women positive for human immunodeficiency virus in the era of highly active antiretroviral therapy. Curr Opin Oncol 15(5):382–388
146. Barbaro G, Barbarini G (2007) HIV infection and cancer in the era of highly active antiretroviral therapy (review). Oncol Rep 17(5):1121–1126
147. Adler DH (2010) The impact of HAART on HPV-related cervical disease. Curr HIV Res 8 (7):493–497
148. Bratcher LF, Sahasrabuddhe VV (2010) The impact of antiretroviral therapy on HPV and cervical intraepithelial neoplasia: current evidence and directions for future research. Infect Agent Cancer 5:8
149. Rohner E, Sengayi M, Goeieman B, Michelow P, Firnhaber C, Maskew M et al (2017) Cervical cancer risk and impact of Pap-based screening in HIV-positive women on antiretroviral therapy in Johannesburg. South Africa. Int J Cancer. 141(3):488–496
150. World Health Organization. Countries using hpv vaccine 2013. Available from: http://www.who.int/immunization/diseases/hpv/decision_implementation/en/

151. Medeiros LR, Rosa DD, da Rosa MI, Bozzetti MC, Zanini RR (2009) Efficacy of human papillomavirus vaccines: a systematic quantitative review. Int. J. Gynecol. cancer Official J Int. Gynecol Cancer Society 19(7):1166–1176
152. Munoz N, Kjaer SK, Sigurdsson K, Iversen OE, Hernandez-Avila M, Wheeler CM et al (2010) Impact of human papillomavirus (HPV)-6/11/16/18 vaccine on all HPV-associated genital diseases in young women. J Natl Cancer Inst 102(5):325–339
153. Skinner SR, Apter D, De Carvalho N, Harper DM, Konno R, Paavonen J et al (2016) Human papillomavirus (HPV)-16/18 AS04-adjuvanted vaccine for the prevention of cervical cancer and HPV-related diseases. Expert Rev Vaccines. 15(3):367–387
154. Garland SM, Kjaer SK, Munoz N, Block SL, Brown DR, DiNubile MJ et al (2016) Impact and effectiveness of the quadrivalent human papillomavirus vaccine: a systematic review of 10 years of real-world experience. Clin Infect Dis 63(4):519–527
155. Hofstetter AM, Ompad DC, Stockwell MS, Rosenthal SL, Soren K (2016) Human papillomavirus vaccination and cervical cytology outcomes among urban low-income minority females. JAMA pediatrics. 170(5):445–452
156. Kim J, Bell C, Sun M, Kliewer G, Xu L, McInerney M, et al (2016) Effect of human papillomavirus vaccination on cervical cancer screening in Alberta. CMAJ: Can. Med. Assoc. J. = J. de l'Assoc. Med.Can. 188(12):E281–8
157. Levin MJ, Moscicki AB, Song LY, Fenton T, Meyer WA 3rd, Read JS et al (2010) Safety and immunogenicity of a quadrivalent human papillomavirus (types 6, 11, 16, and 18) vaccine in HIV-infected children 7 to 12 years old. J Acquir Immune Defic Syndr 55 (2):197–204
158. Weinberg A, Song LY, Saah A, Brown M, Moscicki AB, Meyer WA 3rd et al (2012) Humoral, mucosal, and cell-mediated immunity against vaccine and nonvaccine genotypes after administration of quadrivalent human papillomavirus vaccine to HIV-infected children. J Infect Dis 206(8):1309–1318
159. Denny L, Hendricks B, Gordon C, Thomas F, Hezareh M, Dobbelaere K et al (2013) Safety and immunogenicity of the HPV-16/18 AS04-adjuvanted vaccine in HIV-positive women in South Africa: a partially-blind randomised placebo-controlled study. Vaccine 31(48):5745–5753
160. Kahn JA, Xu J, Kapogiannis BG, Rudy B, Gonin R, Liu N et al (2013) Immunogenicity and safety of the human papillomavirus 6, 11, 16, 18 vaccine in HIV-infected young women. Clin Infect Dis 57(5):735–744
161. Kojic EM, Kang M, Cespedes MS, Umbleja T, Godfrey C, Allen RT et al (2014) Immunogenicity and safety of the quadrivalent human papillomavirus vaccine in HIV-1-infected women. Clin Infect Dis 59(1):127–135
162. Kojic EM, Rana AI, Cu-Uvin S (2016) Human papillomavirus vaccination in HIV-infected women: need for increased coverage. Expert Rev Vaccines. 15(1):105–117
163. Mofenson LM, Brady MT, Danner SP, Dominguez KL, Hazra R, Handelsman E, et al (2009) Guidelines for the prevention and treatment of opportunistic infections among HIV-exposed and HIV-infected children: recommendations from CDC, the National Institutes of Health, the HIV medicine association of the infectious diseases society of America, the pediatric infectious diseases society, and the American academy of pediatrics. MMWR Recomm Rep 58(Rr-11):1–166
164. World Health Organization. Human Papillomavirus (HPV) position paper 2017 [Available from: http://www.who.int/immunization/policy/position_papers/hpv/en/
165. Firnhaber C, Wilkin T (2012) Human papillomavirus vaccines: where do they fit in HIV-infectedi? Curr HIV/AIDS Rep. 9(3):278–286
166. Crum-Cianflone NF, Sullivan E (2017) Vaccinations for the HIV-infected adult: a review of the current recommendations. Part I, Infectious diseases and therapy
167. Gertig DM, Brotherton JM, Budd AC, Drennan K, Chappell G, Saville AM (2013) Impact of a population-based HPV vaccination program on cervical abnormalities: a data linkage study. BMC Med 11:227

168. Firnhaber C, Evans D, Friedman-Khalili R, Willliams S, Michelow P, Matlhagela K et al (2011) Seroprevalence of HPV vaccine types 6, 11, 16 and 18 in HIV-infected women from South Africa, Brazil and Botswana. J Clin Virol 52(3):265–268
169. Moyer VA. Screening for cervical cancer: U.S. Preventive Services Task Force recommendation statement. Ann Intern Med. 2012;156(12):880–91, w312
170. Saslow D, Solomon D, Lawson HW, Killackey M, Kulasingam SL, Cain J, et al (2012) American cancer society, American society for colposcopy and cervical pathology, and American society for clinical pathology screening guidelines for the prevention and early detection of cervical cancer. CA: A Cancer J. for Clinicians 62(3):147–72
171. Huh WK, Ault KA, Chelmow D, Davey DD, Goulart RA, Garcia FA et al (2015) Use of primary high-risk human papillomavirus testing for cervical cancer screening: interim clinical guidance. Obstet Gynecol 125(2):330–337
172. U.S. Preventive Services Task Force. Draft Recommendation Statement Cervical Cancer: Screening 2017 [Available from: https://www.uspreventiveservicestaskforce.org/Page/Document/draft-recommendation-statement/cervical-cancer-screening2
173. Keller MJ, Burk RD, Massad LS, Eltoum IE, Hessol NA, Castle PE et al (2015) Cervical precancer risk in HIV-infected women who test positive for oncogenic human papillomavirus despite a normal pap test. Clin Infect Dis 61(10):1573–1581
174. Robbins HA, Strickler HD, Massad LS, Pierce CB, Darragh TM, Minkoff H et al (2017) Cervical cancer screening intervals and management for women living with HIV: a risk benchmarking approach. Aids 31(7):1035–1044
175. Massad LS, Fazzari MJ, Anastos K, Klein RS, Minkoff H, Jamieson DJ et al (2007) Outcomes after treatment of cervical intraepithelial neoplasia among women with HIV. J Low Genit Tract Dis. 11(2):90–97
176. Howlader N NA, Krapcho M, Miller D, Bishop K, Kosary CL, Yu M, Ruhl J, Tatalovich Z, Mariotto A, Lewis DR, Chen HS, Feuer EJ, Cronin KA (eds).MD 2017 SEER Cancer Statistics Review, 1975–2014. National Cancer Institute, Bethesda, [Available from: https://seer.cancer.gov/csr/1975_2014/
177. Bailey H, Thorne C, Semenenko I, Malyuta R, Tereschenko R, Adeyanova I et al (2012) Cervical screening within HIV care: findings from an HIV-positive cohort in Ukraine. PLoS ONE 7(4):e34706
178. Thorsteinsson K, Ladelund S, Jensen-Fangel S, Katzenstein TL, Johansen IS, Pedersen G et al (2014) Adherence to the cervical cancer screening program in women living with HIV in Denmark: comparison with the general population. BMC Infect Dis 14:256
179. Orang'o O, Liu T, Christoffersen-Deb A, Itsura P, Oguda J, Washington S, et al (2016) Use of VIA, pap smear, or HR-HPV testing in women living with HIV/AIDS for post-treatment cervical cancer screening: same tests, different priorities. Aids

HPV-Associated Oropharyngeal Cancer in the HIV/AIDS Patient

6

Jennifer E. Cameron and Michael Hagensee

Contents

J. E. Cameron
Department of Microbiology, Immunology and Parasitology,
Louisiana State University Health Sciences Center, 1901 Perdido
St. Medical Education Building, Room 6243, New Orleans, LA 70112, USA
e-mail: jcame2@lsuhsc.edu

M. Hagensee (✉)
Department of Medicine, Section of Infectious Diseases,
1542 Tulane Avenue, Box T4M-1, New Orleans, LA, USA
e-mail: mhagen@lsuhsc.edu

C. Meyers (ed.), *HIV/AIDS-Associated Viral Oncogenesis*, Cancer Treatment and Research 177, https://doi.org/10.1007/978-3-030-03502-0_6

Abstract

Since their discovery as the etiologic agents of cervical cancer in the mid-1970s, human papillomaviruses (HPVs) have been linked with a growing number of epithelial-derived tumors, including head and neck squamous cell carcinomas. HPV demonstrates a particular predilection for causing tumors of the oropharynx, with the majority of cases involving infection with high-oncogenic risk HPV-16. People living with HIV are at increased risk of infection with HPV- and HPV-related oral complications even with adequate control of their HIV infection with antiretroviral therapy. In this chapter, we discuss the molecular mechanisms that underlie HPV-mediated oncogenesis in the oropharynx. We also describe the progress that has been made in understanding the epidemiology of oral HPV infection and the determinants of oral HPV-related pathology. Finally, we examine what can be done to treat and prevent oral HPV infection, benign lesions, and cancer, particularly in the context of the HIV-positive patient.

6.1 Introduction

Since their discovery as the etiologic agents of cervical cancer in the mid-1970s, human papillomaviruses (HPVs) have been linked with a growing number of epithelial-derived tumors, including head and neck squamous cell carcinomas (HNSCC, see Table 6.1 for a list of abbreviations). In the head and neck region, the virus demonstrates a particular predilection for causing tumors of the oropharynx. The presence of HPV in oropharyngeal tumors has significant implications for the prognosis of the patient and the treatment regimen indicated. The incidence of HPV-associated HNSCC is reportedly on the rise in several regions globally, including the United States, Europe, and Australia. In all regions, the predominant HPV genotype associated with head and neck tumors is HPV-16. HPV infection, related oral benign lesions, and HNSCC are all increased in the HIV-seropositive population. It is not clear that highly active antiretroviral therapy (HAART) is reducing these rates by improving immune control. HAART may be potentially adding to the increased rates either directly due to damage to the oral epithelium or indirectly due to an increase in lifespan of the HIV-infected individual.

In this chapter, we discuss the findings of investigations that have begun to reveal the molecular mechanisms that underlie HPV-mediated oncogenesis in the oropharynx. We also describe the progress that has been made in determining the prevalence of oral HPV infection and the risks associated with acquiring oral HPV infections. Finally, we examine what can be done to prevent oral HPV infection, benign lesions, and cancer, particularly in the context of the HIV-positive patient.

6.1.1 Classical HPV-Mediated Oncogenesis

Papillomaviruses are ubiquitous, small DNA viruses that commonly infect squamous epithelium. Over 200 genotypes of human papillomavirus have been described. The viruses can be classified according to their predilection for infecting skin or mucosal tissues, with those that preferentially infect the mucosal epithelium showing little to no mucosal site restriction. Another useful way to classify the viruses is by their propensity to associate with clinical disease, with "low-risk" viruses typically causing benign papillomas (warts), "intermediate-risk" viruses causing rare cases of cancer, and "high-risk" viruses causing the majority of cases of HPV-positive tumors. Perhaps the most objective way to classify the viruses is by the relatedness of their genetic sequence. Phylogenetic analysis classifies the human papillomaviruses into alpha, beta, gamma, mu, and nu subgroups. Within this classification, the viruses that infect the mucosal epithelium belong to the alpha-papillomavirus group. The alpha-papillomavirus group contains HPV-16, 18,

31, 33, and 45, which account for more than 90% of HPV-associated cervical, vaginal, penile, and anal malignancies. While all of these alpha group viruses appear capable of establishing oral infections, oropharyngeal cancers appear to derive predominantly from infection with HPV-16. Other members of the alpha-papillomavirus group include the anogenital wart-associated genotypes-6 and -11 and the oral-specific genotypes-7, -13, and -32, all of which are associated with warts on the mucosal surfaces of the oral cavity.

Individual HPV genotypes are numbered according to the order in which they were discovered and subsequently confirmed as a novel virus by genetic sequence. In order to qualify as a unique HPV genotype, the putative virus must demonstrate a greater than 10% genetic variation from known HPV genotypes in three specific regions of the genome: the early genes E6 and E7, and the late gene L1. These regions of the virus genome were selected based on their importance to clinical disease; the E6 and E7 genes encode the oncogenic proteins of the virus and the L1 gene encodes the major capsid protein, which is the primary target of the serological antibody response to HPV infection and the antigen used in papillomavirus vaccines. This genomic definition of viral subtypes generally coincides with unique L1-specific serum antibody responses, such that serological assays can identify exposure to specific HPV genotypes. The serum antibody response to the HPV vaccine is similarly genotype-restricted, and therefore multiple genotypes are included in current vaccines to afford broad protection against the most clinically relevant HPV infections.

Papillomaviruses are believed to initiate infection by gaining access to the basal keratinocyte progenitor cells, either through micro-abrasions of the epithelial tissue or at sites where the epithelial layers are thin, such as the squamo-columnar junction of the endocervix. After successful infection of the basal keratinocytes, papillomaviruses express "early" genes, including the E2, E6, and E7 genes. The E6 protein interacts with many host cell nuclear proteins, the most important of which is the tumor suppressor p53. When in complex with p53, the E6 protein facilitates ubiquitination and subsequent proteasome-mediated degradation of p53 [1]. Loss of p53 leads to failed cell cycle checkpoint control, unrestricted cell proliferation, and resistance to apoptotic stimuli. Similarly, the E7 protein interacts with many host cell nuclear proteins, the most important of which is the retinoblastoma tumor suppressor protein, pRb. High-risk HPV E7 protein mediates degradation of pRB [2, 3], which in turn releases E2F family transcription activators for importation into the nucleus where they promote cell cycle progression to S-phase [4, 5]. These effects largely account for the oncogenic potential of papillomaviruses. During a productive papillomavirus infection, however, the function of these proteins is restricted by the papillomavirus E2 protein. The E2 protein functions as a transcriptional suppressor and specifically down-modulates the expression of E6 and E7. This allows the virus to successfully complete its life cycle without excessive damage to the host.

During productive viral infection, papillomaviruses complete their life cycle with the expression of the late genes L1 and L2 in the upper layers of the keratinized squamous epithelium. Together, the protein products of the L1 and L2 genes make up the viral capsid. The L1 protein predominates in the assembled capsid and is the immunodominant antigen of the serum antibody response to papillomaviruses. Serological responses to HPV L1 are type-specific with limited cross-reactivity between types. Ectopic expression of L1 leads to spontaneous assembly of empty viral particles. These L1 viral particles are used in papillomavirus vaccines to induce protective HPV type-specific immunity. Papillomavirus type-specific serum antibody responses correlate with exposure to the corresponding HPV genotype but are poor predictors of HPV-related disease or disease outcomes.

The vast majority of HPV infections resolve without long-term effects on the host. During the normal productive HPV infection, the viral genome is maintained extra-chromosomally as circular viral episomes capable of expressing viral transcripts depending on the differentiation state of the cell. Under conditions that are poorly understood, however, the viral genome may stably integrate into the host genome. Integration of the HPV genome is a hallmark of HPV-mediated anogenital cancers. In cancer cases, the HPV E2 transcriptional suppressor open reading frame is frequently disrupted, resulting in the loss of E2 expression and unchecked expression of the viral oncoproteins E6 and E7. The abundance of E6 and E7 effectively immortalize the host cell. The immortalized cell is capable of proliferation but loses the ability to differentiate, thus limiting L1 and L2 expression. Although the correlates of immune-mediated clearance of HPV infection are poorly understood, it is likely that this loss of L1 and L2 expression compromises the host's ability to clear the infected cells. Thus, integration of HPV into the host genome and overexpression of E6 and E7 oncoproteins are the predominant features of HPV-mediated anogenital cancers. Several features of oropharyngeal HPV infection suggest that the virus behaves similarly in this tissue. Notably, the molecular biology of HPV-associated head and neck tumors has not been studied in the context of HIV infection; however, it seems highly likely that the molecular mechanisms of HPV-induced carcinogenesis are no different in patients with HIV than they are in patients without HIV.

6.1.2 Canonical Cell Signaling Pathways Involved in HPV-Mediated HNSCC

HPV was first proposed as an etiologic agent of head and neck tumors in the early 1980s. Syrjanen et al. provided circumstantial evidence suggesting the presence of HPV in oral squamous cell carcinomas based on the identification of HPV-associated morphological characteristics in tissue adjacent to the tumor tissue [6]. Detection of high-risk HPV DNA in head and neck tumors followed, revealing a strong association between HPV infection and oropharyngeal cancers, particularly those cancers arising from the lingual and palatine tonsils of the oropharynx [7–15]. The tissue of the tonsillar crypts consists of reticulated squamous epithelium punctuated with gaps in the basement membrane that allow lymphoid derived cell lineages to traverse from the stroma to the apical surface of the epithelium [16]. This loose network of cells is believed to provide infectious viral particles ready access to the basal keratinocytes. Studies comparing HPV detection in oropharyngeal cancer cases and matched controls confirmed the association between HPV and oropharyngeal tumors [17–20]. Further investigation provided strong evidence of the involvement of HPV in the genesis of oropharyngeal tumors by demonstrating HPV integration and expression of high-risk HPV E6 and E7 transcripts in the tumors [19, 21–27].

HPV-associated head and neck tumors demonstrate molecular signatures that are distinct from that of HPV-negative head and neck tumors. HPV-negative head and neck tumors are frequently characterized by genomic mutations in TP53, the gene encoding the tumor suppressor protein p53. In contrast, HPV-positive head and neck tumors often contain wild-type TP53 [10, 21, 25–32]. Likewise, the E7-mediated degradation of pRb in HPV-positive tumors is distinct from the largely intact pRb protein expression seen in HPV-negative tumors [26–28]. The lack of somatic mutations in TP53 and the absence of pRB protein in HPV-positive HNSCC suggest that the HPV E6 and E7 proteins are functionally active in these tumors, similar to their anogenital counterparts. While HNSCC tumors typically demonstrate either TP53 mutations or E6/E7 expression but rarely both, HPV DNA detection and TP53 mutation can coexist [26], suggesting that in some cases the virus may be a bystander rather than the etiologic agent of the tumor. The frequency of wild-type TP53/high E6/E7 expression phenotype tumors is greatest for tumors derived from the oropharynx [26], consistent with the predilection for HPV to associate with oropharyngeal squamous cell carcinoma (OPSCC).

A significant consequence of E7-mediated loss of pRb expression is the activation and nuclear translocation of the E2F transcription factor. Activation of E2F transcription results in the induction of the tumor suppressor protein cyclin-dependent kinase inhibitor 2A gene (CDKN2A) which encodes the $p16^{INK4A}$ protein [33]. In the presence of intact pRB $p16^{INK4A}$ promotes cell cycle arrest, but in the absence of pRB $p16^{INK4A}$ disrupts D1:CDK4/6 complexes and promotes cell cycle progression [34]. In HPV-negative tumors, $p16^{INK4A}$ is frequently silenced due to mutation or epigenetic modification of the gene [35, 36]. Thus, HPV-positive tumors can generally be distinguished from HPV-negative tumors based on the detection of $p16^{INK4A}$ protein [7, 26, 30, 35, 37]. The detection of $p16^{INK4A}$ has also been shown to have diagnostic utility for distinguishing "true" HPV-associated HNSCC from bystander HPV infection [38]. Importantly, the reduction in cyclin D1 subsequent to $p16^{INK4A}$ activity impairs the function of RAD51, a mediator of homologous recombination-directed DNA damage repair and is thought to contribute to the radiosensitivity associated with HPV-positive HNSCC tumors [39, 40].

6.1.3 Integration of HPV into the Host Genome in HNSCC

It is generally believed that integration of HPV into the host genome is an essential step in cervical cancer carcinogenesis. The virus is determined to be integrated if fusion events containing both viral and host genomic elements can be detected within the tumor tissue. Integration of HPV into the host genome affects the gene expression capabilities of the virus. The consequences of integration classically include disruption of the regulatory E2/E4 region of the genome, which in turn leads to induction of the viral oncogenes E6 and E7. Integration may also disrupt the E1 open reading frame (ORF), which may promote DNA damage and growth arrest as shown in in vitro studies [41]. Integration may facilitate host-viral fusion transcripts that are more stable than viral transcripts alone [42]. A viral super-enhancer may be created by integration near regulatory element repeats [42], and integration at sites of host super-enhancers has been shown to boost E6/E7 expression levels [43]. Likely each individual integration event is independent and has unique consequences. Viral breakpoints are not conserved or predictable and can occur anywhere, although one report noted an increased incidence of viral linearization breakpoint in the E1 ORF [44]. Studies primarily focused on tumor-derived cell lines or patient-derived cancer tissues demonstrate that the entire genome is not always present in the integration, but E6 and E7 are frequently intact and highly expressed at the transcript level [45, 46]. This is likely because tumors and resulting cell lines have selected for this type of integration event and these findings may not be reflective of all integration events in vivo. Importantly, while integration may be one event that enhances expression of the oncogenes E6 and E7, integration does not predict E6/E7 expression level or HPV viral load in HNSCC [47]. This indicates that integration is not an absolute requirement to achieve high E6 and E7 expression levels and tumor phenotype. This observation is supported by studies of HPV-positive head and neck tumors that failed to reveal fusion events in 30–60% of the tumors [24, 26, 44, 47]. In fact, one study demonstrated larger tumor size at diagnosis in patients with HPV-positive HNSCC tumors in which the virus remained episomal (extrachromosomal) compared to patients with integrated HPV [24].

When HPV integration does occur, it is rarely a single integration event; viral–host fusion events may be detected at multiple sites in the host genome [44]. In tumor cell lines, HPV integration events are often identified at host genomic fragile sites, for example, areas of the genome that are prone to amplification, deletion, and chromosomal translocations or rearrangements [45]. Integration at or near cancer-related genes is also a common feature of HPV-positive HNSCC-derived cell lines [48]. In patient-derived tumor tissues, HPV integration tends to occur at common fragile sites, areas that are highly transcriptionally active, and areas that contain short sequences of viral–host nucleic acid homology [49–51]. Host genomic copy number variations often co-localize to sites of HPV integration [50], though it is not clear whether the integration of HPV facilitates genomic instability and copy number variation, or if integration of HPV and copy number variation are both consequences of genomic instability [52]. Integration at sites of "micro-homology" suggests that viral genome may have been mistaken for host by the DNA repair machinery during microhomology-mediated end joining repair [49, 51].

Integration of HPV into the host genome may or may not result in phenotypic consequences to the host. If integration occurs in genes that are epigenetically silenced, in intergenic regions, or in intron regions, there is unlikely to be any appreciable change in the cell's function or phenotype. Likewise, disruption of a host coding sequence by viral genome sequence may have no functional consequences as long as the second allele remains intact and compensatory expression of a functional protein is achieved. These types of silent integration events may occur with relative frequency but go undetected due to their benign effect on the cell. In patient-derived HNSCC tissues, integration into an annotated gene occurred 54% of the time, and 71% of all tumors harbored integrated HPV [44]. Amplification, enhanced oncogene expression, loss of function of tumor suppressors and loss- or gain-of-function fusion transcripts are all potential deleterious effects of HPV integration that can result in cancer phenotypes. For example, integration of HPV upstream of NR4A2, a transcription factor often activated in cancers [53], led to 250-fold amplification of the downstream region and induction of NR4A2 [44] (again see Table 6.1 for the definition of abbreviations and putative gene functions). In another example, HPV integration 150 base pairs upstream of a translocation junction involving chromosomes 3 and 13 led to the induction of the oncogenes KLF5, TP63, and TPRG1 [44]. In addition to determining cancer phenotype, defined integration events may also drive treatment responses, with integration in intergenic regions correlating with positive treatment outcomes and integration into or near cancer-related genes predicting recurrence [48]. While the functional consequences of HPV integration can be revealed through modern genomic and phenotypic studies, the mechanisms that drive HPV integration are not well understood.

6.1.4 Epigenetic Regulation of the Host Genome in HPV-Associated HNSCC: Aberrant DNA Methylation

Recent studies have begun to reveal the complex mechanisms of carcinogenesis in HPV-associated epithelial neoplasms in molecular detail. These tumors demonstrate genetic and epigenetic changes that promote tumor phenotypes via many of the same cell signaling pathways that have been described in a variety of HPV-negative tumors. By far, the most common epigenetic modification of the human genome is the addition of a methyl group to the carbon-5 position of cytosine nucleotides, predominantly those that immediately precede a guanosine nucleotide (CpG dinucleotides, or CpG islands). Aberrant methylation and subsequent chromatin remodeling of promoter regions and first exons of coding genes often results in silencing of critical genes in HNSCC, for example, the cell cycle/cell fate regulators CDKN2A and DAPK and the DNA repair genes MGMT and MLH1 [54]. Cases of HNSCC that test positive for HPV are almost three times more differentially methylated at CpG loci than HPV-negative cases of HNSCC when comparing tumor to adjacent normal tissue [55]. In HNSCC-derived cell lines, greater frequency of gene methylation is seen in HPV-positive lines than in HPV-negative lines [56]. Expression of DNA methyltransferases such as DNMT1 and DNMT3A may be upregulated in HPV-positive OPSCC [56, 57]. Ectopic expression of HPV-16 E7 and infection of keratinocytes in vitro confirmed induction of both DNMT1 and DNMT3A [58]. Anayannis et al. [59] proposed that this effect may be due to the activity of E7 via release of E2F (subsequent to pRB degradation) that stimulates DNMT1 transcription [60]. Anayannis et al. also proposed that E6 may contribute to induction of DNMT1 via release of Sp1 transcription factor from inactivating complexes with p53 [61]. The E7 protein may interact directly with

DNMT1 via its zinc finger domain and promote methyltransferase activity (in in vitro studies) [62]. Expression of DNMT1 is also induced in cervical cancer [63], supporting the role of the HPV oncoproteins in mediating DNA methylation. Methylation patterns in OPSCC are sufficiently consistent such that analysis of methylation of a panel of 22 CpG loci can distinguish HPV-positive tumors from HPV-negative tumors [57]. Four of these loci are found in the CDKN2A locus, downstream of the transcription start site of p16. Hypermethylation of this region correlated with increased expression of $p14^{ARF}$ and $p16^{INK4A}$ proteins [57].

HPV-positive OPSCC has higher levels of gene promoter methylation than HPV-negative OPSCC [64]. Promoter methylation signatures associated with HPV infection in OPSCC can be identified and include three cadherins of the polycomb group target genes [65]. The predicted consequence of this modification is silencing of cadherin expression, leading to disruption of cell-cell adhesion and dysregulation of tissue morphogenesis. Other promoters found to be hypermethylated in HPV-positive HNSCC include cell cycle regulators CCNA1 (cyclin A1) [66] and TP73 [67]; invasion and metastasis mediators CADM1, CDH13, TIMP3 [67] and IGSF4 [68]; mediator of WNT signaling SFRP4 [69]; signaling mediator receptors ESR1 and RARβ [64]; decider of cell fate APC [67]; and proapoptotic DAPK [67]. The latter promoter hypermethylation events (ESR1, RARβ, APC, and DAPK) have also been identified frequently in HPV-negative OPSCC, suggesting that perturbation of these pathways is important for the tumor phenotype in oropharyngeal tissue. These hypermethylated promoters have all been reported in isolated studies and without empiric data to solidify their mechanistic roles in the tumors. However, specific gene promoter methylation signatures have been shown to predict prognosis of HPV-positive OPSCC, for example, ALDH1A2, GATA4, GFR4, IRX4, and OSR2 [70], suggesting that they may have a mechanistic role in promoting tumor progression. Finally, in addition to promoter methylation, host gene expression may be regulated via the interactions of HPV E6 and E7 with cellular p300 and MYC, two major activators of promoter enhancers [71–74].

6.1.5 Epigenetic Regulation of the Host Genome in HPV-Associated HNSCC: Dysregulation of Gene Transcription via Host Chromatin Modification

Methylation and acetylation of histones modify the structure of host chromatin and regulate transcriptional activity of host genes (Fig. 6.1). Histone acetyltransferases (HATs) produce an open chromatin structure to allow transcription factors and coactivators to bind and induce transcription; histone deacetylases (HDACs) promote condensed and inactive chromatin. Histone methyltransferases (HMTs) methylate H3 and H4 histone tails at arginine and lysine residues. Both acetylation and methylation are reversible; however, acetylation always results in open chromatin whereas methylation can result in either activation or repression depending on the specific residue that is methylated. Demethylation occurs through the activity of histone demethylases (HDMs). Therefore, transcriptionally active areas of the

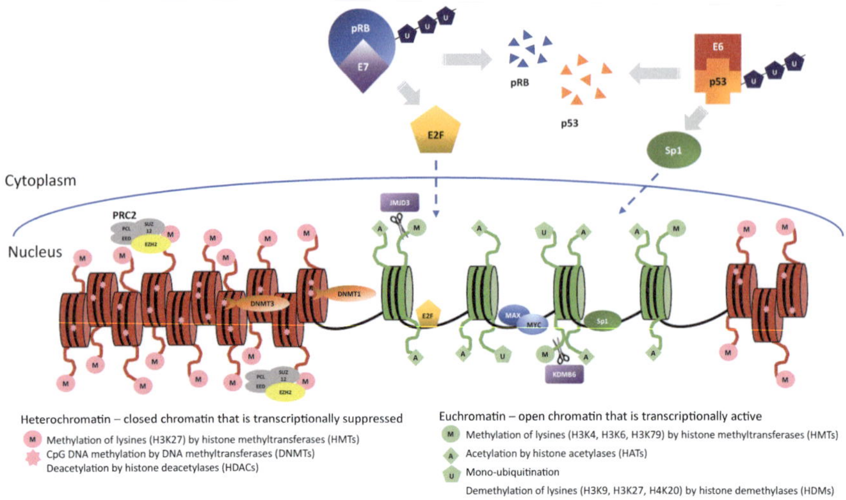

Fig. 6.1 Epigenetic modifications associated with HPV-mediated oncogenesis. Degradation of p53 by HPV E6 and pRB by HPV E7 releases transcription factors such as Sp1 and E2F for nuclear importation and transcription initiation. The HPV oncoproteins promote the activity of DNA methyltransferases DNMT1 and DNMT3, resulting in silencing of host tumor suppressors. Induction of EZH2 promotes activity of polycomb repressor complex 2 (PRC2) to repress chromatin via trimethylation of lysine 27 on histone 3 (H3K27me3). E7 also activates histone acetyltransferases (HATs), inactivates histone deacetylases (HDACs) and promotes the activity of histone demethylases such as KDM6B and JMJD3 to promote open chromatin and host gene transcription. Interactions between E6/E7 and the promoter enhancer activators p300 and MYC also promote host gene transcription. Diagram not drawn to scale

genome are unmethylated at promoter CpG islands and have loose chromatin structure to allow transcription factors and regulatory proteins to bind. Loose chromatin is characterized by acetylation of histones at lysine residues and flanking nucleosomes that are trimethylated at lysine 4 on histone 3 (designated H3K4me3; reviewed by Baylin and Jones [75]). Transcriptionally inactive areas generally are in complex with inhibitory polycomb group complex proteins, for example, EZH2, a histone-lysine *N*-methyltransferase, which catalyzes the trimethylation of lysine 27 on histone 3 (H3K27me3), resulting in a repressive phenotype.

Host histone modification has been shown to be mediated by the HPV E7 protein. The E7-mediated activation of histone acetyltransferases and inactivation of histone deacetylases leads to acetylation of histone tails and chromatin opening [76]. This is consistent with the functional role of E7 in dysregulating cell cycle progression and growth, which requires activation of gene transcription. For instance, the E7-mediated induction of the lysine demethylase KDM6B leads to removal of repressive H3K27me3 marks from the CDKN2A promoter, allowing transcriptional activation of the gene locus [77]. Alternatively, the JMJD3 demethylase activates CDKN2A via RAS [78]. The result of these activation events is the production of p16^{INK4A}, a tumor suppressor protein that is frequently upregulated in cancer and is an established marker of HPV-associated dysplastic tissue, as noted above. Conversely, the E7-mediated release of E2F promotes transcription of the polycomb repressor complex (PRC2) methyltransferase EZH2 in cervical cancer cell lines [79]. In HPV-positive HNSCC, signatures indicative of EZH2 activity include global elevation of H3K27me3 marker of transcriptional repression [80] and hypermethylation of the PRC2 promoter in HNSCC cell lines [56]. Clinically, HDAC inhibitors may be promising adjunctive therapeutics for HNSCC [81–85] but their role in HPV-positive OPSCC specifically is not known.

6.1.6 Epigenetic Regulation of the Host Genome in HPV-Associated HNSCC: Dysregulation of MicroRNA Expression

A third epigenetic mechanism by which HPV may drive host cell tumor phenotypes is through dysregulation of cellular microRNAs (miRNAs). These small noncoding RNA molecules regulate protein production by mediating interactions between translationally suppressive RNA-induced silencing complexes (RISC) and cognate sequences encoded within the 3-prime untranslated region of mRNA transcripts. MicroRNA expression signatures can differentiate HPV-positive from HPV-negative HNSCC [86, 87]. Further, miRNA dysregulation appears to be similar in HPV-positive HNSCC and cervical cancer, suggesting the involvement of HPV and likely E6/E7 in the dysregulation of host miRNAs [88]. Common features between HPV-positive HNSCC and cervical cancer include the induction of miR-15a and miR-16 and suppression of miR-195 and miR-497 [86]. Several miRNAs have been reported to be dysregulated in HPV-associated HNSCC by more than one group, including induction of miR-9 [87, 89, 90], miR-20b [87, 90], and miR-363 [87, 91] and down-regulation of miR-126, miR-143, miR-145, and miR-199a/b [87, 91]. In addition to viral-mediated miRNA dysregulation, single nucleotide polymorphisms in the precursor sequences of miRNAs may increase the risk of OPSCC by altering their ability to be cleaved into mature, functional miRNAs. Examples include miR-146, miR-149, miR-196, and miR-499 [92].

The functional consequence of aberrant miRNA expression is complex to unravel, with multiple miRNAs often targeting a single transcript and multiple transcripts being targeted by a single miRNA; however, some research has shed light into possible oncogenic mechanisms resulting from miRNA dysregulation. For instance, the oncogenic miRNA miR-21, which is consistently induced in HPV-positive OPSCC, targets the tumor suppressors PTEN, TPM1, and Bcl-2 [93]. Expression of the tumor suppressor PDCD4 is lost in the majority of tonsil-derived tumors and may be a consequence of translational suppression by miR-21 and miR-499 working in concert [94]. Induction of miR-363 is purportedly mediated by HPV E6 [95] and may promote a less-aggressive tumor phenotype via reduction of MYO1B (myosin 1B), a protein that promotes cell migration and invasion in vitro [96].

MicroRNA expression may be an important prognosticator of patient outcomes. Low levels of let-7d, miR-205 [97], and miR-375 [98] were shown to predict prognosis of HNSCC independently of HPV status. Other studies showed miRNA profiles that may predict prognosis but no consistent prognostic miRNA profile has emerged, largely due to heterogeneity in populations, variable inclusion of cases (all HNSCC sites vs. OPSCC only), and inconsistent methodologies [89, 90, 99–102]. Still, with careful parsing of patient subgroups and application of standardized techniques, this area shows promise for future clinical utility. Of note, loss of miR-375 expression is associated with both HPV-positivity of HNSCC and poor prognosis [98]. The loss of miR-375 was demonstrated to be the result of epigenetic silencing via E6-mediated induction of DNMT1 in HPV-16-positive cervical cancer cells [103]. Ectopic expression of miR-375 directly targeted E6/E7 transcripts, activated p21 and suppressed telomerase activity in HNSCC and cervical cancer models [104]. An important target of miR-375 is MALAT1, a long noncoding RNA associated with tumor phenotypes [103]. In the absence of miR-375 MALAT1 is overexpressed. Ectopic expression of HPV16 E6 in oral keratinocytes led to increased expression of MALAT1 and this effect was associated with loss of p53 [105]. Conversely, knockdown of E6/E7 expression in cervical cancer cells resulted in a reduction in MALAT1 expression [106]. The consequences of overexpression of MALAT1 include invasive and epithelial–mesenchymal transition (EMT) tumor phenotypes [98, 103, 107] and may explain the association between loss of miR-375 expression and poor patient prognosis.

6.1.7 Epigenetic Modification of the Viral Genome in HPV-Mediated HNSCC

Epigenetic modifications may promote oncogenesis not only through changes in host gene expression but also by changes in viral gene expression. Papillomaviruses do not encode any proteins with methyltransferase activity, but the HPV genome does have CpG dinucleotides within conserved palindromic sequences [108]. The viral genome is often methylated when flanking host DNA is methylated [109]. Even though this may be a bystander effect, it could still have an impact on viral gene expression. Late gene regions (at the boundary of L1 and L2) may be methylated in cervical cancer [110, 111] and HPV-positive HNSCC [65, 112], but HPV-16 LCR/E6/E7 promoter regions are generally unmethylated in HPV-positive HNSCC [65, 112]. Methylation of the viral long control regions may be a strategy used by the host to control the virus or by the virus to control viral gene expression. For instance, demethylation of the HPV-16 LCR in an OPSCC cell line using 5-aza-2′-deoxycytidine caused repression of E6/E7 transcript expression followed by cell cycle arrest at G2/M checkpoint [113], indicating that methylation of the HPV LCR led to high E6/E7 expression and tumor phenotype. Further, HPV-positive OPSCC with intact HPV E2 sequences (either episomal virus or integrated concatemers) had partial to complete methylation of the E2 binding sites 3 and 4 [114]. High methylation at those sites corresponded to high E6/E7 transcript expression and poor prognosis [114] and may explain cases of OPSCC in which HPV remains episomal with an intact E2 open reading frame and high E6/E7 expression.

6.1.8 Noncanonical Cell Signaling Pathways Involved in HPV-Mediated HNSCC

While the oncoproteins of HPV are thought to play a major role in promoting sustained cell proliferation, somatic mutations of the host genome may be crucial to achieving a fully transformed phenotype. A proportion of the somatic mutations uncovered in HPV-associated HNSCC may be the consequence of off-target damage to the host genome resulting from a potent host cell response to viral infection. The class of enzymes known as apolipoprotein B mRNA editing catalytic polypeptide-like (APOBEC) consists of cytosine deaminases that restrict viral replication by mutating viral DNA (reviewed by Harris and Dudley [115]). Bystander mutation of host DNA occurs as a consequence of sustained APOBEC activity, and therefore APOBEC-mediated mutations are common events in virally induced cancers [116]. These host somatic mutations are marked by cytosine-to-thymine or guanine point mutations at TpC sites [116]. Infection with HPV appears to induce APOBEC3A which may improve clearance of virus [117]. Infection with high-risk HPV also induces APOBEC3B via the activity of E6 and expression of E6 may exacerbate APOBEC mutagenesis [118], with the consequence of increased off-target mutagenesis of host DNA.

The activity of APOBEC can result in C-to-T mutations in two hotspots in the PIK3CA gene. These mutations cause amino acid changes that result in gain of function and constitutive activation of the PIK3CA gene product, the p110α catalytic subunit of phosphoinositol 3-kinase (PI3K) [119]. PIK3CA is a component of the phosphatidylinositol 3-kinase/protein kinase B (AKT)/mammalian target of rapamycin (mTOR) pathway, which is often dysregulated in cancers and is involved in cell growth, proliferation, differentiation, glucose metabolism, protein synthesis, and apoptosis [120–123]. PIK3CA is considered to be an oncogene since activating mutations and duplications of PIK3CA result in unchecked growth, invasion and metastasis [124]. Mutation of PIK3CA occurs in 22–56% of HPV-positive HNSCC [50, 125–127]. Hotspots for PIK3CA mutation in HPV-positive HNSCC include E542K and E545K mutations in the helical domain. These mutations are a result of C-to-T changes in TCW genomic regions, consistent with APOBEC induced mutagenesis [119, 128–130]. The proposed mechanism by which these two amino acid changes evoke constitutive activation of PI3K is via attenuation of the binding capacity of the negative regulator p85α unit to p110α [131]. The importance of the role of PIK3CA in HPV-associated carcinogenesis is further supported by the frequent finding of PIK3CA gene amplification in up to 52% of all HPV-positive HNSCC [50, 125] and PIK3CA mutation in up to 42% of cervical cancers [126, 132–137]. To date, PIK3CA is the most frequently mutated gene reported in HPV-positive cancers.

Activation of PI3K signaling is further exacerbated by mutation and subsequent loss of expression of the tumor suppressor and cell cycle regulator PTEN. The protein product of PTEN regulates PI3K signaling via dephosphorylation of the PI3K target PIP3, which in turn interacts with PDK1 to activate AKT signaling. Somatic mutation of PTEN has been described in HPV-positive HNSCC [50, 125], with 24–56% of HPV-positive HNSCC demonstrating both PIK3CA and PTEN mutations [122, 138]. Remarkably, more than 80% of anal cancers with PIK3CA mutations also harbored PTEN mutations [122] indicating the importance of the combined gain of function of PIK3CA and loss of function of PTEN and sustained activation of the PI3K/AKT/mTOR pathway in HPV-mediated tumorigenesis. Using murine xenograft model systems, treatment of anal cancer with the mTOR inhibitor rapamycin resulted in reduced growth [139], suggesting that this pathway may be a tractable target for adjunctive HPV-associated cancer therapy.

Somatic mutations that promote signaling via PI3K activation and the RAS/MEK/ERK signaling axis have been identified in multiple receptor tyrosine kinases in HPV-positive HNSCC. Activating mutations have been described in FGFR2 and/or FGFR3 kinases in 10–17% of HPV-positive HNSCC [50, 125], KRAS in 6% [127], and HRAS in 1–12% [125]. Inactivating somatic mutation in NF1, a negative regulator of the RAS signaling pathway, was seen in up to 14% of HPV-positive HNSCC [50, 125]. Activation of these signaling pathways promotes cell cycle progression, proliferation, and survival. Kinase inhibitors may be effective in defined subsets of patients with these mutations [127], though this remains under investigation.

In addition to the above receptor tyrosine kinases, the receptor tyrosine kinases of the epidermal growth factor receptor family, ERBB2/HER2 and ERBB3/HER3, may also be induced by HPV E6 and E7 expression. Members of the ERBB family promote cell cycle progression and proliferation and are frequently activated in solid organ tumors. Expression of HER2 and HER3 and receptor complexes indicative of signaling activation were shown to be elevated in HPV-positive tumors [140]. Silencing of E6 and E7 in tumor cell lines mitigated HER3 expression [141], linking the activity of growth factor receptor tyrosine kinases with HPV oncogene expression. Clinically this suggests that targeted therapy with Afatinib, an anti-ErbB family small molecule inhibitor, may be an effective therapeutic approach. However, recent results from a multicenter Phase-III clinical trial reporting efficacy of Afatinib in HNSCC patients demonstrated little to no benefit for patients with HPV-positive tumors [142]. It is possible that better definition of tumor phenotypes will reveal a subset of patients who will respond favorably to ERBB family blockers.

Genetic changes in immune response related genes may also facilitate HPV-associated tumor development. Presentation of antigen in MHC class I molecules (human leukocyte antigens, HLA) facilitate immune clearance of viral infections. Germline variants in the HLA-A/B genes are associated with HPV-positive HNSCC risk [143]. Somatic mutations in HLA-A/B genes are also identified in 11% of HPV-positive HNSCC [127]. Similar findings have been described in cervical cancer [144, 145]. Activation of receptor signaling on immune cells is often mediated via NF-κB signal transduction, and constitutive activation of this pathway promotes cell proliferation and immortalization. Inactivating somatic mutations in two negative regulators of NF-κB signaling, TRAF3, and CYLD, have been described in

HPV-positive HNSCC [50, 146]. Interestingly, while TRAF3 deletions (14%) or truncating somatic mutations (8%) were common in HPV-positive HNSCC [50], TRAF3 does not appear to be commonly mutated in cervical cancers [132].

In summary, HPV-mediated oncogenesis in the oropharynx is similar to that in the anogenital regions, with a few key differences. Elevated expression of the viral oncoproteins E6 and E7 appears to play a critical role in promoting the tumor phenotype, with less of a reliance on integration of the viral genome into the host genome than what is typically observed in anogenital tract tumors. Progression of tumor formation is clearly a multistep process involving changes to the host genome. Some of these changes, such as epigenetic modification of the host chromatin, may be reversible; others, such as APOBEC-mediated somatic mutations and integration-associated disruption of the host genome, are irreversible. Improved understanding of the specific cell signaling pathways that are frequently altered during HPV-mediated tumorigenesis continues to reveal opportunities for personalized adjunctive therapies to treat patients with HPV-associated HNSCC.

6.2 Genotype Prevalence and Risk Factors for Oral HPV Infection in the HIV-Seropositive Individual

6.2.1 Oral HPV Infection Is More Common in HIV-Infected Individuals

The majority of oral HPV infections are asymptomatic, and the development of oral HPV-associated cancer is a rare event that presumably occurs only after persistent infection. Therefore, epidemiological studies often examine the risk factors for asymptomatic detection of oral HPV infection. Only a handful of studies compared the rates of HPV infection in HIV-positive and HIV-negative individuals. Oral HPV infection is associated with markers of sexual risk, such as homosexuality, unprotected oral sex, and a history of previous sexually transmitted disease (STD), indicating that HPV is likely transmitted to the oral cavity through sexual contact.

Early studies examined relatively small cohorts (n = 1 to 300) comparing infection in HIV-positive and negative individuals [147–152]. These studies showed a significant increase in prevalent oral HPV infections of 3–7-fold in HIV-positive individuals as compared to HIV-negative individuals. Low CD4+ T cell counts, increased number of oral sex partners and evidence of other STDs (chlamydia and herpes simplex virus, HSV) were associated with oral HPV infection in the HIV-positive cohort. Interestingly, other studies have not shown an association with lower CD4+ T cell counts [152]. Risk factor analysis for oral HPV infection in a study of HIV-positive individuals from New Orleans noted increases in Caucasians (57% vs. 24%) and males (68% vs. 25%) [147]. It is of interest that neither CD4+ T cell count nor HIV viral load correlated with the presence of oral HPV [147].

In the U.S., two large cohort studies have examined the prevalence and risk factors for oral HPV infection in HIV-positive individuals. The Woman's Interagency HIV Study (WIHS) enrolled 2794 HIV-seropositive women and 972 high-risk HIV seronegative women in two installments (from 1994 to 1995 and from 2001 to 2002) at clinics in New York City, Boston, Washington D.C., Chicago, Los Angeles, and San Francisco [149]. The Multicenter AIDS Cohort Study (MACS) recruited 2963 HIV-positive men and 4124 HIV-negative men from Baltimore, Chicago, Pittsburgh, and Los Angeles since 1987 [153]. These large cohort studies confirmed the early work showing a significantly higher prevalence of oral HPV infection (25–40%) in HIV-positive individuals as compared to HIV-negative subjects (9–25%). The overall risk factors for HPV oral infection included smoking, lower CD4+ T cell count, and higher lifetime number of sex partners. Lower CD4+ T cell counts at sampling predicted oral HPV risk better than nadir CD4+ T cell count. Studies from Europe and Australia, where HPV-associated oropharyngeal cancer incidence is rising as it is in North America, also found a 2–3-fold increased prevalence of oral HPV infection in HIV-positive subjects [154–156]. In these cohorts, an increased reported lifetime number of oral sex partners were associated with the presence of an oral HPV infection. Finally, a recent meta-analysis on oral infection in men who have sex with men (MSMs) substantiated the increase in high-risk HPV infections in HIV-positive men as compared to similarly risked HIV-negative men [157].

One striking observation in HIV-negative populations is that oral HPV infection is more common in males, which parallels the increased incidence of HPV-positive HNSCC predominantly affecting men [151, 152]. This male predominance generally seems to be found in the HIV-positive population as well, with 20–68% of men and 20–38% of women testing positive for oral HPV infection [147]. Interestingly, both subclinical HPV infection and OPSCC appear to be more common in heterosexual men than in men who have sex with men (MSMs) [153]. Although the data did not reach statistical significance, this observation was consistent among three different studies [150, 158, 159]. Whether this observation can be attributed to differences in sexual behaviors, independency of sexual networks, or gender-specific differences in transmission efficiency between mucosal sites is not known. In contrast, a recent large international study found a higher proportion of OPSCC cases tested positive for HPV in women than in men (see details below), suggesting that regional environmental or behavioral factors may be more important than de facto gender in terms of oral HPV acquisition and OPSCC [38]. A study focusing on a cohort of women who have sex with women would be of interest to better delineate the role of gender in the risk of oral HPV infection and OPSCC.

6.2.2 Genotypes in Oral HPV Infection in HIV-Infected Individuals

In addition to defining the rates of oral infection, an important consideration is the genotype(s) of HPV that are found in the oral cavity. HPV-16 is detected in 95% of

OPSCC [9, 151] and therefore many studies have determined the rates of asymptomatic HPV-16 oral infection. The other high-risk HPV genotypes included in screening tests vary greatly among studies. Most studies showed that the prevalence of HPV-16 varied from 0.6 to 6% in HIV-positive individuals and HPV-16 made up between 1.4 and 18% of the detectable oral HPV infections [147, 155, 156, 160–162]. Prevalence of HPV-33 was reportedly between 0.6 and 10% in a few studies and it accounted for 6–29% of detected oral HPV infections [155, 160–163]. The prevalence of other high-risk genotypes is more difficult to assess since studies did not consistently test for the same range of genotypes.

Low-risk HPV genotypes-6 and -11 are classically associated with genital warts and are expected to be found in the oral cavity since oral sex is believed to facilitate transmission between mucosal sites. The prevalence of HPV-6 was up to 4% of HIV-positive individuals and it accounted for up to 9% of oral HPV infections [147, 155, 160–163]. Similarly, the prevalence of HPV-11 was 0.6–6% in the HIV-infected population and made up 1.4–18% of the oral HPV infections detected [155, 160, 162]. Other HPV genotypes that were commonly found were HPV-55 and HPV-83, with prevalence similar to HPV-6 and -11 (up to 5% of the subjects and up to 15% of the detected oral HPV infections) [147, 162]. It is not clear how often HPV-55 is seen in oral pathology, but its classification as a high-risk HPV genotype and its relative frequency of detection may warrant further investigation [147, 161, 163]. The most complete HPV genotyping study published to date utilized a novel approach to simultaneously detect the common anogenital HPV genotypes (alpha) as well as the cutaneous beta and gamma genotypes of HPV [164]. By this more comprehensive approach, 87% of a cohort of 52 HIV-positive individuals was positive for any genotype of HPV. The oncogenic alpha genotypes (-16, -18, and others) were found in 23% and the non-oncogenic alpha types in 40% of the participants, which is comparable to the above studies. Of note, only coinfection with hepatitis C was seen as a risk factor for oncogenic oral HPV infections in this study.

Curiously, genotypes which show a predilection for the oral cavity (HPV-7, -13, and -32) have rarely been studied. Pursuant to an outbreak of HPV-32-positive oral warts in HIV-seropositive patients initiating combination antiretroviral therapy, Cameron and Hagensee found HPV-32 in 9.5% of all HIV subjects examined (Fig. 6.2) [147] from a cohort enrolled from 2000 to 2004. In a follow-up study in HIV-positive individuals (2013–2017), the prevalence of oral HPV-32 infection was only 4.8%, which is a reduction of 49%. The reduction in subclinical detection of HPV-32 infection coincides with clinical evidence that the prevalence of oral warts in HIV-positive individuals had declined (M. Hagensee, unpublished observation). The prevalence of oral HPV-16 did not decline as drastically in the same time span (from 1.3% to 1.1, 15% reduction), suggesting that the reduction is not due to general HPV acquisition or clearance.

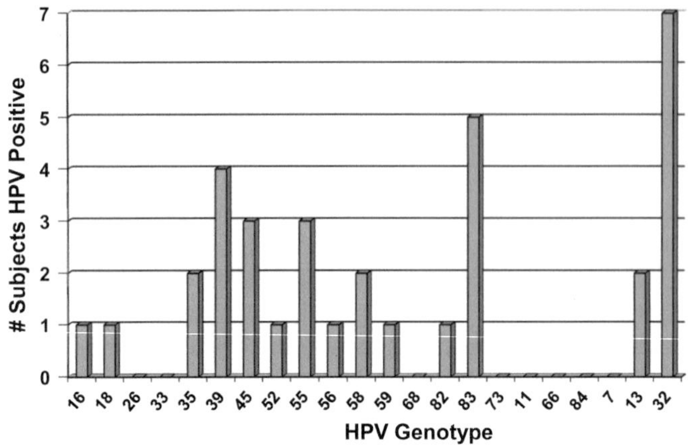

Fig. 6.2 The HPV genotypes found in the oral cavity of HIV-positive individuals

6.2.3 Comparison of Oral and Genital HPV Genotype Prevalence

In studies that compared oral HPV infection to HPV infection of the anogenital tract in the same individual, oral HPV infection was detected less frequently than anogenital infection. The most common site of HPV infection in HIV-positive men was the anus, with 73–97% testing positive, followed by 25–44% testing positive at the penis and 16–44% testing positive in the oral cavity [155, 163, 165, 166]. High-oncogenic risk HPV genotypes were also found less often in the oral cavity (11–27% HPV-positive) than the anus (24–86% HPV-positive) or the penis (23–41% HPV-positive) [155, 163, 165, 167]. A similar trend was observed in women, with oral HPV prevalence between 20 and 25% and prevalence at the cervix between 76 and 97% [150, 168, 169]. High-risk HPV genotypes were seen in the oral cavity in 3–15% of those screened as compared to 58–63% HPV-positive at the cervix [150, 168, 169]. There was a notable lack of HPV genotype concordance between the oral cavity and genital tract [155, 168]. It is logical to predict that the HPV genotypes found in the oral cavity of an individual might better reflect those detected in his or her recent sex partner. Prospective studies in monogamous couples would be of interest to examine the kinetics of HPV transmission via various forms of sexual contact.

6.2.4 Oral HPV Incidence, Persistence, and Resolution

The natural history of HPV infection in the female genital tract has been well described. Cervical HPV infection is typically cleared naturally without clinical intervention, but the virus often establishes infection for more than a year before being cleared. Continued persistence of the virus is essential to malignant transformation and therefore persistent infection puts the host at risk for dysplasia that can progress to cancer. The natural history of oral HPV infection has been more challenging to investigate because of the difficulty in pinpointing the exact site of infection for prospective sampling, the lack of well-defined precancerous pathology in the oropharynx, and the relatively lower frequency of HPV infection in the oral cavity. Additionally, there are no consensus definitions for HPV acquisition or persistence and clearance, and these outcomes are dependent on sampling frequency and duration of prospective follow-up. Despite inconsistencies in study design and outcome definitions across published prospective studies, a picture of the natural history of oral HPV infection is beginning to emerge.

Defining HPV incidence as the detection of any new HPV genotype at a follow-up visit that was not detected at the previous visit or baseline, incidence ranged between 4.8 and 24 per 100 person-years for any HPV, 3.2–9.0 per 100 person-years for any high-risk HPV, and 0.8–1.2 per 100 person-years for HPV-16 in HIV infected individuals [153, 157, 170]. The highest incidence rates were seen for HPV-16 and -18 [171]. Incident detection of HPV was associated with lower CD4+ T cell count and increased numbers of oral sex partners [153]. When persistence was defined as an HPV genotype seen at baseline and at the next consecutive visit, 35–75% of infections met the criteria for persistence [153, 155, 157, 170, 172]. Risk factors for increased persistence included male gender, smoking, lower CD4+ T cell count, increased duration of HIV infection, taking ART, and increased time on ART [153, 155, 172]. Interestingly, one study showed increased HPV persistence in those who have had a tonsillectomy [170].

Conversely, 44–83% of HPV infections were considered to have cleared when clearance was defined as the absence of detection of a HPV genotype seen at baseline at the subsequent follow-up visit [153, 155, 170, 172]. Sampling error leads to false-negative HPV tests, which can artificially inflate clearance estimates. Using the more stringent criteria of two consecutive HPV-negative visits after a positive test, clearance decreased to 35–53% of infections detected at baseline [153, 157, 166, 171]. Factors associated with clearance included female gender, shorter duration of HIV infection and no previous history of sexually transmitted infections [166]. However, increasing number of oral sex partners and the presence of anal warts may also predict the clearance of an oral HPV infection [171]. Finally, incident infections detected during prospective follow-up cleared more quickly than prevalent infections detected at baseline.

Using a different approach, Lam et al. attempted to define a serum immunological marker that would predict those who would clear an oral HPV infection [173]. A cohort of 1601 adults (75% with HIV infection) had oral rinse samples collected and tested for HPV status at 6-month intervals for 4 years. From these, a multiplex cytokine assay from paired serum samples was performed in those with incident HPV infection, prevalent infection and no HPV infection (roughly 300 in each group). Serum TNF-alpha, IL-8, IFN-gamma, IL-10, and IL-2 levels were most highly elevated in those with prevalent HPV infection and were moderately elevated in those with incident infection compared to the HPV-negative group. Higher TNF-alpha levels were seen in those with persistent oral HPV infection in both men and women with higher IL-2 levels seen with persistent infection in men.

6.2.5 Oral Warts in the HIV-Positive Individual—The Benign Tumors Caused by HPV

Papillomas due to HPV infection can occur on virtually all oral mucosal surfaces [174]. While the majority of papillomas occur on the labial mucosa, they can also occur on the buccal mucosa, the tongue, the soft palate, and the gingiva. While the histopathology of oral warts almost invariably demonstrates poorly differentiated, large, vacuolated koilocytic cells, the gross appearance varies greatly. Often the clinical appearance of lesions is reflective of the specific HPV genotype causing the lesion. For instance, HPV genotypes 6 and 11 tend to cause soft, sessile, and cauliflower-like lesions (condyloma accuminatum) in the oral cavity. HPV genotypes 1, 2, and 7, which are associated with cutaneous warts, cause firm, sessile, and oral common warts (verruca vulgaris). HPV genotypes 13 and 32, which have been described exclusively in the oral cavity, are the cause of oral focal epithelial hyperplasia (FEH), a dysplastic lesion characterized by multiple small, flat papules generally found on the lower lip. While there is some degree of HPV genotype-specific clinical presentation, unusual manifestations of oral HPV disease in the HIV-positive patient frequently occur [174]. Examinations of oral wart biopsies from HIV-positive individuals prior to the routine use of highly active antiretroviral therapy (HAART) contain a range of HPV genotypes, including cutaneous type 2; genital types 6, 11, 16, and 18; and oral type 13 [174]. However, the most common HPV genotypes identified in HIV-associated oral warts are the oral-specific HPV type 32 and the cutaneous HPV type 7 [174]. A study by Cameron and Hagensee examined the genotypes of HPV involved in oral warts in the HIV-positive individual in the setting of routine use of HAART. The vast majority of these samples were HPV-32 positive with other HPVs detected being 6, 7, 53, 73, and 84 [147]. Additional studies investigating the HPV genotypes found in warts and other oral lesions found HPV types 6, 11, 13, and 32 to be the most prevalent [175]. Cumulatively, pooling the limited data from these studies, HPV-13 and -32 are the most common HPV genotypes seen in oral warts in HIV-positive individuals (Fig. 6.3) [147, 174, 175].

6.2.6 Treatment and Prognosis of HIV-Associated Oral Warts

The treatment of oral warts in the HIV-positive patient is difficult due to both the wide distribution of lesions throughout the oral mucosa and the high recurrence rate. Although these lesions are generally painless, they can become traumatized. They can mechanically interfere with eating and talking and look unsightly when present externally on the lips. Treatments utilized include both medical and surgical modalities, depending upon the site of the wart, the characteristics of the wart, and the

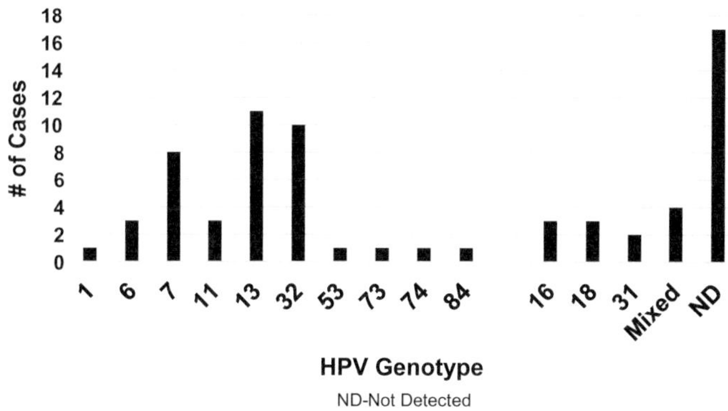

Fig. 6.3 The HPV genotypes found in biopsies of oral warts in HIV-positive individuals. Data combined from three studies [147, 174, 175]. ND, no HPV detected

number of lesions. Surgical techniques include excision, electrosurgery, cryosurgery, and CO_2 laser, whereas medical modalities include podophyllin resin and interferon-α injections [176]. Treatment with Imiquimod is indicated for cutaneous and genital warts, but this topical medicine is not approved for use in the oral cavity.

Surgical excision is difficult when multiple warts cover a large area. However, surgical or electrosurgical debulking of the wart is useful prior to use of topical agents or the CO_2 laser. The use of the CO_2 laser has proven problematic because the dispersal of HPV in the laser plume can lead to nasal warts in either operator or patient [176]. The use of surgical techniques does not often lead to postoperative scarring of the intraoral mucosa, but scarring of the lips can occur, leading to stricture and diminished opening. The use of podophyllin resin (25%) as a topical agent has proven disappointing [176]. Case reports showed promising results of the use of interferon-α as a topical and systemic combination with weekly intralesional injections in addition to twice-weekly subcutaneous injections. Follow-up ranged from 12 months to three years, with no recurrence of warts at the site of treatment [177]. Topical 1% Cidofovir gel was reportedly successful in treating a case of recalcitrant oral warts in a HIV-positive individual [178], but data from controlled clinical trials are lacking.

Trials focusing on nongenital warts may provide some new treatment options. Taking oral zinc sulfate 600 mg/day led to resolution of 50% of nongenital (skin) warts after two months of treatment [179]. In addition, oral isotretinoin was shown superior to topical isotretinoin in the treatment of nongenital warts [180] with a 69% remission rate. It would be of interest to determine the efficacy of these medications for the treatment of oral warts.

6.2.7 Is a Wart a Cancer in Disguise?

Up to one-half of genital warts may harbor concomitant infections with both low- and high-risk HPV genotypes [181]. The clinical concern is that a lesion that appears benign may actually have the potential to transform into a malignant lesion. A published case report indicated that a clinically diagnosed oral wart showed an epithelial neoplasm upon histological examination of the biopsy [182]. In situ hybridization demonstrated focal positivity for HPV-16 and -18. This suggests that high-risk HPV genotypes can masquerade as low-risk genotypes and cause warts in the oral cavity [182]. It appears, however, that this is a rare occurrence since reports utilizing sensitive real-time PCR assays and p16 staining have failed to demonstrate high-risk HPV genotypes in oral wart tissues [181, 183]. Importantly, these studies did not investigate HIV-infected individuals that are prone to infections with multiple HPV genotypes and the possibility that warts in this population might mask concomitant infections with high-risk HPV genotypes.

6.2.8 The Role of HPV Infection in HNSCC in HIV-Positive and HIV-Negative Individuals

Long associated with cancers and precancerous lesions of the anogenital mucosa, HPV is also involved in the etiology of a subset of head and neck squamous cell carcinomas [151, 184]. Over 50% of tumors arising from the lingual and palatine tonsils (oropharynx) contain HPV DNA with over 90% of this harboring HPV-16. These tumors arise mostly in upper income, younger (2–5 years) white men than those that are HPV-negative. This is likely due to sexual exposure (i.e., oral sex) [151, 184]. These tumors present at an earlier tumor stage but a more advanced nodal stage with a nonkeratinized basaloid histopathology. Surprisingly, they also respond better to conventional chemotherapy and radiotherapy [184, 185].

Parameters of HPV infection are highly correlated with the development of OPSCC. Seropositivity to HPV-16 viral capsid proteins (L1) confers a 32-fold increased risk of OPSCC [151]. Those with an oral HPV-16 infection as determined by PCR detection have a 50–200-fold higher likelihood of developing OPSCC [175, 186]. Since oral HPV infection is more common in HIV-positive individuals, it is of concern that HNSCC due to HPV will also become more common as this population ages. Tobacco and alcohol use and abuse are also more common in HIV-positive individuals and are classic risk factors for HNSCC. It has been controversial whether these traditional risk factors may also increase the risk of HPV-related OPSCC with some studies showing that tobacco and alcohol use increased the risk of HPV-related OPSCC [186] and others indicating that HPV-associated and substance-associated HNSCC are two distinct diseases [151]. Most studies have shown a male predominance in HNSCC overall and in OPSCC specifically but many of these did not test the tumors for the presence of HPV. In a large international study, a large male predominance for OPSCC (83%) was found and HPV DNA was detected in 25% of these tumors using a sophisticated and sensitive assay [38]. In this large cohort, 32% of the tumors in women were HPV-16-positive as compared to only 19% of the tumors in men [38]. Any high-risk HPV detection was also higher in women (40%) versus 22% in men. This observation is provocative but needs to be corroborated in other cohorts. Multiple studies have shown an increase in HPV-related OPSCC over the past few decades that are attributed to a birth cohort effect [184]. This could be due to changes in attitudes regarding sex that occurred in the 1960s and may have led to increased oral HPV-16 infection via sexual practices. It is tempting to also associate the HIV epidemic with this increase in HPV-related oral cancers but this has not borne out in large epidemiological studies likely due to the relatively low prevalence of HIV/AIDS in the general population.

6.2.9 HPV-Related OPSCC Is More Common in the HIV-Seropositive Population

HIV-positive individuals are at increased risk for all HPV-associated cancers. Cervical cancer rates are increased 2.9–5.4-fold in HIV-positive women. Increases are also seen in anal (7.8× for females, 60× for males) vaginal/vulvar (3.9×) and penile (5.4–6.9×) cancers [150, 187]. The data for oral cancers is complicated due to some studies failing to distinguish OPSCC from HNSCC or lacked testing for HPV. Most studies show that the occurrence of OPSCC in HIV-positive individuals is increased by 1.6- to 2.6-fold over the HIV-negative population [150, 158, 187–190]. These studies represented individuals from North America, Australia, and Europe and spanned cohorts from the mid-1990s to the mid-2000s. A few studies showed even higher rates in the HIV-positive population. An early Swiss cohort study showed a higher rate of OPSCC at 4.1-fold in HIV-positive individuals as compared to the general population [159] and a recent large cohort from America showed an increased frequency (3.2-fold) in HIV-positive individuals [191]. In all these studies, HPV-positive cancers were defined simply by anatomical locations (oropharynx) and not by testing of biopsy samples for the virus. Furthermore, many studies matched HIV/AIDS registries with cancer databases to provide this insight [158, 187, 189, 191].

With the increased life expectancy of the HIV-infected individual due to HAART, there is a potential risk for the HPV-related oral malignancies to increase in prevalence over time. Indeed, early studies noted an increase in head and neck cancer in HIV-positive individuals over time [192]. In more formal analyses, two studies examined the rates of OPSCC in HIV-positive individuals over time. Both studies noted an increase of 60–80% in oral cancer prevalence from the mid-1990s to the mid-2000s in HIV-positive individuals [189, 193]. These cancers tend to occur at a younger age than in HIV-negative individuals. For example, the age of diagnosis of tongue cancer was 45.6 years in HIV-positive and 58.9 years in HIV-negative individuals ($p = 0.03$) [194]. Immune suppression as indicated by diminishing peripheral blood CD4+ T cell count does not reliably predict OPSCC risk [158, 187, 191].

A few recent studies have examined the role of smoking in the prevalence of OPSCC in HIV-positive individuals. Chew et al. revealed an OPSCC incidence rate of 23.2/100,000 among a cohort of U.S. military veterans with HIV [195]. Risk factors for OPSCC in this cohort were age over 50, CD4+ T cell count less than 200 and relatively poor HIV control. Smoking history was not associated with OPSCC but a large proportion of this group (84%) smoked. Finally, Silverberg et al. examined the large Kaiser Permanente cohort of about 20,000 HIV-positive and 210,000 HIV-negative adults [196] and demonstrated a modest increase of OPSCC (Relative Risk [RR], 1.9) in HIV-positive individuals after adjustment for age, race/ethnicity, and gender. However, this was no longer statistically significant (RR, 1.4) after adjusting for smoking, BMI, and alcohol and drug use. In conclusion, HIV-positive people are at increased risk of OPSCC but the exact roles of HPV, tobacco and alcohol exposure are not well defined.

6.2.10 Examination of the Site of HPV-Related HNSCC in HIV-Positive Individuals

To help distinguish between smoking, alcohol and HPV infection as risk factors, Picard et al. examined 40 HIV-positive individuals with HNSCC and performed immunohistochemistry for p16, in situ hybridization for 12 high-risk HPV genotypes, and PCR (L1 targeted) for 15 high-risk HPV genotypes [197]. Thirty of the 47 identified patients had tumors in the oropharynx or oral cavity (64%). Twelve of the 40 tested for HPV were positive with 42% of those with oropharyngeal cancer testing positive and 33% of those with oral cancer testing positive. HPV-16 was detected in 50% of the HPV-positive cancers. Median survival was increased in those with HPV-related tumors (Hazard Ratio 2.9, 0.9–10.1). McLemore et al. examined 12 HIV-positive patients with oropharynx or oral cavity carcinoma [198]. They found 5/12 positive for HPV by L1 PCR with HPV-16 in 4 out of 5 cases. A case-control study of 41 HIV-positive HNSCC patients showed 28% being positive for HPV by L1 PCR. Of the HPV-positive tumors, 50% occurred in the orophaynx and 50% occurred at other sites, and 7/12 was positive for HPV-16 [46]. In sum, these small studies demonstrated HPV in 33–50% of HNSCC in HIV-seropositive patients with higher HPV-positive frequency in the oropharynx. Clearly more studies focused on determining the role of HPV in OPSCC in HIV-positive individuals need to be undertaken.

6.2.11 Treatment of OPSCC in the HIV-Positive Individual

Many studies have shown that treatment of HPV-related OPSCC has a favorable outcome [199, 200]. Due to this favorable outcome in HPV-related OPSCC, there has been a discussion about potentially reducing the amount of radiation used to treat this condition [201]. There are only a few reports focusing on the success rate of treating the HIV-positive person with HPV-related OPSCC. A very early pilot study showed worse outcomes (death) in 5 of 6 cases of HNSCC in HIV-positive individuals as compared to 0 of 4 who were HIV-negative [202]. None of these HIV-positive patients were on ART. Later studies have shown more favorable results with local control and 5-year survival of 55–80% and in some cases no difference in outcomes between HIV-positive and HIV-negative individuals [203–205]. Although a number of clinical trials are focusing on dose reduction for HPV-related OPSCC, there are none that focus on the HIV-positive person. At this time, dose reduction for HIV-positive individuals with OPSCC cannot be recommended. Finally, there is no data on screening for recurrence in the HIV-positive individual to support any deviation from the screening guidelines for the HIV-negative population.

Although many investigators have examined ways to screen for OPSCC using HPV diagnostic testing or an oral Pap smear, no test is clinically approved and no studies have focused on the HIV-positive individual. An interesting approach utilized an bead-based multiplex serological assay for numerous HPV genotypes and proteins [206]. Antibodies to HPV-16 E1, E2, E6, E7, and L1 proteins were detected in those with OPSCC. The most robust association was seen with E6 in which 35% of those with OPSCC had HPV-16 E6 antibodies as compared to 0.6% of the controls (odds ratio [95% confidence interval], 274 [110-6981]). These antibodies were not seen in individuals with cancers at other oral sites and in some cases, these antibodies were detectible 10 years prior to diagnosis. It is not clear how often this panel of serum antibodies are detected in the general population and if screening for them would be a cost-effective approach to detect oral or perhaps any HPV-related cancers. These findings need to be examined in an HIV-positive population.

6.2.12 The Role of Antiretroviral Therapy on Oral HPV Infection and Warts

The widespread administration of highly active antiretroviral therapy (HAART) in the United States has had a profound impact on the incidence of HIV-associated oral-opportunistic infections (OI) and death from these OIs. A number of studies from the early 2000s showed a decreased incidence of oral lesions, especially oral candidiasis (thrush) and oral hairy leukoplakia (OHL) [207–210] as compared to the cohorts examined in the early to mid-90s. The prevalence of oral candidiasis generally dropped from 32–43% to 14–19%, OHL from 8% to 4–6%, and necrotizing periodontal or gum disease from 4–5% to 2%. Interestingly, regimens containing non-nucleoside reverse transcriptase inhibitors (NNRTIs) may reduce the risk of OIs to a greater extent than regimens containing protease inhibitors (PIs), despite similar levels of immune control [208]. This implies that the use of different ART combinations could have differential effects on oral infection and disease. It is less clear what impact HAART has had on HPV-related oral lesions and chronic HPV infection since this has rarely been reported. Contradictory reports have shown both positive and negative impact of HAART on the prevalence of oral warts [207, 209].

In the year 2000, Dr. Janet Leigh, Director of the HIV Dental Clinic at the Medical Center of Louisiana, New Orleans, reported a striking clinical observation that the incidence of oral warts in HIV-seropositive individuals had increased substantially since the introduction of HAART [211]. Shortly thereafter, John and Deborah Greenspan reported a significant three–four-fold increase in the prevalence of oral warts in a San Francisco based cohort of HIV-seropositive people during the decade of the 1990s [207]. The associated risk factors for this paradoxical increase in oral warts was the use of ART, male gender, and seropositivity for hepatitis B virus and having more than a one-log drop in HIV viral load in the year prior to wart diagnosis [207, 212]. Finally, Cameron et al. [147] reported that there was a sixfold increase in oral HPV detected in those prescribed HAART (71% vs. 28%) but only in the Caucasian HIV-positive population [147]. There was not a significant increase in oral HPV detection in African-American members of the cohort who were prescribed HAART. The study was unable to evaluate patient compliance with the prescribed regimen, but did report an association between oral HPV detection and significant drop in HIV viral load, suggesting that those with oral HPV were taking their medications as prescribed.

These initial studies implied a lack of reduction in HPV oral infection and warts in those on ART implicating either a lack of restoration of oral immunity against HPV or some direct augmentation of oral HPV infection due to ART. A few recent studies add further insight into this paradox. A cross-sectional study from Mexico City reported a 6.9% prevalence of HPV-related benign oral lesions with most of these cases diagnosed as either papillomas or multifocal epithelial hyperplasia [175]. In comparison to a cohort without HPV oral lesions, the patients with lesions were more likely to be older (over age 40, $p = 0.002$). In addition, those on ART for more than 12 months ($p < 0.001$) were more likely to have an oral HPV-related lesion, implicating an effect of ART on HPV itself or on the oral epithelium. Additional insight comes from a study focused on the best time to start ART [213]. Two cohorts of HIV-positive individuals were enrolled, one that started ART immediately following HIV diagnosis and a second group that delayed the start of therapy until they experienced a drop in CD4+ T cell count below 200 cells/ml or acquired an opportunistic infection. The delayed group had an incident rate of oral lesions of 17% versus 4% in the immediate group ($p < 0.01$) with increases seen in oral candidiasis, OHL and herpes simplex virus infections. There was no significant difference in HPV-related lesions in these cohorts; however, the incidence of oral warts was increased after ART was initiated in the delayed group as compared to before starting ART (4.3% vs. 0.97%, $p < 0.01$). The AIDS Clinical Trial Group (ACTG) protocol A5272 study set out to determine if oral HPV infection or warts increased during prospective follow-up after starting ART [161]. This study followed 388 HIV-positive individuals as they started ART and collected oral wash specimens pre-ART and after 16 and 24 weeks of therapy. At baseline, 18% of participants had at least one HPV genotype present as compared to 24% at the next follow-up visit. HPV infection at baseline was more likely to persist than to clear.

Finally, those who acquired a new HPV genotype had a larger increase in CD4+ T cell count and no significant change in HIV viral load as compared to those who did not acquire a new HPV infection. There were no changes in the incidence of warts but the follow-up was only 24 weeks. These studies all point to a potential detrimental effect of ART on controlling oral HPV infection with implications for future rates of HPV-related oral disease.

Studies from the Hagensee laboratory also imply a role of ART in the pathogenesis of oral wart formation. We performed a retrospective chart review on 21 HIV-positive patients with oral warts seeking dental care at the Medical Center of Louisiana HIV outpatient clinic (Cameron and Hagensee, unpublished). Interestingly, 13 (62%) patients had started ART in the year prior to wart diagnosis. There was no clear association with current CD4+ T cell count, nor was there a clear indication of recent change in CD4+ T cell count with the majority (14, 67%) having stable CD4+ T cell counts (±100 cells/ml) in the previous year prior to diagnosis. Strikingly, 76% of these individuals had a four-log drop in HIV viral load within 6 months prior to wart development despite no observed change in CD4+ T cell count. Thus, one could hypothesize that wart development is an unconventional form of immune reconstitution disease following initiation of antiretroviral therapy for HIV.

To investigate the possibility that HPV-associated oral warts represent an immune reconstitution syndrome Lilly et al. [214] examined biopsy specimens from HIV-positive individuals for evidence of immune cell infiltration and activity. Biopsies of oral lesions were taken along with control biopsies of tissue adjacent to the lesion. They studied a total of 12 patients with oral warts and found no difference in a large panel of inflammatory marker mRNA profiles (Th1, Th2, and inflammatory) as well as quantities of CD3+, CD4+, and CD8+ cells. Thus, consistent with the lack of association with appreciable differences in CD4+ T cell counts, restoration of immunity in itself does not seem to promote wart formation. This may imply an interaction of the ART agents and the oral epithelium that increases HPV oral infection and/or persistence and subsequent lesion development. A publication from the Meyers laboratory adds an interesting in vitro correlate of these clinical observations. Danaher et al. noted a marked growth inhibition of organotypic raft cultures with a number of commonly used protease inhibitors (PIs) at concentrations that approximated the levels achieved in the oral cavity [215]. The most profound inhibition of growth was observed with Nelfinavir followed by Lopinavir and Saquinivir. Indeed, PIs have been proposed as new cancer chemotherapy drugs due to their profound activity on apoptosis [216, 217]. Thus, toxic effects of PIs on the oral epithelium may promote the development of HPV-associated lesions.

6.2.13 Antiretroviral Therapy and Risk of OPSCC

As noted above, HPV-associated OPSCC has been well described in the HIV-negative population with rates that are increasing over the past years [184]. It is not clear if the widespread use of ART increases the risk of OPSCC in HIV-positive individuals. Insight may be found by examining other patient populations that are immune suppressed. Rates of OPSCC are increased in solid organ transplant patients 2.2–5.3-fold but the HPV status of these tumors are unknown [188, 218]. The highest rates of OPSCC are in those undergoing liver transplantation. Interestingly, the reason for liver transplant was mostly alcoholic liver disease implying interaction of traditional substance use risk factors, immune suppression and potentially HPV infection in the increased risk of OPSCC.

The study by Powles et al. examined the rates of many non-AIDS defining cancers in the pre-ART (1983–1995), early ART (1996–2001) and established ART eras (2002–2007). They found a nonsignificant increase in head and neck cancers in those classified as on established ART (2.66-fold) as compared to those in the early ART (1.75-fold) and pre-ART (1.34-fold) eras [219]. A similar study using the Italian linkage registry also showed a slight increase in head and neck cancers in the ART period (1997–2002, Standardized Incidence Ratio = 1.8) as compared to the pre-ART era (prior to 1997, SIR = 1.4) [220]. These studies do not clearly demonstrate an increase in OPSCC over time in the HIV-positive population; however, it is clear that these rates are not declining to HIV-negative populations despite improved immunity. If these cancers were significantly immune-related then it would be expected that the rates would decline as is seen for Kaposi's sarcoma. One explanation is that the lesion(s) started development while immune suppressed and ART has increased the lifespan of the individual so that these lesions can be detected (survivor bias). Conversely, the improved immunity may be offset by toxicity of the ART on the oral epithelium.

6.2.14 Projected Impact of HPV Vaccination for Prevention of Oropharyngeal Cancer

Primary prevention of oncogenic HPV infection has been made possible through the advent of the HPV vaccine. The available vaccines are recombinant protein vaccines formulated with empty virus-like particles (VLPs) consisting of HPV L1 capsid protein. The VLPs induce serum antibody responses resembling the natural response to HPV infection. This serum antibody response is HPV type-specific with little to no cross-protection against other HPV genotypes; therefore, multiple HPV genotypes are included in the vaccine to promote broad-spectrum protection. To date the most broad-spectrum HPV vaccine available, Gardasil-9 (Merck), provides protection against seven oncogenic HPV genotypes (HPV-16, -18, -31, -33, -45, -52, and -58) and the two most common low-risk (wart-associated) genotypes (HPV-6 and -11). In the U.S., this vaccine is currently approved for both males and females ages 9–45, with a recommended two-dose vaccine regimen (0, 6–12 months) for adolescents ages 9–14 and three-dose regimen (0, 2, and 6 months) for recipients ages 15 and up.

Given the slow development of HPV-associated cancer following the acquisition of HPV infection, HPV vaccine clinical trials were designed with genotype-specific infection and early precancer endpoints as surrogate outcomes for cancer. The vaccine was found to be safe and highly effective at preventing infection with the genotypes covered by the vaccine. Likewise, there were no incident diagnoses of vaccine genotype positive cervical dysplasia among women receiving the vaccine in large-scale clinical trials [221]. In practice, populations with high rates of vaccination reported remarkably rapid declines in genital wart incidence [222], and vaccine efficacy remains high (96% protection from infection and 100% protection from cervical intraepithelial neoplasia) up to ten years after the initial dose [223]. These findings support the current projections that the nonavalent HPV vaccine will prevent up to 90% of HPV-associated cervical cancer, assuming adequate uptake of the vaccine.

Although empirical evidence is scant, experts believe that the HPV vaccine will prove to be an effective primary prevention strategy for HPV-associated oropharyngeal cancer. Studies to determine the HPV genotype distribution in OPSCC have consistently shown a predominance of HPV-16 (80–90% of HPV-positive cases) followed by HPV-33 (3–10% of HPV-positive cases), with HPV-18 making up an additional 2–5% of HPV-positive cases [38, 224, 225]. Therefore, assuming that parenteral vaccination generates sufficient protective immunity in the oral cavity, the nonavalent vaccine should provide adequate protection to prevent more than 90% of HPV-related OPSCC. The low rates of oral HPV infection, the lack of a readily identifiable dysplastic precursor lesion in the oropharynx, and the relatively rare incidence of oropharyngeal cancer (estimated 4.5 cases per 100,000 population in the U.S. [226]) preclude clinical trials to demonstrate vaccine efficacy for prevention of HPV-associated oropharyngeal cancer. Nevertheless, there is evidence to support the inference that the vaccine will be effective at preventing HPV-associated OPSCC. First, antibodies against HPV can be detected in the oral cavity following HPV infection [227, 228] or HPV vaccination [229, 230]. Second, vaccination appears to prevent the acquisition of oral HPV infection. In a large cohort of Costa Rican women, the prevalence of oral HPV-16 and -18 infections four years after vaccination was 0.03% in those receiving the HPV vaccine compared to 0.5% in those in the control arm of the trial for a HPV vaccine efficacy estimate of 62.5–99.7% [231]. Similarly, data from the National Health and Nutrition Examination Survey (NHANES) cohort, including both males and females, demonstrated an 88% reduction in the prevalence of oral infections with HPV-6, -11, -16, and -18 in young adults who received at least one dose of the HPV vaccine [232]. While definitive conclusions from these reports are not possible due to the low prevalence of oral HPV infection and limited observation periods, the data thus far favors a positive impact for the vaccine as a primary prevention strategy for HPV-associated OPSCC.

People living with HIV are likely to derive benefit from the vaccine similar to the general population. While no published efficacy trials have specifically targeted HIV-infected populations, phase 1 clinical studies in HIV-seropositive men and children have demonstrated good safety and immunogenicity of the quadrivalent HPV vaccine [233, 234]. Importantly, individuals with HIV infection appear capable of generating anti-HPV antibodies in oral fluids [227], suggesting that vaccine-mediated protection from OPSCC is feasible in this population. The U.S. Centers for Disease Control recommends three-dose HPV vaccination for immunocompromised adolescents and young adults, including those living with HIV, on the same schedule as recommended for the general population [235].

6.3 Summary

In summary, oral HPV infection is more prevalent in HIV-positive individuals (13–40%) than in HIV-negative individuals (4–25%). Those with greater lifetime sexual partners, particularly those reporting a history of oral sex and those with compromised immunity are the most at risk for oral HPV infection. OPSCC diagnoses are on the rise in the general population in the U.S., Europe, and Australia, and in HIV-infected individuals, restoration of immunity and reduction of HIV viral load does not appear to reduce the risk of developing OPSCC. Based on the established natural history of HPV infection at anogenital mucosal sites, it is believed that chronic HPV infection of the oral epithelium will lead to precancer and cancerous lesions. Natural history studies are needed to prove this as well as to establish additional risk factors for HPV-related oral disease development such as warts or cancer. Partner studies focused on oral HPV acquisition are also needed to better establish the mode of transmission. The mechanism by which HPV causes OPSCC also needs further study in order to reveal targeted diagnostic and therapeutic opportunities for cancer prevention. In contrast to HPV-associated disease in the genital tract, HPV-associated head and neck tumors are overwhelmingly dominated by HPV-16 infection, and they may not require viral integration into the host chromosome in order to promote tumor formation. Both reversible and irreversible genetic and epigenetic changes to the host genome contribute to head and neck tumor formation. Finally, HPV vaccine implementation needs to be improved to increase primary prevention of the oral-related HPV diseases.

Appendix

Table 6.1 List of abbreviations used in this chapter

Abbreviations	Definitions
HPV	Human papillomavirus
E2	HPV early gene encoding a protein that regulates viral transcription. Also refers to the protein product of the gene
E6	HPV early gene encoding an oncoprotein that interacts with cellular p53. Also refers to the protein product of the gene
E7	HPV early gene encoding an oncoprotein that interacts with cellular pRb. Also refers to the protein product of the gene

(continued)

Table 6.1 (continued)

Abbreviations	Definitions
L1	HPV late gene that encodes the major capsid protein of the virus. Also refers to the protein product of the gene
L2	HPV late gene that encodes the minor capsid protein of the virus. Also refers to the protein product of the gene
p53	Tumor suppressor protein that promotes apoptosis
pRb	Retinoblastoma protein; tumor suppressor protein that regulates cell cycle
E2F	Transcription factor involved in cell cycle progression and proliferation
TP53	Gene encoding the p53 protein
OPSCC	Oropharyngeal squamous cell carcinoma, a subset of head and neck cancers
HNSCC	Head and neck squamous cell carcinoma
$P16^{INK4A}$	Tumor suppressor encoded by CKN2A. Inhibits cell cycle progression via inhibition of CDK4 kinase. Biomarker of HPV infection
CDKN2A	Cyclin-dependent kinase inhibitor 2A. Gene encodes several transcript variants, the protein products of which are tumor suppressors that regulate cell cycle through inhibition of CDK4 kinase
D1	Cyclin D1. Promotes cell cycle progression
CDK4/6	Cyclin-dependent kinase 4, 6. Serine/threonine kinases that promote cell cycle progression. These kinases phosphorylate and inactivate pRb
RAD51	RAD51 Recombinase. Mediates homologous recombination and DNA damage repair
ORF	Open reading frame, or predicted coding regions of the viral genome
NR4A2	Nuclear receptor subfamily 4 group A member 2. Putative transcription factor regulating differentiation during neuronal development
KLF5	Kruppel like factor 5. Zinc finger protein that functions as a transcriptional activator
TP63	Tumor protein p63. Member of p53 transcription factor family
TPRG1	Tumor protein p63 regulated
CpG	Cytosine-Guanine dinucleotides that can be methylated at the carbon-5 position of cytosine
DAPK	Death-associated protein kinase. Mediates gamma-interferon inducible apoptosis
MGMT	O-6-methylguanine-DNA methyltransferase. DNA repair protein that protects genome from alkylating agent-induced mutagenesis
MLH1	MutL homolog 1. Component of the DNA mismatch repair machinery
DNMT1	DNA methyltransferase 1. Mediates methylation of genomic CpG dinucleotides. Involved in maintenance of methylation following DNA replication
DNMT3A	DNA methyltransferase 3 alpha. Responsible for de novo methylation of CpG dinucleotides
Sp1	Specificity protein 1. Zinc finger transcription factor involved in many cell processes. Posttranslational modifications determine its function. Can act as an activator or a repressor

(continued)

Table 6.1 (continued)

Abbreviations	Definitions
P14^{ARF}	Tumor suppressor encoded by CKN2A. Inhibits cell cycle progression via inhibition of CDK4 kinase
CCNA1	Gene that encodes cyclin A1. Coordinates cell division via regulation of cyclin-dependent kinases. Interacts with Rb family proteins, E2F and p21 family proteins
TP73	Tumor protein p73. Member of p53 transcription factor family. Involved in cell stress response
CADM1	Cell adhesion molecule 1. Also called IGSF4A. Mediates cell-cell adhesion. Involved in activating natural killer cell cytotoxicity and interferon-gamma production by CD8+ T cells
CDH13	Cadherin-13. Appears to protect vascular endothelial cells from oxidative stress-induced apoptosis
TIMP3	Tissue inhibitor of metalloproteinases 3, TIMP Metallopeptidase inhibitor 3. Prevent degradation of extracellular matrix via inhibition of matrix metalloproteinases
IGSF4	See CADM1
WNT	Wingless-type MMTV integration site family member. Secreted signaling protein involved in developmental processes. Promotes beta-catenin-mediated transcriptional activation
SFRP4	Secreted frizzled-related protein 4. Soluble ligand of WNT family proteins. Important in bone morphogenesis
ESR1	Estrogen receptor 1. Hormone-responsive receptor that also functions in the nucleus as a transcription factor
RARβ	Retinoic acid receptor beta. Steroid-thyroid hormone receptor that binds retinoic acid. Modulates cell growth
APC	Adenomatous polyposis coli tumor suppressor. Inhibitor of WNT signaling
ALDH1A2	Aldehyde dehydrogenase family 1 member A2. Enzyme that catalyzes synthesis of retinoic acid from precursor retinaldehyde
GATA4	GATA binding protein 4. Zinc finger transcription factor that recognizes GATA promoter motifs. Important in developmental processes
GFRA4	GDNF family receptor alpha 4. Receptor for persephin. Mediates activation of RET tyrosine kinase receptor
IRX4	Iroquois homeobox 4. Mediator of cardiac tissue development
OSR2	Odd-skipped related transcription factor 2. Likely involved in developmental processes
P300	E1A binding protein p300. Transcriptional co-activator protein. Plays a role in activation of hypoxia response genes
MYC	MYC proto-oncogene, BHLH transcription factor. Works in concert with MAX to initiate gene transcription. Mediates cell growth, apoptosis, transformation and angiogenesis
HAT	Histone acetyltransferase. Catalyzes transfer of acetyl group to lysine residues on histones to promote open chromatin and transcriptional activity
HDAC	Histone deacetylase. Removes acetyl groups from lysine residues of histones to regulate gene expression

(continued)

Table 6.1 (continued)

Abbreviations	Definitions
HMT	Histone methyltransferase. Catalyze transfer of up to three methyl groups to lysine and arginine residues of histones. Regulates gene expression
HDM	Histone demethylase
H3K4me3	Histone 3 Lysine 4 methyl 3. Denotes trimethylation of the fourth lysine on histone H3. Marker of active transcription
EZH2	Enhancer of zeste 2 polycomb repressive complex subunit 2. In complex with PRC2, catalyzes the methylation of lysines 9 and 27 on histone 3 to suppress gene transcription
H3K27me3	Histone 3 Lysine 27 methyl 3. Trimethylation of lysine 27 of histone H3. Marker of repressed chromatin
JMJD3	Jumonji domain containing 3. Also known as lysine demethylase 6B (KDM6B). Removes methyl groups from lysine 27 of histone H3, thereby promoting transcriptional activity
RAS	Family of small GTPases that transmit intracellular signals. Promotes cell growth, differentiation, and survival
PRC2	Polycomb repressive complex 2. Protein involved in transfer of methyl groups to lysine 27 of histone H3, promoting transcriptional repression. Responsible for silencing of chromatin during embryonic development
RISC	RNA-induced silencing complex. A complex of proteins that suppress mRNA translation. Specificity is conferred by miRNAs in complex with RISC
miRNA	microRNA. Short noncoding RNA species that suppress translation of proteins by directing RISC complexes to cognate target nucleotide sequences on the mRNA transcript 3′ untranslated region
PTEN	Phosphatase and tensin homolog. Dephosphorylates phosphoinositide substrates. Inhibitor of AKT signaling pathway
TPM1	Tropomyosin 1. Actin-binding protein that stabilizes actin filaments. Functions in muscle contraction
Bcl-2	Apoptosis regulator. Protein located on outer membrane of mitochondria. Promotes survival of cells, particularly lymphocytes
PDCD4	Programmed cell death 4. Inhibits translation by binding and interfering with eukaryotic translation initiation factor 4A1
MYO1B	Myosin 1B. Motor protein that facilitates cell migration
MALAT1	Metastasis-associated lung adenocarcinoma transcript 1. Long noncoding RNA that may function as a scaffold for ribonucleoprotein complexes in the nucleus. Regulates gene transcription and promotes cell migration and metastasis
EMT	Epithelial-mesenchymal transition. Process in which epithelial cells lose phenotypic properties such as polarity and cell-cell adhesion, and gain properties that resemble pluripotent mesenchymal stem cells, including migratory capacity. Functions in wound healing
LCR	Long control region; portion of viral genome that contains regulatory nucleotide sequences
G2/M	Cell cycle checkpoint important for repair of DNA damage

(continued)

Table 6.1 (continued)

Abbreviations	Definitions
PIK3CA	Phosphatidylinositol-4,5-bisphosphate 3-kinase catalytic subunit alpha. Gene encodes catalytic alpha subunit of phosphoinositide-3-kinase (PI3K), p110. The kinase initiates signaling cascades that promote cell proliferation, survival, and migration. Activates AKT signaling
AKT	AKT serine/threonine kinase. Following activation of AKT by PI3K cascade AKT phosphorylates a broad array of proteins to promote growth, survival, migration, and angiogenesis
PI3K	Phosphoinositide-3-kinase. Enzyme that phosphorylates the hydroxyl group at the third position in the inositol ring of phosphatidylinositol. Promotes signal transduction that results in a number of pro-tumorigenic processes
mTOR	Mammalian target of rapamycin. Serine/threonine kinase that phosphorylates a wide range of proteins involved in pro-tumorigenic processes. Activated by PI3K/AKT signaling
APOBEC	Apolipoprotein B mRNA editing enzyme, catalytic polypeptide-like. Cytidine deaminase. Bind RNA or single-stranded DNA. Converts cytidine to uridine, thymidine or guanine
TpC	Thymidine-Cytidine dinucleotides. Motif recognized by APOBEC cytidine deaminases
APOBEC3A	Apolipoprotein B mRNA editing enzyme, catalytic polypeptide-like 3A. Cytidine deaminase. Plays an antiviral role by editing viral genome sequences. Active primarily in cytoplasm of monocytes/macrophages
APOBEC3B	Apolipoprotein B mRNA editing enzyme catalytic subunit 3B. Cytidine deaminase. May play an anti-viral role in host defense. Localizes to nucleus and therefore can cause off-target mutagenesis of host DNA
TCW	Thymidine-Cytidine-(Adenosine or Thymidine) nucleotide motif that is targeted by APOBEC cytidine deaminases
PIP_3	Phosphatidylinositol (3,4,5)-trisphosphate. Cell membrane-associated phospholipid created by phosphorylation of phosphatidylinositol (4,5)-bisphosphate (PIP_2) by PI3K. Mediates AKT signaling
PDK1	Pyruvate dehydrogenase kinase 1. Phosphorylates and deactivates pyruvate dehydrogenase, a mitochondrial enzyme involved in regulation of carbohydrate metabolism. Promotes survival during hypoxia and oxidative stress
MEK	MAPK/ERK kinase 1; Mitogen-activated protein kinase kinase 1 (MAP2K1). Phosphorylates and activates ERK/MAPK
ERK	Extracellular signal-regulated kinase, mitogen-activated protein kinase (MAPK). Part of receptor–ligand-mediated signaling cascade that activates downstream transcription factors
FGFR2	Fibroblast growth factor receptor 2. Tyrosine-protein kinase and cell surface receptor that promotes cell signaling via RAS/MEK/ERK or PI3K/AKT/mTOR pathways
FGFR3	Fibroblast growth receptor 3. Tyrosine-protein kinase and cell surface receptor that promotes cell signaling via RAS/MEK/ERK or PI3K/AKT/mTOR pathways

(continued)

Table 6.1 (continued)

Abbreviations	Definitions
KRAS	KRAS proto-oncogene, GTPase. Protein binds GDP/GTP and has GTPase activity. Mediates cell proliferation and induces silencing of tumor suppressor genes
HRAS	HRAS proto-oncogene, GTPase. Protein binds GDP/GTP and has GTPase activity. Mediates cell proliferation and induces silencing of tumor suppressor genes
ERBB	Epidermal growth factor receptor (EGFR); HER1. Transmembrane glycoprotein cell surface receptor tyrosine kinase. Induces cell proliferation upon interaction with epidermal growth factor ligand
HER2	Human epidermal growth factor receptor 2; Erb-B2 receptor tyrosine kinase 2 (ERBB2). Lacks a ligand binding domain but heterodimerizes with other epidermal growth factor receptors and enhances kinase activity and signal transduction
HER3	Human epidermal growth factor receptor 3; Erb-B2 receptor tyrosine kinase 3 (ERBB3). Receptor binding site for neuregulins but lacks active kinase domain. Functions by forming heterodimers with other epidermal growth factor receptors that have kinase domains
HLA-A/B	Human leukocyte antigens A and B; major histocompatibility complex (MHC). Cell surface receptor that presents short polypeptides to immune cells
NF-κB	Nuclear factor kappa-light-chain-enhancer of activated B cells. Transcription factor protein complex that initiates gene transcription in response to danger stimuli (cell stress, infection, toxins, etc.)
TRAF3	Tumor necrosis factor (TNF) receptor-associated factor 3. Signal transduction mediator of TNF receptor signaling that leads to activation of NF-κB
CYLD	CYLD lysine 63 deubiquitinase. Cleaves lysine-63-linked polyubiquitin chains. Contributes to NF-κB activation
MSM	Men who have sex with men
HAART	Highly active antiretroviral therapy
OHL	Oral hairy leukoplakia
OPC	Oral pharyngeal cancer

References

1. Scheffner M, Werness BA, Huibregtse JM, Levine AJ, Howley PM (1990) The E6 oncoprotein encoded by human papillomavirus types 16 and 18 promotes the degradation of p53. Cell 63:1129–1136
2. Boyer SN, Wazer DE, Band V (1996) E7 protein of human papilloma virus-16 induces degradation of retinoblastoma protein through the ubiquitin-proteasome pathway. Can Res 56:4620–4624
3. Gonzalez SL, Stremlau M, He X, Basile JR, Munger K (2001) Degradation of the retinoblastoma tumor suppressor by the human papillomavirus type 16 E7 oncoprotein is important for functional inactivation and is separable from proteasomal degradation of E7. J Virol 75:7583–7591

4. Chellappan S, Kraus VB, Kroger B, Munger K, Howley PM, Phelps WC, Nevins JR (1992) Adenovirus E1A, simian virus 40 tumor antigen, and human papillomavirus E7 protein share the capacity to disrupt the interaction between transcription factor E2F and the retinoblastoma gene product. Proc Natl Acad Sci USA 89:4549–4553
5. Johnson DG, Schwarz JK, Cress WD, Nevins JR (1993) Expression of transcription factor E2F1 induces quiescent cells to enter S phase. Nature 365:349–352
6. Syrjanen K, Syrjanen S, Lamberg M, Pyrhonen S, Nuutinen J (1983) Morphological and immunohistochemical evidence suggesting human papillomavirus (HPV) involvement in oral squamous cell carcinogenesis. Int J Oral Surg 12:418–424
7. El-Mofty SK, Lu DW (2003) Prevalence of human papillomavirus type 16 DNA in squamous cell carcinoma of the palatine tonsil, and not the oral cavity, in young patients: a distinct clinicopathologic and molecular disease entity. Am J Surg Pathol 27:1463–1470
8. Fouret P, Monceaux G, Temam S, Lacourreye L, St Guily JL (1997) Human papillomavirus in head and neck squamous cell carcinomas in nonsmokers. Arch Otolaryngol Head Neck Surg 123:513–516
9. Gillison ML, Koch WM, Capone RB, Spafford M, Westra WH, Wu L, Zahurak ML, Daniel RW, Viglione M, Symer DE, Shah KV, Sidransky D (2000) Evidence for a causal association between human papillomavirus and a subset of head and neck cancers. J Natl Cancer Inst 92:709–720
10. Haraf DJ, Nodzenski E, Brachman D, Mick R, Montag A, Graves D, Vokes EE, Weichselbaum RR (1996) Human papilloma virus and p53 in head and neck cancer: clinical correlates and survival. Clin Cancer Res 2:755–762
11. Klussmann JP, Weissenborn SJ, Wieland U, Dries V, Kolligs J, Jungehuelsing M, Eckel HE, Dienes HP, Pfister HJ, Fuchs PG (2001) Prevalence, distribution, and viral load of human papillomavirus 16 DNA in tonsillar carcinomas. Cancer 92:2875–2884
12. Niedobitek G, Pitteroff S, Herbst H, Shepherd P, Finn T, Anagnostopoulos I, Stein H (1990) Detection of human papillomavirus type 16 DNA in carcinomas of the palatine tonsil. J Clin Pathol 43:918–921
13. Paz IB, Cook N, Odom-Maryon T, Xie Y, Wilczynski SP (1997) Human papillomavirus (HPV) in head and neck cancer. An association of HPV 16 with squamous cell carcinoma of Waldeyer's tonsillar ring. Cancer 79:595–604
14. Ringstrom E, Peters E, Hasegawa M, Posner M, Liu M, Kelsey KT (2002) Human papillomavirus type 16 and squamous cell carcinoma of the head and neck. Clin Cancer Res 8:3187–3192
15. Ritchie JM, Smith EM, Summersgill KF, Hoffman HT, Wang D, Klussmann JP, Turek LP, Haugen TH (2003) Human papillomavirus infection as a prognostic factor in carcinomas of the oral cavity and oropharynx. Int J Cancer 104:336–344
16. Perry ME (1994) The specialised structure of crypt epithelium in the human palatine tonsil and its functional significance. J Anat 185(Pt 1):111–127
17. Brandsma JL, Abramson AL (1989) Association of papillomavirus with cancers of the head and neck. Arch Otolaryngol Head Neck Surg 115:621–625
18. Schwartz SM, Daling JR, Doody DR, Wipf GC, Carter JJ, Madeleine MM, Mao EJ, Fitzgibbons ED, Huang S, Beckmann AM, McDougall JK, Galloway DA (1998) Oral cancer risk in relation to sexual history and evidence of human papillomavirus infection. J Natl Cancer Inst 90:1626–1636
19. Snijders PJ, Cromme FV, van den Brule AJ, Schrijnemakers HF, Snow GB, Meijer CJ, Walboomers JM (1992) Prevalence and expression of human papillomavirus in tonsillar carcinomas, indicating a possible viral etiology. Int J Cancer 51:845–850
20. Strome SE, Savva A, Brissett AE, Gostout BS, Lewis J, Clayton AC, McGovern R, Weaver AL, Persing D, Kasperbauer JL (2002) Squamous cell carcinoma of the tonsils: a molecular analysis of HPV associations. Clin Cancer Res 8:1093–1100 (An official journal of the American Association for Cancer Research)

21. Balz V, Scheckenbach K, Gotte K, Bockmuhl U, Petersen I, Bier H (2003) Is the p53 inactivation frequency in squamous cell carcinomas of the head and neck underestimated? Analysis of p53 exons 2-11 and human papillomavirus 16/18 E6 transcripts in 123 unselected tumor specimens. Can Res 63:1188–1191
22. Begum S, Cao D, Gillison M, Zahurak M, Westra WH (2005) Tissue distribution of human papillomavirus 16 DNA integration in patients with tonsillar carcinoma. Clin Cancer Res 11:5694–5699 (An official journal of the American Association for Cancer Research)
23. Klussmann JP, Gultekin E, Weissenborn SJ, Wieland U, Dries V, Dienes HP, Eckel HE, Pfister HJ, Fuchs PG (2003) Expression of p16 protein identifies a distinct entity of tonsillar carcinomas associated with human papillomavirus. Am J Pathol 162:747–753
24. Koskinen WJ, Chen RW, Leivo I, Makitie A, Back L, Kontio R, Suuronen R, Lindqvist C, Auvinen E, Molijn A, Quint WG, Vaheri A, Aaltonen LM (2003) Prevalence and physical status of human papillomavirus in squamous cell carcinomas of the head and neck. Int J Cancer 107:401–406
25. van Houten VM, Snijders PJ, van den Brekel MW, Kummer JA, Meijer CJ, van Leeuwen B, Denkers F, Smeele LE, Snow GB, Brakenhoff RH (2001) Biological evidence that human papillomaviruses are etiologically involved in a subgroup of head and neck squamous cell carcinomas. Int J Cancer 93:232–235
26. Wiest T, Schwarz E, Enders C, Flechtenmacher C, Bosch FX (2002) Involvement of intact HPV16 E6/E7 gene expression in head and neck cancers with unaltered p 53 status and perturbed pRb cell cycle control. Oncogene 21:1510–1517
27. Wilczynski SP, Lin BT, Xie Y, Paz IB (1998) Detection of human papillomavirus DNA and oncoprotein overexpression are associated with distinct morphological patterns of tonsillar squamous cell carcinoma. Am J Pathol 152:145–156
28. Andl T, Kahn T, Pfuhl A, Nicola T, Erber R, Conradt C, Klein W, Helbig M, Dietz A, Weidauer H, Bosch FX (1998) Etiological involvement of oncogenic human papillomavirus in tonsillar squamous cell carcinomas lacking retinoblastoma cell cycle control. Can Res 58:5–13
29. Brachman DG, Graves D, Vokes E, Beckett M, Haraf D, Montag A, Dunphy E, Mick R, Yandell D, Weichselbaum RR (1992) Occurrence of p53 gene deletions and human papilloma virus infection in human head and neck cancer. Can Res 52:4832–4836
30. Hafkamp HC, Speel EJ, Haesevoets A, Bot FJ, Dinjens WN, Ramaekers FC, Hopman AH, Manni JJ (2003) A subset of head and neck squamous cell carcinomas exhibits integration of HPV 16/18 DNA and overexpression of p16INK4A and p53 in the absence of mutations in p53 exons 5-8. Int J Cancer 107:394–400
31. Sisk EA, Soltys SG, Zhu S, Fisher SG, Carey TE, Bradford CR (2002) Human papillomavirus and p53 mutational status as prognostic factors in head and neck carcinoma. Head Neck 24:841–849
32. Snijders PJ, Steenbergen RD, Top B, Scott SD, Meijer CJ, Walboomers JM (1994) Analysis of p53 status in tonsillar carcinomas associated with human papillomavirus. J Gen Virol 75 (Pt 10):2769–2775
33. Li Y, Nichols MA, Shay JW, Xiong Y (1994) Transcriptional repression of the D-type cyclin-dependent kinase inhibitor p16 by the retinoblastoma susceptibility gene product pRb. Can Res 54:6078–6082
34. Ohtani N, Yamakoshi K, Takahashi A, Hara E (2004) The p16INK4a-RB pathway: molecular link between cellular senescence and tumor suppression. J Med Invest 51:146–153
35. Olshan AF, Weissler MC, Pei H, Conway K, Anderson S, Fried DB, Yarbrough WG (1997) Alterations of the p16 gene in head and neck cancer: frequency and association with p53, PRAD-1 and HPV. Oncogene 14:811–818
36. Shintani S, Nakahara Y, Mihara M, Ueyama Y, Matsumura T (2001) Inactivation of the p14 (ARF), p15(INK4B) and p16(INK4A) genes is a frequent event in human oral squamous cell carcinomas. Oral Oncol 37:498–504

37. Begum S, Gillison ML, Ansari-Lari MA, Shah K, Westra WH (2003) Detection of human papillomavirus in cervical lymph nodes: a highly effective strategy for localizing site of tumor origin. Clin Cancer Res 9:6469–6475 (An official journal of the American Association for Cancer Research)
38. Castellsague X, Alemany L, Quer M, Halec G, Quiros B, Tous S, Clavero O, Alos L, Biegner T, Szafarowski T, Alejo M, Holzinger D, Cadena E, Claros E, Hall G, Laco J, Poljak M, Benevolo M, Kasamatsu E, Mehanna H, Ndiaye C, Guimera N, Lloveras B, Leon X, Ruiz-Cabezas JC, Alvarado-Cabrero I, Kang CS, Oh JK, Garcia-Rojo M, Iljazovic E, Ajayi OF, Duarte F, Nessa A, Tinoco L, Duran-Padilla MA, Pirog EC, Viarheichyk H, Morales H, Costes V, Felix A, Germar MJ, Mena M, Ruacan A, Jain A, Mehrotra R, Goodman MT, Lombardi LE, Ferrera A, Malami S, Albanesi EI, Dabed P, Molina C, Lopez-Revilla R, Mandys V, Gonzalez ME, Velasco J, Bravo IG, Quint W, Pawlita M, Munoz N, de Sanjose S, Xavier Bosch F (2016) HPV involvement in head and neck cancers: comprehensive assessment of biomarkers in 3680 patients. J Natl Cancer Inst 108:djv403
39. Dok R, Kalev P, Van Limbergen EJ, Asbagh LA, Vazquez I, Hauben E, Sablina A, Nuyts S (2014) p16INK4a impairs homologous recombination-mediated DNA repair in human papillomavirus-positive head and neck tumors. Can Res 74:1739–1751
40. Jirawatnotai S, Hu Y, Michowski W, Elias JE, Becks L, Bienvenu F, Zagozdzon A, Goswami T, Wang YE, Clark AB, Kunkel TA, van Harn T, Xia B, Correll M, Quackenbush J, Livingston DM, Gygi SP, Sicinski P (2011) A function for cyclin D1 in DNA repair uncovered by protein interactome analyses in human cancers. Nature 474:230–234
41. Sakakibara N, Mitra R, McBride AA (2011) The papillomavirus E1 helicase activates a cellular DNA damage response in viral replication foci. J Virol 85:8981–8995
42. McBride AA, Warburton A (2017) The role of integration in oncogenic progression of HPV-associated cancers. PLoS Pathog 13:e1006211
43. Dooley KE, Warburton A, McBride AA (2016) Tandemly integrated HPV16 can form a Brd4-dependent super-enhancer-like element that drives transcription of viral oncogenes. MBio 7
44. Parfenov M, Pedamallu CS, Gehlenborg N, Freeman SS, Danilova L, Bristow CA, Lee S, Hadjipanayis AG, Ivanova EV, Wilkerson MD, Protopopov A, Yang L, Seth S, Song X, Tang J, Ren X, Zhang J, Pantazi A, Santoso N, Xu AW, Mahadeshwar H, Wheeler DA, Haddad RI, Jung J, Ojesina AI, Issaeva N, Yarbrough WG, Hayes DN, Grandis JR, El-Naggar AK, Meyerson M, Park PJ, Chin L, Seidman JG, Hammerman PS, Kucherlapati R (2014) Characterization of HPV and host genome interactions in primary head and neck cancers. Proc Natl Acad Sci USA 111:15544–15549
45. Akagi K, Li J, Broutian TR, Padilla-Nash H, Xiao W, Jiang B, Rocco JW, Teknos TN, Kumar B, Wangsa D, He D, Ried T, Symer DE, Gillison ML (2014) Genome-wide analysis of HPV integration in human cancers reveals recurrent, focal genomic instability. Genome Res 24:185–199
46. Walline HM, Carey TE, Goudsmit CM, Bellile EL, D'Souza G, Peterson LA, McHugh JB, Pai SI, Lee JJ, Shin DM, Ferris RL (2017) High-risk HPV, biomarkers, and outcome in matched cohorts of head and neck cancer patients positive and negative for HIV. Mol Cancer Res MCR 15:179–188
47. Olthof NC, Speel EJ, Kolligs J, Haesevoets A, Henfling M, Ramaekers FC, Preuss SF, Drebber U, Wieland U, Silling S, Lam WL, Vucic EA, Kremer B, Klussmann JP, Huebbers CU (2014) Comprehensive analysis of HPV16 integration in OSCC reveals no significant impact of physical status on viral oncogene and virally disrupted human gene expression. PLoS ONE 9:e88718
48. Walline HM, Goudsmit CM, McHugh JB, Tang AL, Owen JH, Teh BT, McKean E, Glover TW, Graham MP, Prince ME, Chepeha DB, Chinn SB, Ferris RL, Gollin SM, Hoffmann TK, Bier H, Brakenhoff R, Bradford CR, Carey TE, University of Michigan H,

Neck Specialized Program of Research Excellence P (2017) Integration of high-risk human papillomavirus into cellular cancer-related genes in head and neck cancer cell lines. Head Neck 39:840–852

49. Hu Z, Zhu D, Wang W, Li W, Jia W, Zeng X, Ding W, Yu L, Wang X, Wang L, Shen H, Zhang C, Liu H, Liu X, Zhao Y, Fang X, Li S, Chen W, Tang T, Fu A, Wang Z, Chen G, Gao Q, Li S, Xi L, Wang C, Liao S, Ma X, Wu P, Li K, Wang S, Zhou J, Wang J, Xu X, Wang H, Ma D (2015) Genome-wide profiling of HPV integration in cervical cancer identifies clustered genomic hot spots and a potential microhomology-mediated integration mechanism. Nat Genet 47:158–163
50. Network CGA (2015) Comprehensive genomic characterization of head and neck squamous cell carcinomas. Nature 517:576–582
51. Wentzensen N, Vinokurova S, von Knebel Doeberitz M (2004) Systematic review of genomic integration sites of human papillomavirus genomes in epithelial dysplasia and invasive cancer of the female lower genital tract. Can Res 64:3878–3884
52. Bodelon C, Vinokurova S, Sampson JN, den Boon JA, Walker JL, Horswill MA, Korthauer K, Schiffman M, Sherman ME, Zuna RE, Mitchell J, Zhang X, Boland JF, Chaturvedi AK, Dunn ST, Newton MA, Ahlquist P, Wang SS, Wentzensen N (2016) Chromosomal copy number alterations and HPV integration in cervical precancer and invasive cancer. Carcinogenesis 37:188–196
53. Safe S, Jin UH, Hedrick E, Reeder A, Lee SO (2014) Minireview: role of orphan nuclear receptors in cancer and potential as drug targets. Mol Endocrinol 28:157–172
54. Maruya S, Issa JP, Weber RS, Rosenthal DI, Haviland JC, Lotan R, El-Naggar AK (2004) Differential methylation status of tumor-associated genes in head and neck squamous carcinoma: incidence and potential implications. Clin Cancer Res 10:3825–3830 (An official journal of the American Association for Cancer Research)
55. Lleras RA, Smith RV, Adrien LR, Schlecht NF, Burk RD, Harris TM, Childs G, Prystowsky MB, Belbin TJ (2013) Unique DNA methylation loci distinguish anatomic site and HPV status in head and neck squamous cell carcinoma. Clin Cancer Res 19:5444–5455 (An official journal of the American Association for Cancer Research)
56. Sartor MA, Dolinoy DC, Jones TR, Colacino JA, Prince ME, Carey TE, Rozek LS (2011) Genome-wide methylation and expression differences in HPV(+) and HPV(-) squamous cell carcinoma cell lines are consistent with divergent mechanisms of carcinogenesis. Epigenetics 6:777–787
57. Schlecht NF, Ben-Dayan M, Anayannis N, Lleras RA, Thomas C, Wang Y, Smith RV, Burk RD, Harris TM, Childs G, Ow TJ, Prystowsky MB, Belbin TJ (2015) Epigenetic changes in the CDKN2A locus are associated with differential expression of P16INK4A and P14ARF in HPV-positive oropharyngeal squamous cell carcinoma. Cancer Med 4:342–353
58. Laurson J, Khan S, Chung R, Cross K, Raj K (2010) Epigenetic repression of E-cadherin by human papillomavirus 16 E7 protein. Carcinogenesis 31:918–926
59. Anayannis NV, Schlecht NF, Belbin TJ (2015) Epigenetic mechanisms of human papillomavirus-associated head and neck cancer. Arch Pathol Lab Med 139:1373–1378
60. Kimura H, Nakamura T, Ogawa T, Tanaka S, Shiota K (2003) Transcription of mouse DNA methyltransferase 1 (Dnmt1) is regulated by both E2F-Rb-HDAC-dependent and -independent pathways. Nucleic Acids Res 31:3101–3113
61. Koutsodontis G, Tentes I, Papakosta P, Moustakas A, Kardassis D (2001) Sp1 plays a critical role in the transcriptional activation of the human cyclin-dependent kinase inhibitor p21(WAF1/Cip1) gene by the p53 tumor suppressor protein. J Biol Chem 276:29116–29125
62. Burgers WA, Blanchon L, Pradhan S, de Launoit Y, Kouzarides T, Fuks F (2007) Viral oncoproteins target the DNA methyltransferases. Oncogene 26:1650–1655
63. Sawada M, Kanai Y, Arai E, Ushijima S, Ojima H, Hirohashi S (2007) Increased expression of DNA methyltransferase 1 (DNMT1) protein in uterine cervix squamous cell carcinoma and its precursor lesion. Cancer Lett 251:211–219

64. van Kempen PM, Noorlag R, Braunius WW, Stegeman I, Willems SM, Grolman W (2014) Differences in methylation profiles between HPV-positive and HPV-negative oropharynx squamous cell carcinoma: a systematic review. Epigenetics 9:194–203
65. Wilson GA, Lechner M, Koferle A, Caren H, Butcher LM, Feber A, Fenton T, Jay A, Boshoff C, Beck S (2013) Integrated virus-host methylome analysis in head and neck squamous cell carcinoma. Epigenetics 8:953–961
66. Weiss D, Basel T, Sachse F, Braeuninger A, Rudack C (2011) Promoter methylation of cyclin A1 is associated with human papillomavirus 16 induced head and neck squamous cell carcinoma independently of p53 mutation. Mol Carcinog 50:680–688
67. van Kempen PM, van Bockel L, Braunius WW, Moelans CB, van Olst M, de Jong R, Stegeman I, van Diest PJ, Grolman W, Willems SM (2014) HPV-positive oropharyngeal squamous cell carcinoma is associated with TIMP3 and CADM1 promoter hypermethylation. Cancer Med 3:1185–1196
68. Chen KM, Stephen JK, Havard S, Mahan M, Divine G, Worsham MJ (2015) IGSF4 methylation as an independent marker of human papillomavirus-positive oropharyngeal squamous cell carcinoma. JAMA Otolaryngol Head Neck Surg 141:257–263
69. Marsit CJ, McClean MD, Furniss CS, Kelsey KT (2006) Epigenetic inactivation of the SFRP genes is associated with drinking, smoking and HPV in head and neck squamous cell carcinoma. Int J Cancer 119:1761–1766
70. Kostareli E, Holzinger D, Bogatyrova O, Hielscher T, Wichmann G, Keck M, Lahrmann B, Grabe N, Flechtenmacher C, Schmidt CR, Seiwert T, Dyckhoff G, Dietz A, Hofler D, Pawlita M, Benner A, Bosch FX, Plinkert P, Plass C, Weichenhan D, Hess J (2013) HPV-related methylation signature predicts survival in oropharyngeal squamous cell carcinomas. J Clin Investig 123:2488–2501
71. Miller DL, Puricelli MD, Stack MS (2012) Virology and molecular pathogenesis of HPV (human papillomavirus)-associated oropharyngeal squamous cell carcinoma. Biochem J 443:339–353
72. Patel D, Huang SM, Baglia LA, McCance DJ (1999) The E6 protein of human papillomavirus type 16 binds to and inhibits co-activation by CBP and p300. EMBO J 18:5061–5072
73. Veldman T, Liu X, Yuan H, Schlegel R (2003) Human papillomavirus E6 and Myc proteins associate in vivo and bind to and cooperatively activate the telomerase reverse transcriptase promoter. Proc Natl Acad Sci USA 100:8211–8216
74. Xie X, Piao L, Bullock BN, Smith A, Su T, Zhang M, Teknos TN, Arora PS, Pan Q (2014) Targeting HPV16 E6-p300 interaction reactivates p53 and inhibits the tumorigenicity of HPV-positive head and neck squamous cell carcinoma. Oncogene 33:1037–1046
75. Baylin SB, Jones PA (2011) A decade of exploring the cancer epigenome - biological and translational implications. Nat Rev Cancer 11:726–734
76. Bodily JM, Mehta KP, Laimins LA (2011) Human papillomavirus E7 enhances hypoxia-inducible factor 1-mediated transcription by inhibiting binding of histone deacetylases. Can Res 71:1187–1195
77. McLaughlin-Drubin ME, Crum CP, Munger K (2011) Human papillomavirus E7 oncoprotein induces KDM6A and KDM6B histone demethylase expression and causes epigenetic reprogramming. Proc Natl Acad Sci USA 108:2130–2135
78. Barradas M, Anderton E, Acosta JC, Li S, Banito A, Rodriguez-Niedenfuhr M, Maertens G, Banck M, Zhou MM, Walsh MJ, Peters G, Gil J (2009) Histone demethylase JMJD3 contributes to epigenetic control of INK4a/ARF by oncogenic RAS. Genes Dev 23:1177–1182
79. Holland D, Hoppe-Seyler K, Schuller B, Lohrey C, Maroldt J, Durst M, Hoppe-Seyler F (2008) Activation of the enhancer of zeste homologue 2 gene by the human papillomavirus E7 oncoprotein. Can Res 68:9964–9972
80. Biron VL, Mohamed A, Hendzel MJ, Alan Underhill D, Seikaly H (2012) Epigenetic differences between human papillomavirus-positive and -negative oropharyngeal squamous

cell carcinomas. J Otolaryngol—Head Neck Surg = Le Journal d'oto-rhino-laryngologie et de chirurgie cervico-faciale 41(1):S65–S70

81. Brieger J, Mann SA, Pongsapich W, Koutsimpelas D, Fruth K, Mann WJ (2012) Pharmacological genome demethylation increases radiosensitivity of head and neck squamous carcinoma cells. Int J Mol Med 29:505–509
82. Chung YL, Lee MY, Pui NN (2009) Epigenetic therapy using the histone deacetylase inhibitor for increasing therapeutic gain in oral cancer: prevention of radiation-induced oral mucositis and inhibition of chemical-induced oral carcinogenesis. Carcinogenesis 30:1387–1397
83. Diyabalanage HV, Granda ML, Hooker JM (2013) Combination therapy: histone deacetylase inhibitors and platinum-based chemotherapeutics for cancer. Cancer Lett 329:1–8
84. Gasche JA, Goel A (2012) Epigenetic mechanisms in oral carcinogenesis. Future Oncol 8:1407–1425
85. Viet CT, Dang D, Achdjian S, Ye Y, Katz SG, Schmidt BL (2014) Decitabine rescues cisplatin resistance in head and neck squamous cell carcinoma. PLoS ONE 9:e112880
86. Lajer CB, Garnaes E, Friis-Hansen L, Norrild B, Therkildsen MH, Glud M, Rossing M, Lajer H, Svane D, Skotte L, Specht L, Buchwald C, Nielsen FC (2012) The role of miRNAs in human papilloma virus (HPV)-associated cancers: bridging between HPV-related head and neck cancer and cervical cancer. Br J Cancer 106:1526–1534
87. Miller DL, Davis JW, Taylor KH, Johnson J, Shi Z, Williams R, Atasoy U, Lewis JS Jr, Stack MS (2015) Identification of a human papillomavirus-associated oncogenic miRNA panel in human oropharyngeal squamous cell carcinoma validated by bioinformatics analysis of the Cancer Genome Atlas. Am J Pathol 185:679–692
88. Zheng ZM, Wang X (2011) Regulation of cellular miRNA expression by human papillomaviruses. Biochem Biophys Acta 1809:668–677
89. Gao G, Gay HA, Chernock RD, Zhang TR, Luo J, Thorstad WL, Lewis JS Jr, Wang X (2013) A microRNA expression signature for the prognosis of oropharyngeal squamous cell carcinoma. Cancer 119:72–80
90. Hui AB, Lin A, Xu W, Waldron L, Perez-Ordonez B, Weinreb I, Shi W, Bruce J, Huang SH, O'Sullivan B, Waldron J, Gullane P, Irish JC, Chan K, Liu FF (2013) Potentially prognostic miRNAs in HPV-associated oropharyngeal carcinoma. Clin Cancer Res 19:2154–2162 (An official journal of the American Association for Cancer Research)
91. Lajer CB, Nielsen FC, Friis-Hansen L, Norrild B, Borup R, Garnaes E, Rossing M, Specht L, Therkildsen MH, Nauntofte B, Dabelsteen S, von Buchwald C (2011) Different miRNA signatures of oral and pharyngeal squamous cell carcinomas: a prospective translational study. Br J Cancer 104:830–840
92. Manasa VG, Kannan S (2017) Impact of microRNA dynamics on cancer hallmarks: an oral cancer scenario. Tumour Biol 39:1010428317695920
93. Hui AB, Lenarduzzi M, Krushel T, Waldron L, Pintilie M, Shi W, Perez-Ordonez B, Jurisica I, O'Sullivan B, Waldron J, Gullane P, Cummings B, Liu FF (2010) Comprehensive MicroRNA profiling for head and neck squamous cell carcinomas. Clin Cancer Res 16:1129–1139 (An official journal of the American Association for Cancer Research)
94. Zhang X, Gee H, Rose B, Lee CS, Clark J, Elliott M, Gamble JR, Cairns MJ, Harris A, Khoury S, Tran N (2016) Regulation of the tumour suppressor PDCD4 by miR-499 and miR-21 in oropharyngeal cancers. BMC Cancer 16:86
95. Wald AI, Hoskins EE, Wells SI, Ferris RL, Khan SA (2011) Alteration of microRNA profiles in squamous cell carcinoma of the head and neck cell lines by human papillomavirus. Head Neck 33:504–512
96. Chapman BV, Wald AI, Akhtar P, Munko AC, Xu J, Gibson SP, Grandis JR, Ferris RL, Khan SA (2015) MicroRNA-363 targets myosin 1B to reduce cellular migration in head and neck cancer. BMC Cancer 15:861
97. Childs G, Fazzari M, Kung G, Kawachi N, Brandwein-Gensler M, McLemore M, Chen Q, Burk RD, Smith RV, Prystowsky MB, Belbin TJ, Schlecht NF (2009) Low-level expression

of microRNAs let-7d and miR-205 are prognostic markers of head and neck squamous cell carcinoma. Am J Pathol 174:736–745
98. Harris T, Jimenez L, Kawachi N, Fan JB, Chen J, Belbin T, Ramnauth A, Loudig O, Keller CE, Smith R, Prystowsky MB, Schlecht NF, Segall JE, Childs G (2012) Low-level expression of miR-375 correlates with poor outcome and metastasis while altering the invasive properties of head and neck squamous cell carcinomas. Am J Pathol 180:917–928
99. Avissar M, McClean MD, Kelsey KT, Marsit CJ (2009) MicroRNA expression in head and neck cancer associates with alcohol consumption and survival. Carcinogenesis 30:2059–2063
100. Chen X, Sturgis EM, Wang C, Cao X, Li Y, Wei Q, Li G (2016) Significance of microRNA-related variants in susceptibility to recurrence of oropharyngeal cancer patients after definitive radiotherapy. Oncotarget 7:35015–35025
101. Firmino N, Martinez VD, Rowbotham DA, Enfield KSS, Bennewith KL, Lam WL (2016) HPV status is associated with altered PIWI-interacting RNA expression pattern in head and neck cancer. Oral Oncol 55:43–48
102. Wong N, Khwaja SS, Baker CM, Gay HA, Thorstad WL, Daly MD, Lewis JS Jr, Wang X (2016) Prognostic microRNA signatures derived from the cancer genome atlas for head and neck squamous cell carcinomas. Cancer Med 5:1619–1628
103. Liu S, Song L, Yao H, Zhang L, Xu D, Gao F, Li Q (2016) MiR-375 is epigenetically downregulated by HPV-16 E6 mediated DNMT1 upregulation and modulates EMT of cervical cancer cells by suppressing lncRNA MALAT1. PLoS ONE 11:e0163460
104. Jung HM, Phillips BL, Chan EK (2014) miR-375 activates p21 and suppresses telomerase activity by coordinately regulating HPV E6/E7, E6AP, CIP2A, and 14-3-3zeta. Mol Cancer 13:80
105. Jeffers LK, Duan K, Ellies LG, Seaman WT, Burger-Calderon RA, Diatchenko LB, Webster-Cyriaque J (2013) Correlation of transcription of MALAT-1, a novel noncoding RNA, with deregulated expression of tumor suppressor p 53 in small DNA tumor virus models. J Cancer Ther 4(3)
106. Jiang Y, Li Y, Fang S, Jiang B, Qin C, Xie P, Zhou G, Li G (2014) The role of MALAT1 correlates with HPV in cervical cancer. Oncol Lett 7:2135–2141
107. Jimenez L, Jayakar SK, Ow TJ, Segall JE (2015) Mechanisms of invasion in head and neck cancer. Arch Pathol Lab Med 139:1334–1348
108. Galvan SC, Martinez-Salazar M, Galvan VM, Mendez R, Diaz-Contreras GT, Alvarado-Hermida M, Alcantara-Silva R, Garcia-Carranca A (2011) Analysis of CpG methylation sites and CGI among human papillomavirus DNA genomes. BMC Genom 12:580
109. Hatano T, Sano D, Takahashi H, Hyakusoku H, Isono Y, Shimada S, Sawakuma K, Takada K, Oikawa R, Watanabe Y, Yamamoto H, Itoh F, Myers JN, Oridate N (2017) Identification of human papillomavirus (HPV) 16 DNA integration and the ensuing patterns of methylation in HPV-associated head and neck squamous cell carcinoma cell lines. Int J Cancer 140:1571–1580
110. Fernandez AF, Rosales C, Lopez-Nieva P, Grana O, Ballestar E, Ropero S, Espada J, Melo SA, Lujambio A, Fraga MF, Pino I, Javierre B, Carmona FJ, Acquadro F, Steenbergen RD, Snijders PJ, Meijer CJ, Pineau P, Dejean A, Lloveras B, Capella G, Quer J, Buti M, Esteban JI, Allende H, Rodriguez-Frias F, Castellsague X, Minarovits J, Ponce J, Capello D, Gaidano G, Cigudosa JC, Gomez-Lopez G, Pisano DG, Valencia A, Piris MA, Bosch FX, Cahir-McFarland E, Kieff E, Esteller M (2009) The dynamic DNA methylomes of double-stranded DNA viruses associated with human cancer. Genome Res 19:438–451
111. Mirabello L, Frimer M, Harari A, McAndrew T, Smith B, Chen Z, Wentzensen N, Wacholder S, Castle PE, Raine-Bennett T, Schiffman M, Burk RD (2015) HPV16 methyl-haplotypes determined by a novel next-generation sequencing method are associated with cervical precancer. Int J Cancer 136:E146–E153

112. Park IS, Chang X, Loyo M, Wu G, Chuang A, Kim MS, Chae YK, Lyford-Pike S, Westra WH, Saunders JR, Sidransky D, Pai SI (2011) Characterization of the methylation patterns in human papillomavirus type 16 viral DNA in head and neck cancers. Cancer Prev Res (Phila) 4:207–217
113. Zhang C, Deng Z, Pan X, Uehara T, Suzuki M, Xie M (2015) Effects of methylation status of CpG sites within the HPV16 long control region on HPV16-Positive head and neck cancer cells. PLoS ONE 10:e0141245
114. Reuschenbach M, Huebbers CU, Prigge ES, Bermejo JL, Kalteis MS, Preuss SF, Seuthe IM, Kolligs J, Speel EJ, Olthof N, Kremer B, Wagner S, Klussmann JP, Vinokurova S, von Knebel Doeberitz M (2015) Methylation status of HPV16 E2-binding sites classifies subtypes of HPV-associated oropharyngeal cancers. Cancer 121:1966–1976
115. Harris RS, Dudley JP (2015) APOBECs and virus restriction. Virology 479–480:131–145
116. Rebhandl S, Huemer M, Greil R, Geisberger R (2015) AID/APOBEC deaminases and cancer. Oncoscience 2:320–333
117. Warren CJ, Xu T, Guo K, Griffin LM, Westrich JA, Lee D, Lambert PF, Santiago ML, Pyeon D (2015) APOBEC3A functions as a restriction factor of human papillomavirus. J Virol 89:688–702
118. Vieira VC, Leonard B, White EA, Starrett GJ, Temiz NA, Lorenz LD, Lee D, Soares MA, Lambert PF, Howley PM, Harris RS (2014) Human papillomavirus E6 triggers upregulation of the antiviral and cancer genomic DNA deaminase APOBEC3B. MBio 5
119. Henderson S, Chakravarthy A, Su X, Boshoff C, Fenton TR (2014) APOBEC-mediated cytosine deamination links PIK3CA helical domain mutations to human papillomavirus-driven tumor development. Cell Rep 7:1833–1841
120. Kennedy SG, Wagner AJ, Conzen SD, Jordan J, Bellacosa A, Tsichlis PN, Hay N (1997) The PI 3-kinase/Akt signaling pathway delivers an anti-apoptotic signal. Genes Dev 11:701–713
121. Klippel A, Escobedo MA, Wachowicz MS, Apell G, Brown TW, Giedlin MA, Kavanaugh WM, Williams LT (1998) Activation of phosphatidylinositol 3-kinase is sufficient for cell cycle entry and promotes cellular changes characteristic of oncogenic transformation. Mol Cell Biol 18:5699–5711
122. Koncar RF, Feldman R, Bahassi EM, Hashemi Sadraei N (2017) Comparative molecular profiling of HPV-induced squamous cell carcinomas. Cancer Med 6:1673–1685
123. Ma YY, Wei SJ, Lin YC, Lung JC, Chang TC, Whang-Peng J, Liu JM, Yang DM, Yang WK, Shen CY (2000) PIK3CA as an oncogene in cervical cancer. Oncogene 19:2739–2744
124. Samuels Y, Ericson K (2006) Oncogenic PI3K and its role in cancer. Curr Opin Oncol 18:77–82
125. Chung CH, Guthrie VB, Masica DL, Tokheim C, Kang H, Richmon J, Agrawal N, Fakhry C, Quon H, Subramaniam RM, Zuo Z, Seiwert T, Chalmers ZR, Frampton GM, Ali SM, Yelensky R, Stephens PJ, Miller VA, Karchin R, Bishop JA (2015) Genomic alterations in head and neck squamous cell carcinoma determined by cancer gene-targeted sequencing. Ann Oncol 26:1216–1223
126. Ojesina AI, Lichtenstein L, Freeman SS, Pedamallu CS, Imaz-Rosshandler I, Pugh TJ, Cherniack AD, Ambrogio L, Cibulskis K, Bertelsen B, Romero-Cordoba S, Trevino V, Vazquez-Santillan K, Guadarrama AS, Wright AA, Rosenberg MW, Duke F, Kaplan B, Wang R, Nickerson E, Walline HM, Lawrence MS, Stewart C, Carter SL, McKenna A, Rodriguez-Sanchez IP, Espinosa-Castilla M, Woie K, Bjorge L, Wik E, Halle MK, Hoivik EA, Krakstad C, Gabino NB, Gomez-Macias GS, Valdez-Chapa LD, Garza-Rodriguez ML, Maytorena G, Vazquez J, Rodea C, Cravioto A, Cortes ML, Greulich H, Crum CP, Neuberg DS, Hidalgo-Miranda A, Escareno CR, Akslen LA, Carey TE, Vintermyr OK, Gabriel SB, Barrera-Saldana HA, Melendez-Zajgla J, Getz G, Salvesen HB, Meyerson M (2014) Landscape of genomic alterations in cervical carcinomas. Nature 506:371–375
127. Seiwert TY, Zuo Z, Keck MK, Khattri A, Pedamallu CS, Stricker T, Brown C, Pugh TJ, Stojanov P, Cho J, Lawrence MS, Getz G, Bragelmann J, DeBoer R, Weichselbaum RR,

Langerman A, Portugal L, Blair E, Stenson K, Lingen MW, Cohen EE, Vokes EE, White KP, Hammerman PS (2015) Integrative and comparative genomic analysis of HPV-positive and HPV-negative head and neck squamous cell carcinomas. Clin Cancer Res 21:632–641 (An official journal of the American Association for Cancer Research)

128. Lechner M, Frampton GM, Fenton T, Feber A, Palmer G, Jay A, Pillay N, Forster M, Cronin MT, Lipson D, Miller VA, Brennan TA, Henderson S, Vaz F, O'Flynn P, Kalavrezos N, Yelensky R, Beck S, Stephens PJ, Boshoff C (2013) Targeted next-generation sequencing of head and neck squamous cell carcinoma identifies novel genetic alterations in HPV+ and HPV- tumors. Genome Med 5:49
129. Lui VW, Hedberg ML, Li H, Vangara BS, Pendleton K, Zeng Y, Lu Y, Zhang Q, Du Y, Gilbert BR, Freilino M, Sauerwein S, Peyser ND, Xiao D, Diergaarde B, Wang L, Chiosea S, Seethala R, Johnson JT, Kim S, Duvvuri U, Ferris RL, Romkes M, Nukui T, Kwok-Shing Ng P, Garraway LA, Hammerman PS, Mills GB, Grandis JR (2013) Frequent mutation of the PI3K pathway in head and neck cancer defines predictive biomarkers. Cancer Discov 3:761–769
130. Vogt PK, Kang S, Elsliger MA, Gymnopoulos M (2007) Cancer-specific mutations in phosphatidylinositol 3-kinase. Trends Biochem Sci 32:342–349
131. Liu P, Cheng H, Roberts TM, Zhao JJ (2009) Targeting the phosphoinositide 3-kinase pathway in cancer. Nat Rev Drug Discov 8:627–644
132. Cancer Genome Atlas Research, N., M. Albert Einstein College of, S. Analytical Biological, H. Barretos Cancer, M. Baylor College of, H. Beckman Research Institute of City of, A. Buck Institute for Research on, C. Canada's Michael Smith Genome Sciences, S. Harvard Medical, F. G. C. C. Helen, S. Research Institute at Christiana Care Health, B. HudsonAlpha Institute for, L. L. C. Ilsbio, M. Indiana University School of, V. Institute of Human, B. Institute for Systems, C. International Genomics, B. Leidos, H. Massachusetts General, U. McDonnell Genome Institute at Washington, W. Medical College of, C. Medical University of South, C. Memorial Sloan Kettering Cancer, C. Montefiore Medical, NantOmics, I. National Cancer, A. N. National Hospital, I. National Human Genome Research, S. National Institute of Environmental Health, D. National Institute on, D. Other Communication, L. H. S. C. Ontario Tumour Bank, O. I. f. C. R. Ontario Tumour Bank, T. O. H. Ontario Tumour Bank, H. Oregon, U. Science, C.-S. M. C. Samuel Oschin Comprehensive Cancer Institute, S. R. A. International, S. St Joseph's Candler Health, Eli, L. B. I. o. M. I. o. T. Edythe, U. Harvard, H. Research Institute at Nationwide Children's, U. Sidney Kimmel Comprehensive Cancer Center at Johns Hopkins, B. University of, M. D. A. C. C. University of Texas, H. University of Abuja Teaching, B. University of Alabama at, I. University of California, C. University of California Santa, C. University of Kansas Medical, L. University of, C. University of New Mexico Health Sciences, H. University of North Carolina at Chapel, C. University of Oklahoma Health Sciences, P. University of, R. a. P. M. S. University of Sao Paulo, C. University of Southern, W. University of, M. University of Wisconsin School of, H. Public, I. Van Andel Research, and L. Washington University in St (2017) Integrated genomic and molecular characterization of cervical cancer. Nature 543:378–384
133. Cui B, Zheng B, Zhang X, Stendahl U, Andersson S, Wallin KL (2009) Mutation of PIK3CA: possible risk factor for cervical carcinogenesis in older women. Int J Oncol 34:409–416
134. Lou H, Villagran G, Boland JF, Im KM, Polo S, Zhou W, Odey U, Juarez-Torres E, Medina-Martinez I, Roman-Basaure E, Mitchell J, Roberson D, Sawitzke J, Garland L, Rodriguez-Herrera M, Wells D, Troyer J, Pinto FC, Bass S, Zhang X, Castillo M, Gold B, Morales H, Yeager M, Berumen J, Alvirez E, Gharzouzi E, Dean M (2015) Genome analysis of latin American cervical cancer: frequent activation of the PIK3CA pathway. Clin Cancer Res 21:5360–5370 (An official journal of the American Association for Cancer Research)
135. McIntyre JB, Wu JS, Craighead PS, Phan T, Kobel M, Lees-Miller SP, Ghatage P, Magliocco AM, Doll CM (2013) PIK3CA mutational status and overall survival in patients with cervical cancer treated with radical chemoradiotherapy. Gynecol Oncol 128:409–414

136. Tornesello ML, Annunziata C, Buonaguro L, Losito S, Greggi S, Buonaguro FM (2014) TP53 and PIK3CA gene mutations in adenocarcinoma, squamous cell carcinoma and high-grade intraepithelial neoplasia of the cervix. J Transl Med 12:255
137. Wright AA, Howitt BE, Myers AP, Dahlberg SE, Palescandolo E, Van Hummelen P, MacConaill LE, Shoni M, Wagle N, Jones RT, Quick CM, Laury A, Katz IT, Hahn WC, Matulonis UA, Hirsch MS (2013) Oncogenic mutations in cervical cancer: genomic differences between adenocarcinomas and squamous cell carcinomas of the cervix. Cancer 119:3776–3783
138. Feldman R, Gatalica Z, Knezetic J, Reddy S, Nathan CA, Javadi N, Teknos T (2016) Molecular profiling of head and neck squamous cell carcinoma. Head Neck 38(Suppl 1): E1625–E1638
139. Stelzer MK, Pitot HC, Liem A, Lee D, Kennedy GD, Lambert PF (2010) Rapamycin inhibits anal carcinogenesis in two preclinical animal models. Cancer Prev Res (Phila) 3:1542–1551
140. Pollock NI, Wang L, Wallweber G, Gooding WE, Huang W, Chenna A, Winslow J, Sen M, DeGrave KA, Li H, Zeng Y, Grandis JR (2015) Increased expression of HER2, HER3, and HER2:HER3 heterodimers in HPV-positive HNSCC using a novel proximity-based assay: implications for targeted therapies. Clinical Cancer Research 21:4597–4606 (An official journal of the American Association for Cancer Research)
141. Brand TM, Hartmann S, Bhola NE, Peyser ND, Li H, Zeng Y, Isaacson Wechsler E, Ranall MV, Bandyopadhyay S, Duvvuri U, LaVallee TM, Jordan RCK, Johnson DE, Grandis JR (2017) Human Papillomavirus regulates HER3 expression in head and neck cancer: implications for targeted HER3 therapy in HPV(+) patients. Clin Cancer Res 23:3072–3083 (An official journal of the American Association for Cancer Research)
142. Machiels JP, Haddad RI, Fayette J, Licitra LF, Tahara M, Vermorken JB, Clement PM, Gauler T, Cupissol D, Grau JJ, Guigay J, Caponigro F, de Castro G, Jr., de Souza Viana L, Keilholz U, Del Campo JM, Cong XJ, Ehrnrooth E, Cohen EE, Lux H, Investigators N (2015) Afatinib versus methotrexate as second-line treatment in patients with recurrent or metastatic squamous-cell carcinoma of the head and neck progressing on or after platinum-based therapy (LUX-Head & Neck 1): an open-label, randomised phase 3 trial. Lancet Oncol 16:583–594
143. Lesseur C, Diergaarde B, Olshan AF, Wunsch-Filho V, Ness AR, Liu G, Lacko M, Eluf-Neto J, Franceschi S, Lagiou P, Macfarlane GJ, Richiardi L, Boccia S, Polesel J, Kjaerheim K, Zaridze D, Johansson M, Menezes AM, Curado MP, Robinson M, Ahrens W, Canova C, Znaor A, Castellsague X, Conway DI, Holcatova I, Mates D, Vilensky M, Healy CM, Szeszenia-Dabrowska N, Fabianova E, Lissowska J, Grandis JR, Weissler MC, Tajara EH, Nunes FD, de Carvalho MB, Thomas S, Hung RJ, Peters WH, Herrero R, Cadoni G, Bueno-de-Mesquita HB, Steffen A, Agudo A, Shangina O, Xiao X, Gaborieau V, Chabrier A, Anantharaman D, Boffetta P, Amos CI, McKay JD, Brennan P (2016) Genome-wide association analyses identify new susceptibility loci for oral cavity and pharyngeal cancer. Nat Genet 48:1544–1550
144. Chen D, Gyllensten U (2015) Lessons and implications from association studies and post-GWAS analyses of cervical cancer. Trends Genet 31:41–54
145. Martinez-Nava GA, Fernandez-Nino JA, Madrid-Marina V, Torres-Poveda K (2016) Cervical cancer genetic susceptibility: a systematic review and meta-analyses of recent evidence. PLoS ONE 11:e0157344
146. Hajek M, Sewell A, Kaech S, Burtness B, Yarbrough WG, Issaeva N (2017) TRAF3/CYLD mutations identify a distinct subset of human papillomavirus-associated head and neck squamous cell carcinoma. Cancer 123:1778–1790
147. Cameron JE, Mercante D, O'Brien M, Gaffga AM, Leigh JE, Fidel PL Jr, Hagensee ME (2005) The impact of highly active antiretroviral therapy and immunodeficiency on human papillomavirus infection of the oral cavity of human immunodeficiency virus-seropositive adults. Sex Transm Dis 32:703–709

148. Coutlee F, Trottier AM, Ghattas G, Leduc R, Toma E, Sanche G, Rodrigues I, Turmel B, Allaire G, Ghadirian P (1997) Risk factors for oral human papillomavirus in adults infected and not infected with human immunodeficiency virus. Sex Transm Dis 24:23–31
149. Fakhry C, D'Souza G, Sugar E, Weber K, Goshu E, Minkoff H, Wright R, Seaberg E, Gillison M (2006) Relationship between prevalent oral and cervical human papillomavirus infections in human immunodeficiency virus-positive and -negative women. J Clin Microbiol 44:4479–4485
150. Frisch M, Biggar RJ, Goedert JJ (2000) Human papillomavirus-associated cancers in patients with human immunodeficiency virus infection and acquired immunodeficiency syndrome. J Natl Cancer Inst 92:1500–1510
151. Gillison ML, D'Souza G, Westra W, Sugar E, Xiao W, Begum S, Viscidi R (2008) Distinct risk factor profiles for human papillomavirus type 16-positive and human papillomavirus type 16-negative head and neck cancers. J Natl Cancer Inst 100:407–420
152. Kreimer AR, Alberg AJ, Daniel R, Gravitt PE, Viscidi R, Garrett ES, Shah KV, Gillison ML (2004) Oral human papillomavirus infection in adults is associated with sexual behavior and HIV serostatus. J Infect Dis 189:686–698
153. Beachler DC, Sugar EA, Margolick JB, Weber KM, Strickler HD, Wiley DJ, Cranston RD, Burk RD, Minkoff H, Reddy S, Xiao W, Guo Y, Gillison ML, D'Souza G (2015) Risk factors for acquisition and clearance of oral human papillomavirus infection among HIV-infected and HIV-uninfected adults. Am J Epidemiol 181:40–53
154. Mooij SH, van der Klis FR, van der Sande MA, Schepp RM, Speksnijder AG, Bogaards JA, de Melker HE, de Vries HJ, Snijders PJ, van der Loeff MF (2013) Seroepidemiology of high-risk HPV in HIV-negative and HIV-infected MSM: the H2 M study. Cancer Epidemiol Biomark Prev 22:1698–1708 (A publication of the American Association for Cancer Research, cosponsored by the American Society of Preventive Oncology)
155. Parisi SG, Cruciani M, Scaggiante R, Boldrin C, Andreis S, Dal Bello F, Pagni S, Barelli A, Sattin A, Mengoli C, Palu G (2011) Anal and oral human papillomavirus (HPV) infection in HIV-infected subjects in northern Italy: a longitudinal cohort study among men who have sex with men. BMC Infect Dis 11:150
156. Read TR, Hocking JS, Vodstrcil LA, Tabrizi SN, McCullough MJ, Grulich AE, Garland SM, Bradshaw CS, Chen MY, Fairley CK (2012) Oral human papillomavirus in men having sex with men: risk-factors and sampling. PLoS ONE 7:e49324
157. King EM, Oomeer S, Gilson R, Copas A, Beddows S, Soldan K, Jit M, Edmunds WJ, Sonnenberg P (2016) Oral human papillomavirus infection in men who have sex with men: a systematic review and meta-analysis. PLoS ONE 11:e0157976
158. Chaturvedi AK, Madeleine MM, Biggar RJ, Engels EA (2009) Risk of human papillomavirus-associated cancers among persons with AIDS. J Natl Cancer Inst 101:1120–1130
159. Clifford GM, Polesel J, Rickenbach M, Dal Maso L, Keiser O, Kofler A, Rapiti E, Levi F, Jundt G, Fisch T, Bordoni A, De Weck D, Franceschi S (2005) Cancer risk in the Swiss HIV cohort study: associations with immunodeficiency, smoking, and highly active antiretroviral therapy. J Natl Cancer Inst 97:425–432
160. Marais DJ, Passmore JA, Denny L, Sampson C, Allan BR, Williamson AL (2008) Cervical and oral human papillomavirus types in HIV-1 positive and negative women with cervical disease in South Africa. J Med Virol 80:953–959
161. Shiboski CH, Lee A, Chen H, Webster-Cyriaque J, Seaman T, Landovitz RJ, John M, Reilly N, Naini L, Palefsky J, Jacobson MA (2016) Human papillomavirus infection in the oral cavity of HIV patients is not reduced by initiating antiretroviral therapy. AIDS 30:1573–1582
162. Steinau M, Reddy D, Sumbry A, Reznik D, Gunthel CJ, Del Rio C, Lennox JL, Unger ER, Nguyen ML (2012) Oral sampling and human papillomavirus genotyping in HIV-infected patients. J. Oral Pathol Med 41:288–291 (Official publication of the International Association of Oral Pathologists and the American Academy of Oral Pathology)

163. Sirera G, Videla S, Pinol M, Canadas MP, Llatjos M, Ballesteros AL, Garcia-Cuyas F, Castella E, Guerola R, Tural C, Rey-Joly C, Clotet B (2006) High prevalence of human papillomavirus infection in the anus, penis and mouth in HIV-positive men. AIDS 20:1201–1204
164. Fatahzadeh M, Schlecht NF, Chen Z, Bottalico D, McKinney S, Ostoloza J, Dunne A, Burk RD (2013) Oral human papillomavirus detection in older adults who have human immunodeficiency virus infection. Oral Surg Oral Med Oral Pathol Oral Radiol 115:505–514
165. Blas MM, Brown B, Menacho L, Alva IE, Silva-Santisteban A, Carcamo C (2015) HPV prevalence in multiple anatomical sites among men who have sex with men in Peru. PLoS ONE 10:e0139524
166. Videla S, Darwich L, Canadas MP, Coll J, Pinol M, Garcia-Cuyas F, Molina-Lopez RA, Cobarsi P, Clotet B, Sirera G (2013) Natural history of human papillomavirus infections involving anal, penile, and oral sites among HIV-positive men. Sex Transm Dis 40:3–10
167. van Rijn VM, Mooij SH, Mollers M, Snijders PJ, Speksnijder AG, King AJ, de Vries HJ, van Eeden A, van der Klis FR, de Melker HE, van der Sande MA, van der Loeff MF (2014) Anal, penile, and oral high-risk HPV infections and HPV seropositivity in HIV-positive and HIV-negative men who have sex with men. PLoS ONE 9:e92208
168. Fakhry C, Gillison ML (2006) Clinical implications of human papillomavirus in head and neck cancers. J Clin Oncol 24:2606–2611
169. Richter KL, van Rensburg EJ, van Heerden WF, Boy SC (2008) Human papilloma virus types in the oral and cervical mucosa of HIV-positive South African women prior to antiretroviral therapy. J Oral Pathol Med 37:555–559 (Official publication of the International Association of Oral Pathologists and the American Academy of Oral Pathology)
170. Ong JJ, Read TR, Vodstrcil LA, Walker S, Chen M, Bradshaw CS, Garland SM, Tabrizi SN, Cornall A, Grulich A, Hocking J, Fairley CK (2014) Detection of oral human papillomavirus in HIV-positive men who have sex with men 3 years after baseline: a follow up cross-sectional study. PLoS ONE 9:e102138
171. van Aar F, Mooij SH, van der Sande MA, Meijer CJ, King AJ, Verhagen DW, Heijman T, Coutinho RA, Schim van der Loeff MF (2014) Twelve-month incidence and clearance of oral HPV infection in HIV-negative and HIV-infected men who have sex with men: the H2M cohort study. BMC Infect Dis 14:668
172. D'Souza G, Fakhry C, Sugar EA, Seaberg EC, Weber K, Minkoff HL, Anastos K, Palefsky JM, Gillison ML (2007) Six-month natural history of oral versus cervical human papillomavirus infection. Int J Cancer 121:143–150
173. Lam JO, Bream JH, Sugar EA, Coles CL, Weber KM, Burk RD, Wiley DJ, Cranston RD, Reddy S, Margolick JB, Strickler HD, Wentz A, Jacobson L, Guo Y, Xiao W, Gillison ML, D'Souza G (2016) Association of serum cytokines with oral HPV clearance. Cytokine 83:85–91
174. Greenspan D, de Villiers EM, Greenspan JS, de Souza YG, zur Hausan H (1988) Unusual HPV types in oral warts in association with HIV infection. J Oral Pathol 17:482–488
175. Anaya-Saavedra G, Flores-Moreno B, Garcia-Carranca A, Irigoyen-Camacho E, Guido-Jimenez M, Ramirez-Amador V (2013) HPV oral lesions in HIV-infected patients: the impact of long-term HAART. J Oral Pathol Med 42:443–449 (Official publication of the International Association of Oral Pathologists and the American Academy of Oral Pathology)
176. Drake LA, Ceilley RI, Cornelison RL, Dobes WL, Dorner W, Goltz RW, Lewis CW, Salasche SJ, Turner ML, Lowery BJ (1995) Guidelines of care for warts: human papillomavirus. Committee on Guidelines of Care. J Am Acad Dermatol 32:98–103

177. Lozada-Nur F, Glick M, Schubert M, Silverberg I (2001) Use of intralesional interferon-alpha for the treatment of recalcitrant oral warts in patients with AIDS: a report of 4 cases. Oral Surg Oral Med Oral Pathol Oral Radiol Endod 92:617–622
178. DeRossi SS, Laudenbach J (2004) The management of oral human papillomavirus with topical cidofovir: a case report. Cutis 73:191–193
179. Mun JH, Kim SH, Jung DS, Ko HC, Kim BS, Kwon KS, Kim MB (2011) Oral zinc sulfate treatment for viral warts: an open-label study. J Dermatol 38:541–545
180. Kaur GJ, Brar BK, Kumar S, Brar SK, Singh B (2017) Evaluation of the efficacy and safety of oral isotretinoin versus topical isotretinoin in the treatment of plane warts: a randomized open trial. Int J Dermatol 56:1352–1358
181. Snietura M, Lamch R, Kopec A, Waniczek D, Likus W, Lange D, Markowski J (2017) Oral and oropharyngeal papillomas are not associated with high-risk human papillomavirus infection. Eur Arch Otorhinolaryngol 274:3477–3483
182. Piattelli A, Rubini C, Fioroni M, Iezzi T (2001) Warty carcinoma of the oral mucosa in an HIV+ patient. Oral Oncol 37:665–667
183. Dona MG, Pichi B, Rollo F, Gheit T, Laquintana V, Covello R, Pescarmona E, Spriano G, Pellini R, Giuliani M, Tommasino M, Benevolo M (2017) Mucosal and cutaneous human papillomaviruses in head and neck squamous cell papillomas. Head Neck 39:254–259
184. Chaturvedi AK, Engels EA, Anderson WF, Gillison ML (2008) Incidence trends for human papillomavirus-related and -unrelated oral squamous cell carcinomas in the United States. J Clin Oncol 26:612–619
185. Gillison ML (2009) Oropharyngeal cancer: a potential consequence of concomitant HPV and HIV infection. Curr Opin Oncol 21:439–444
186. D'Souza G, Kreimer AR, Viscidi R, Pawlita M, Fakhry C, Koch WM, Westra WH, Gillison ML (2007) Case-control study of human papillomavirus and oropharyngeal cancer. N Engl J Med 356:1944–1956
187. Engels EA, Biggar RJ, Hall HI, Cross H, Crutchfield A, Finch JL, Grigg R, Hylton T, Pawlish KS, McNeel TS, Goedert JJ (2008) Cancer risk in people infected with human immunodeficiency virus in the United States. Int J Cancer 123:187–194
188. Grulich AE, van Leeuwen MT, Falster MO, Vajdic CM (2007) Incidence of cancers in people with HIV/AIDS compared with immunosuppressed transplant recipients: a meta-analysis. Lancet 370:59–67
189. Patel P, Hanson DL, Sullivan PS, Novak RM, Moorman AC, Tong TC, Holmberg SD, Brooks JT (2008) Incidence of types of cancer among HIV-infected persons compared with the general population in the United States, 1992-2003. Ann Intern Med 148:728–736
190. Shiels MS, Cole SR, Kirk GD, Poole C (2009) A meta-analysis of the incidence of non-AIDS cancers in HIV-infected individuals. J Acquir Immune Defic Syndr 52:611–622
191. Beachler DC, Abraham AG, Silverberg MJ, Jing Y, Fakhry C, Gill MJ, Dubrow R, Kitahata MM, Klein MB, Burchell AN, Korthuis PT, Moore RD, D'Souza G (2014) Incidence and risk factors of HPV-related and HPV-unrelated head and neck squamous cell carcinoma in HIV-infected individuals. Oral Oncol 50:1169–1176
192. Berretta M, Cinelli R, Martellotta F, Spina M, Vaccher E, Tirelli U (2003) Therapeutic approaches to AIDS-related malignancies. Oncogene 22:6646–6659
193. Simard EP, Pfeiffer RM, Engels EA (2010) Spectrum of cancer risk late after AIDS onset in the United States. Arch Intern Med 170:1337–1345
194. Demopoulos BP, Vamvakas E, Ehrlich JE, Demopoulos R (2003) Non-acquired immunodeficiency syndrome-defining malignancies in patients infected with human immunodeficiency virus. Arch Pathol Lab Med 127:589–592
195. Chew EY, Hartman CM, Richardson PA, Zevallos JP, Sikora AG, Kramer JR, Chiao EY (2017) Risk factors for oropharynx cancer in a cohort of HIV-infected veterans. Oral Oncol 68:60–66
196. Silverberg MJ, Chao C, Leyden WA, Xu L, Horberg MA, Klein D, Towner WJ, Dubrow R, Quesenberry CP Jr, Neugebauer RS, Abrams DI (2011) HIV infection, immunodeficiency,

viral replication, and the risk of cancer. Cancer Epidemiol Biomark Prev 20:2551–2559 (A publication of the American Association for Cancer Research, cosponsored by the American Society of Preventive Oncology)

197. Picard A, Badoual C, Hourseau M, Halimi C, Pere H, Dib F, Barry B, Albert S (2016) Human papilloma virus prevalence in HIV patients with head and neck squamous cell carcinoma. AIDS 30:1257–1266
198. McLemore MS, Haigentz M Jr, Smith RV, Nuovo GJ, Alos L, Cardesa A, Brandwein-Gensler M (2010) Head and neck squamous cell carcinomas in HIV-positive patients: a preliminary investigation of viral associations. Head Neck Pathol 4:97–105
199. Fakhry C, Westra WH, Li S, Cmelak A, Ridge JA, Pinto H, Forastiere A, Gillison ML (2008) Improved survival of patients with human papillomavirus-positive head and neck squamous cell carcinoma in a prospective clinical trial. J Natl Cancer Inst 100:261–269
200. Weinberger PM, Yu Z, Haffty BG, Kowalski D, Harigopal M, Brandsma J, Sasaki C, Joe J, Camp RL, Rimm DL, Psyrri A (2006) Molecular classification identifies a subset of human papillomavirus–associated oropharyngeal cancers with favorable prognosis. J Clin Oncol 24:736–747 (Official journal of the American Society of Clinical Oncology)
201. Kimple RJ, Harari PM (2014) Is radiation dose reduction the right answer for HPV-positive head and neck cancer? Oral Oncol 50:560–564
202. Singh B, Sabin S, Rofim O, Shaha A, Har-El G, Lucente FE (1999) Alterations in head and neck cancer occurring in HIV-infected patients–results of a pilot, longitudinal, prospective study. Acta Oncol 38:1047–1050
203. D'Souza G, Carey TE, William WN Jr., Nguyen ML, Ko EC, Riddell JT, Pai SI, Gupta V, Walline HM, Lee JJ, Wolf GT, Shin DM, Grandis JR, Ferris RL (2014) Epidemiology of head and neck squamous cell cancer among HIV-infected patients. J Acquir Immune Defic Syndr 65:603–610
204. Grew DJ, Cooper BT, Nguy S, Halperin J, Sanfilippo NJ (2014) Toxicity and disease-related outcomes after radiotherapy for head and neck cancer in human immunodeficiency virus-positive patients. Front Oncol 4:316
205. Mourad WF, Hu KS, Shasha D, Concert C, Ishihara D, Lin W, Gamez ME, Lukens JN, Shourbaji RA, Ryniak M, Li Z, Culliney BE, Khorsandi AS, Tran T, Jacobson A, Manolidis S, Schantz S, Urken M, Persky MS, Harrison LB (2013) Long-term outcome of seropositive HIV patients with head and neck squamous cell carcinoma treated with radiation therapy and chemotherapy. Anticancer Res 33:5511–5516
206. Kreimer AR, Johansson M, Waterboer T, Kaaks R, Chang-Claude J, Drogen D, Tjonneland A, Overvad K, Quiros JR, Gonzalez CA, Sanchez MJ, Larranaga N, Navarro C, Barricarte A, Travis RC, Khaw KT, Wareham N, Trichopoulou A, Lagiou P, Trichopoulos D, Peeters PH, Panico S, Masala G, Grioni S, Tumino R, Vineis P, Bueno-de-Mesquita HB, Laurell G, Hallmans G, Manjer J, Ekstrom J, Skeie G, Lund E, Weiderpass E, Ferrari P, Byrnes G, Romieu I, Riboli E, Hildesheim A, Boeing H, Pawlita M, Brennan P (2013) Evaluation of human papillomavirus antibodies and risk of subsequent head and neck cancer. J Clin Oncol 31:2708–2715 (Official journal of the American Society of Clinical Oncology)
207. Greenspan D, Canchola AJ, MacPhail LA, Cheikh B, Greenspan JS (2001) Effect of highly active antiretroviral therapy on frequency of oral warts. Lancet 357:1411–1412
208. Ortega KL, Vale DA, Magalhaes MH (2009) Impact of PI and NNRTI HAART-based therapy on oral lesions of Brazilian HIV-infected patients. J Oral Pathol Med 38:489–494 (Official publication of the International Association of Oral Pathologists and the American Academy of Oral Pathology)
209. Patton LL, McKaig R, Strauss R, Rogers D, Eron JJ Jr (2000) Changing prevalence of oral manifestations of human immuno-deficiency virus in the era of protease inhibitor therapy. Oral Surg. Oral Med. Oral Pathol. Oral Radiol. Endod. 89:299–304
210. Tappuni AR, Fleming GJ (2001) The effect of antiretroviral therapy on the prevalence of oral manifestations in HIV-infected patients: a UK study. Oral Surg Oral Med Oral Pathol Oral Radiol Endod 92:623–628

211. Leigh J (2000) Oral warts rise dramatically with use of new agents in HIV. HIV Clin 12:7
212. King MD, Reznik DA, O'Daniels CM, Larsen NM, Osterholt D, Blumberg HM (2002) Human papillomavirus-associated oral warts among human immunodeficiency virus-seropositive patients in the era of highly active antiretroviral therapy: an emerging infection. Clin Infect Dis 34:641–648
213. Batavia AS, Secours R, Espinosa P, Jean Juste MA, Severe P, Pape JW, Fitzgerald DW (2016) Diagnosis of HIV-associated oral lesions in relation to early versus delayed antiretroviral therapy: results from the CIPRA HT001 trial. PLoS ONE 11:e0150656
214. Lilly EA, Cameron JE, Shetty KV, Leigh JE, Hager S, McNulty KM, Cheeks C, Hagensee ME, Fidel PL Jr (2005) Lack of evidence for local immune activity in oral hairy leukoplakia and oral wart lesions. Oral Microbiol Immunol 20:154–162
215. Danaher RJ, Wang C, Roland AT, Kaetzel CS, Greenberg RN, Miller CS (2010) HIV protease inhibitors block oral epithelial cell DNA synthesis. Arch Oral Biol 55:95–100
216. Badley AD (2005) In vitro and in vivo effects of HIV protease inhibitors on apoptosis. Cell Death Differ 12(Suppl 1):924–931
217. Chow WA, Jiang C, Guan M (2009) Anti-HIV drugs for cancer therapeutics: back to the future? The Lancet. Oncology 10:61–71
218. Madeleine MM, Finch JL, Lynch CF, Goodman MT, Engels EA (2013) HPV-related cancers after solid organ transplantation in the United States. Am J Transplant 13:3202–3209 (Official journal of the American Society of Transplantation and the American Society of Transplant Surgeons)
219. Powles T, Robinson D, Stebbing J, Shamash J, Nelson M, Gazzard B, Mandelia S, Moller H, Bower M (2009) Highly active antiretroviral therapy and the incidence of non-AIDS-defining cancers in people with HIV infection. J Clin Oncol 27:884–890
220. Dal Maso L, Polesel J, Serraino D, Lise M, Piselli P, Falcini F, Russo A, Intrieri T, Vercelli M, Zambon P, Tagliabue G, Zanetti R, Federico M, Limina RM, Mangone L, De Lisi V, Stracci F, Ferretti S, Piffer S, Budroni M, Donato A, Giacomin A, Bellu F, Fusco M, Madeddu A, Vitarelli S, Tessandori R, Tumino R, Suligoi B, Franceschi S (2009) Pattern of cancer risk in persons with AIDS in Italy in the HAART era. Br J Cancer 100:840–847
221. Koutsky LA, Ault KA, Wheeler CM, Brown DR, Barr E, Alvarez FB, Chiacchierini LM, Jansen KU (2002) A controlled trial of a human papillomavirus type 16 vaccine. N Engl J Med 347:1645–1651
222. Ali H, Donovan B, Wand H, Read TR, Regan DG, Grulich AE, Fairley CK, Guy RJ (2013) Genital warts in young Australians five years into national human papillomavirus vaccination programme: national surveillance data. BMJ 346:f2032
223. Naud PS, Roteli-Martins CM, De Carvalho NS, Teixeira JC, de Borba PC, Sanchez N, Zahaf T, Catteau G, Geeraerts B, Descamps D (2014) Sustained efficacy, immunogenicity, and safety of the HPV-16/18 AS04-adjuvanted vaccine: final analysis of a long-term follow-up study up to 9.4 years post-vaccination. Hum Vaccines Immunother. 10:2147–2162
224. Kreimer AR, Clifford GM, Boyle P, Franceschi S (2005) Human papillomavirus types in head and neck squamous cell carcinomas worldwide: a systematic review. Cancer Epidemiol Biomark Prev 14:467–475 (A publication of the American Association for Cancer Research, cosponsored by the American Society of Preventive Oncology)
225. Steinau M, Saraiya M, Goodman MT, Peters ES, Watson M, Cleveland JL, Lynch CF, Wilkinson EJ, Hernandez BY, Copeland G, Saber MS, Hopenhayn C, Huang Y, Cozen W, Lyu C, Unger ER (2014) Human papillomavirus prevalence in oropharyngeal cancer before vaccine introduction, United States. Emerg Infect Dis 20:822–828
226. Viens LJ, Henley SJ, Watson M, Markowitz LE, Thomas CC, Thompson TD, Razzaghi H, Saraiya M (2016) Human Papillomavirus-associated cancers - United States, 2008–2012. MMWR Morb Mortal Wkly Rep 65:661–666

227. Cameron JE, Snowhite IV, Chaturvedi AK, Hagensee ME (2003) Human papillomavirus-specific antibody status in oral fluids modestly reflects serum status in human immunodeficiency virus-positive individuals. Clin Diagn Lab Immunol 10:431–438
228. Marais DJ, Best JM, Rose RC, Keating P, Soeters R, Denny L, Dehaeck CM, Nevin J, Kay P, Passmore JA, Williamson AL (2001) Oral antibodies to human papillomavirus type 16 in women with cervical neoplasia. J Med Virol 65:149–154
229. Pinto LA, Kemp TJ, Torres BN, Isaacs-Soriano K, Ingles D, Abrahamsen M, Pan Y, Lazcano-Ponce E, Salmeron J, Giuliano AR (2016) Quadrivalent Human Papillomavirus (HPV) vaccine induces HPV-specific antibodies in the oral cavity: results from the mid-adult male vaccine trial. J Infect Dis 214:1276–1283
230. Rowhani-Rahbar A, Carter JJ, Hawes SE, Hughes JP, Weiss NS, Galloway DA, Koutsky LA (2009) Antibody responses in oral fluid after administration of prophylactic human papillomavirus vaccines. J Infect Dis 200:1452–1455
231. Herrero R, Quint W, Hildesheim A, Gonzalez P, Struijk L, Katki HA, Porras C, Schiffman M, Rodriguez AC, Solomon D, Jimenez S, Schiller JT, Lowy DR, van Doorn LJ, Wacholder S, Kreimer AR (2013) Reduced prevalence of oral human papillomavirus (HPV) 4 years after bivalent HPV vaccination in a randomized clinical trial in Costa Rica. PLoS ONE 8:e68329
232. Hirth JM, Chang M, Resto VA (2017) Prevalence of oral human papillomavirus by vaccination status among young adults (18–30 years old). Vaccine 35:3446–3451
233. Levin MJ, Moscicki AB, Song LY, Fenton T, Meyer WA 3rd, Read JS, Handelsman EL, Nowak B, Sattler CA, Saah A, Radley DR, Esser MT, Weinberg A (2010) Safety and immunogenicity of a quadrivalent human papillomavirus (types 6, 11, 16, and 18) vaccine in HIV-infected children 7 to 12 years old. J Acquir Immune Defic Syndr 55:197–204
234. Wilkin T, Lee JY, Lensing SY, Stier EA, Goldstone SE, Berry JM, Jay N, Aboulafia D, Cohn DL, Einstein MH, Saah A, Mitsuyasu RT, Palefsky JM (2010) Safety and immunogenicity of the quadrivalent human papillomavirus vaccine in HIV-1-infected men. J Infect Dis 202:1246–1253
235. Meites E, Kempe A, Markowitz LE (2016) Use of a 2-dose schedule for human papillomavirus vaccination - updated recommendations of the advisory committee on immunization practices. MMWR Morb Mortal Wkly Rep 65:1405–1408

HPV-Associated Anal Cancer in the HIV/AIDS Patient

7

Chia-Ching J. Wang and Joel M. Palefsky

Contents

C.-C. J. Wang
Division of Hematology/Oncology, Department of Medicine,
Zuckerberg San Francisco General Hospital, San Francisco, CA, USA
e-mail: chia-ching.wang@ucsf.edu

J. M. Palefsky (✉)
Division of Infectious Diseases, Department of Medicine, University of California at San Francisco, San Francisco, CA, USA
e-mail: joel.palefsky@ucsf.edu

C.-C. J. Wang
995 Potrero Avenue, Building 80, 4th Floor, San Francisco, CA 94110, USA

J. M. Palefsky
513 Parnassus Ave, Med Sci Room 420E, Box 0654, San Francisco, CA 94143, USA

C. Meyers (ed.), *HIV/AIDS-Associated Viral Oncogenesis*, Cancer Treatment and Research 177, https://doi.org/10.1007/978-3-030-03502-0_7

Abstract

The prevalence of anal human papillomavirus (HPV) infection and anal high-grade squamous intraepithelial lesion (HSIL) remain high among HIV-infected individuals on effective antiretroviral therapy (ART). The incidence of HPV-related anal cancers has continued to increase since the introduction of ART. Therefore, ART may confer only limited benefit with respect to reducing the risk of anal HSIL and cancer. Efforts are in progress to define the efficacy of secondary prevention programs for prevention of anal cancer. In the modern ART era, anal cancer recurrence and survival outcomes are similar in HIV-infected and HIV-uninfected patients, but HIV-infected patients may experience more toxicities. This article reviews the current literature on HPV-associated anal cancer in the HIV-infected population, including epidemiology, screening, clinical characteristics, and treatment outcomes.

Keywords

HPV · Anal cancer · HIV · Immunosuppression · Vaccination

7.1 Introduction

Improvements in the care of people living with HIV since the introduction of antiretroviral therapy (ART) 20 years ago have led to prolonged survival of this population. For HIV-infected men and women age 20 years who initiated ART between 2008 and 2010 can expect to live, on average, an additional 40 years or more and those who started ART and whose CD4 counts exceeded 350 cells/mm^3 1 year after ART initiation have an estimated life expectancy approaching that of the general population [1]. These advances also reflect the availability of increasingly effective antiretroviral agents, more options for the management of patients developing resistance, fewer drug interactions, better management of opportunistic infections and chronic diseases, and introduction of HIV screening programs with initiation of ART immediately upon HIV diagnosis and at higher CD4 levels than under older guidelines.

There are currently over 1 million people living with HIV/AIDS in the United States [2]. With decreases in infectious deaths, cancer has become a leading cause of morbidity and mortality in this patient population [3]. Cancer is now estimated to be responsible for over one-third of all deaths in HIV-infected individuals [4]. Kaposi sarcoma (KS), certain non-Hodgkin's lymphomas (NHLs) and cervical cancer confer the diagnosis of AIDS in an HIV-infected patient, and are referred to as AIDS-defining malignancies (ADM). Over the years, it has also been recognized that several additional cancers occur more frequently in HIV-infected patients, such as lung cancer, hepatocellular carcinoma (HCC), anal cancer, oropharyngeal cancer, classical Hodgkin lymphoma, and non-melanoma skin cancer [5, 6]. These neoplasms in HIV patients are referred to as non-AIDS-defining malignancies (NADM).

The majority of cancers associated with HIV are linked to co-infection with oncogenic viruses, with human papillomavirus (HPV) being one of the most common. HPV is responsible for 100% of cervical cancers and 88% of anal cancers, with the majority caused by HPV 16 or 18 [7, 8]. The purpose of this article is to present the most recent information on the epidemiology, treatment, and outcomes of anal cancer in the modern ART era.

7.2 HPV Infection and HPV-Related Diseases in HIV-Infected Individuals

The HPV virion contains a double-stranded, circular DNA genome surrounded by a capsid. HPV initially infects cells of the basal layer of squamous epithelium that has been exposed due to microabrasions or other forms of breach in the epithelium. The viral capsid proteins are L1 and L2. The HPV genome is divided into three regions: early, late, and long control or non-coding (Fig. 7.1). The early region contains the regulatory proteins, E1 and E2, and the main oncogenic proteins, E6, and E7. The region also includes the E4 and E5 proteins, with E4 functioning primarily as a structural protein, and E5 as an accessory oncogenic protein that promotes transformation along with E6 and E7 through reducing turnover of cell surface epidermal growth factor receptor. The oncogenic functions of E6 and E7 are complementary; E7 inactivates the host retinoblastoma protein and increases the rate of mutations by enhancing DNA replication. E6 inactivates the host p53 protein and allows these mutations to accumulate by disrupting DNA repair and cell death [9]. The E6 and E7 oncoproteins also enhance cellular proliferation, resulting in increased numbers of infected cells and infectious virions [10].

The anus consists of a mucosa-lined anal canal and a keratinized epithelium-lined perianal area. The anal canal begins where the rectum enters the puborectalis sling at the apex of the anal sphincter complex (palpable as the anorectal ring on digital anorectal examination and approximately 1–2 cm proximal to the dentate line), and ends where the squamous mucosa blends with the perianal skin, which roughly coincides with the palpable intersphincteric groove or the outermost boundary of the internal sphincter muscle. HPV infects basal cells throughout the anal canal and perianal epithelium, but one of the prime targets of HPV is the anal epithelial transformation zone (TZ). In the anus, the TZ, extends proximally from the squamocolumnar junction (SCJ) where the rectal columnar epithelium meets the squamous epithelium of the anus, to the dentate line distally (Fig. 7.2). The TZ is an area of active transition from columnar epithelium to squamous epithelium through the process of squamous metaplasia.

Clinically, mucosal HPVs are classified into low-risk and high-risk types according to the potential of malignant progression of the lesions they cause [11]. Infections with low-risk HPVs are primarily associated with flat low-grade

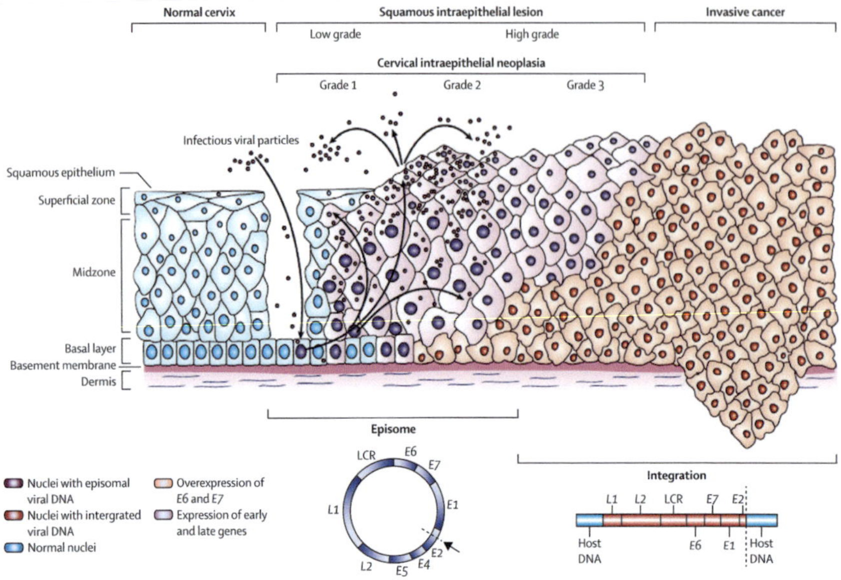

Fig. 7.1 Human papillomavirus lifecycle and organization of its genome [156]. Basal cells in the cervical epithelium rest on the basement membrane, which is supported by the dermis. Human papillomavirus is thought to access the basal cells through microabrasions in the cervical epithelium. After infection, the early human papillomavirus genes *E1*, *E2*, *E4*, *E5*, *E6*, and *E7* are expressed and the viral DNA replicates from episomal DNA. In the upper layers of epithelium (the midzone and superficial zone) the viral genome is replicated further, and the late genes *L1* and *L2*, and *E4* are expressed. *L1* and *L2* encapsulate the viral genomes to form progeny virions in the nucleus. The shed virus can then initiate a new infection. Low-grade intraepithelial lesions support productive viral replication. An unknown number of high-risk human papillomavirus infections progress to high-grade anal intraepithelial neoplasias. The progression of untreated lesions to micro-invasive and invasive cancer is associated with the integration of the human papillomavirus genome into the host chromosomes (red nuclei), with associated loss or disruption of *E2*, and subsequent upregulation of *E6* and *E7* oncogene expression. LCR = long control region. Permission: https://s100.copyright.com/CustomerAdmin/PLF.jsp?ref=bbdd7703-ca29-4109-bed1-4bc1990e7ce0

squamous intraepithelial lesions (LSIL), or genital warts (condyloma acuminata). HPV-6 and HPV-11 are the most abundant low-risk HPVs and cause more than 90% of condylomata acuminate [12]. These lesions are at very low risk for malignant progression and frequently regress spontaneously over time. In rare cases, low-risk HPV infections can cause slow-growing giant condyloma, also known as Buschke–Lowenstein tumor. These lesions are highly destructive to adjacent normal tissue through local spread, and can metastasize [13]. Infections with high-risk HPVs are associated with carcinoma and premalignant lesions, known as high-grade squamous intraepithelial lesions (HSIL), in the cervix, anus, vulva, vagina, and oropharynx.

HPV infection is very common in the perianal region and anal canal in both sexes. The highest anal HPV prevalence (nearly 100%) is found in HIV-infected men having sex with men (MSM). High-risk HPV types can be detected in the

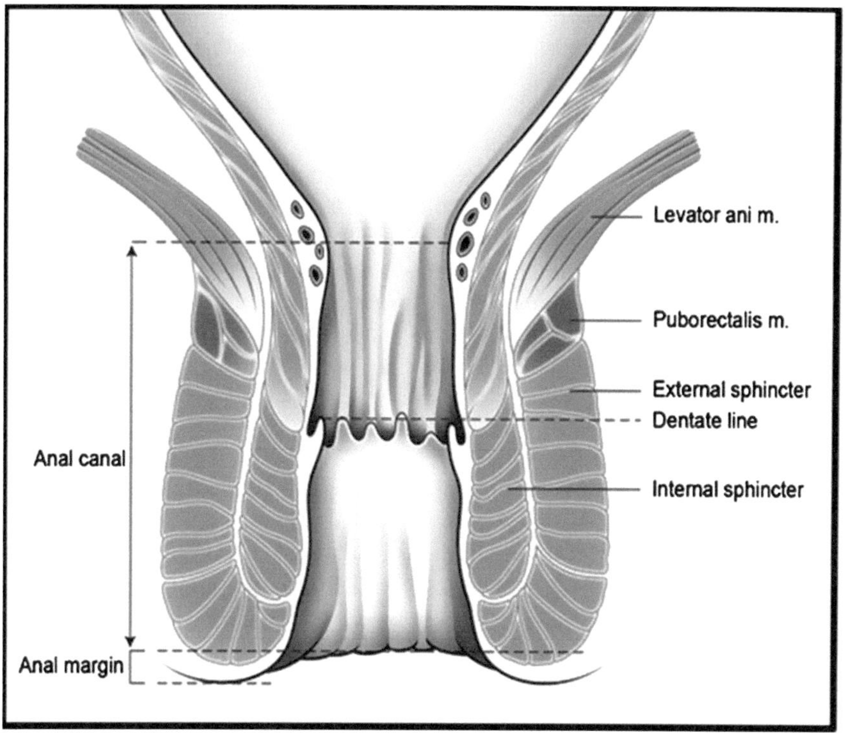

Fig. 7.2 Anatomy of the anal region [157]. The transformation zone, shown in red arrow, extends from the squamocolumnar junction to the dentate line. Permission: https://s100.copyright.com/CustomerAdmin/PLF.jsp?ref=3e25e5ed-bc96-46ec-b88d-2b0643f96fad

majority of HIV-positive MSM (73.5; 95% confidence interval [CI]: 63.7–83.0%) [14]. Multiple HPV genotypes have been associated with HSIL and cancer. HPV-16 accounts for over 50% of cases of HSIL and anal cancer. Other high-risk types include 18, 31, 35, 45, 51, 52, and 58 [15]. High-risk HPV are more prevalent among HIV-infected individuals, which can also contribute to a higher prevalence of HPV-associated malignancies in this population [16].

HIV-infected women are also at very high risk of anal HPV infection and some studies show that it is even more common than cervical HPV infection in this population [17–20]. An early study of HIV-infected and HIV-uninfected women injection drug users reported that anal HPV infection was twice as frequent as cervical HPV infection, and that HPV-associated epithelial abnormalities were associated with lower peripheral blood CD4 cell counts [20]. HIV-infected women were more likely to have the same HPV genotype in the anus and cervix than HIV-uninfected women (18 vs. 3%, $P < 0.001$). This was true for both oncogenic (9 vs. 2%, P.0.003) and non-oncogenic (12 vs. 1%, $P < 0.001$) HPV types. In multivariable analysis of HIV-infected women, CD4 cell count of less than 200 was the strongest factor associated with concomitant oncogenic (OR.4.2) and non-oncogenic (OR.16.5) HPV infection [18]. Anal intercourse is not required for anal HPV infection, and may possibly be spread from the cervix to the anus by wiping after going to the toilet [19]. Likewise, HIV-infected men who have sex with women have a high prevalence of anal HPV infection [17], potentially acquired through autoinoculation.

The terminology for HPV-associated squamous lesions of the lower anogenital tract has a long history marked by confusion caused by the various terminologies employed by pathologists and clinicians from differing specialties to describe the same histopathological entity. The Lower Anogenital Squamous Terminology (LAST) project created a histopathologic nomenclature system that reflects current knowledge of HPV biology. Current data support the two-tiered system of LSIL and HSIL [21], which may be further qualified with the appropriate intraepithelial neoplasia (-IN) terminology for specific location. The biomarker, p16, has the most robust published literature on its utility to help make morphological diagnoses of HPV-associated squamous lesions, particularly intermediate lesions such as intraepithelial neoplasia grade 2 (IN-2) or moderate dysplasia, more objective and reproducible. Negative or non-block positive staining of IN-2 lesions strongly favors an interpretation of LSIL or a non-HPV-associated pathology [22]. Therefore in the anus, LSIL includes condyloma, AIN 1, and p16-negative AIN 2, and these are not considered to be precancerous. In contrast, anal HSIL includes p16-positive AIN 2 and AIN 3. HSIL is considered to be the true cancer precursor [21].

Given the high prevalence of anal HPV in HIV-infected men and women, it is not surprising that anal HSIL is also common in this population. In the Multicenter AIDS Cohort Study, the prevalence of any abnormal anal cytology was 38, 41, and 47% among HIV-infected MSM with current CD4+ T-cell count ≥ 500, 350–499, and <350 cells/mm3 (P < 0001), respectively [23]. A prospective cohort study to assess the natural history of anal HPV infection in HIV-infected MSM in the ART era showed that the incidence of any anal HPV infection and oncogenic anal HPV infection was 21.3/100 and 13.3/100 person-years, respectively [24]. 20% of these men with an incident HPV infection also had more than one new HPV type detected during follow-up [24]. Low CD4 counts are a risk factor for HIV-positive individuals developing anal squamous intraepithelial lesions (ASIL). Palefsky et al. showed that, for HIV-infected men, having CD4 cell counts below 200 cells/mm^3 was associated with more than threefold increased incidence of progression (based on cytology and/or biopsy) of normal or atypical epithelium to ASIL, or from anal LSIL to a higher grade lesion [25]. HIV-infected MSM on effective ART for 24 months or more have also been shown to be have less HPV infection and less anal HSIL, although the reduction in HPV burden is relatively modest (from 100 to 88%) [26, 27].

Consistent with their high prevalence of anal HPV infection, HIV-infected women have a high prevalence of ASIL. Even in the era of effective ART, the prevalence of ASIL has been found to be significantly increased among HIV-infected women (16%) compared with HIV-uninfected women (4%) [28]. The Women's Interagency HIV Study showed a prevalence of 9 and 1% of anal HSIL in HIV-infected and HIV-uninfected women, respectively [28]. More recent data indicate that the prevalence of anal HSIL in HIV-infected women was 28% [29]. In HIV-infected women, progressive immunodeficiency and higher HIV viral load also are associated with an increased rate of cervical HPV carriage, cervical HSIL and cervical cancer [30, 31]. Previously, smaller studies with short follow-up periods did not show a beneficial impact of ART on high-risk HPV infection [32, 33]. More recently, larger prospective cohorts demonstrated that sustained virologic suppression by effective ART can decrease the risk of persistent high-risk HPV infection and lead to more rapid clearance of HPV-related cervical SILs [34–36].

It has been demonstrated that HPV-induced anal HSIL lesions are the direct precursors of anal cancer. In a retrospective review of 138 HIV-infected MSM diagnosed with anal cancer during 1997–2011, anal cancer developed at the previously biopsied site of anal HSIL in 27 men [37]. Sixty six men were diagnosed with anal cancer at their first clinic visit, and they all had HSIL as well. The concordance between the location of the HSIL and the cancer could not be definitively confirmed in 45 men, but most of them had HSIL overlying or immediately adjacent to their cancer. However, the risk and rates of progression from HSIL to anal cancer among HIV-infected patients who were not treated for HSIL are not precisely known. In a recent meta-analysis, Machalek et al. estimated the progression rate from anal HSIL to anal cancer among HIV-infected men in the modern ART era to be one in 377 per year in the absence of treatment for precursor lesions [14]. In a retrospective cohort analysis of HIV-infected patients under care at the University of California at San Diego Owen Clinic, patients with a baseline HSIL anal cytology had an estimated 5-year probability of progression to anal cancer of 1.7% and an estimated annual progression risk of 1 in 263 [38]. This group of investigators also found a high probability of regression of the anal HSIL state (27–62%) at 2 years after initial cytology screening using a 3-state Markov model of clinical pathogenesis [39]. More recently, Dalla Pria et al. reported on the experience of an HIV-infected MSM cohort in which HRA with intervention for HSIL was routinely offered [40]. In this HSIL-treated cohort, the estimated rate of anal cancer from histopathologic diagnosis of anal HSIL ascertained at the first HRA was 6.1 per 1000 person-years (95% CI: 4.2–7.8); this rate corresponds to per person per year rate of 1/164. Of note, these estimated rates of progression for anal HSIL are lower than the risk of progression to cancer for cervical HSIL (approximately 1 in 80 per year) [41]. It appears that on a per-lesion basis, anal HSIL is less likely to lead to anal cancer than cervical HSIL is to cervical cancer. The reasons for the lower susceptibility of the anus to malignant transformation compared with the cervix are unknown, but the hormonal milieu and potentially the different microbiomes of the two sites may be involved.

7.3 Primary Anal Cancer Prevention

Vaccination with the bivalent and quadrivalent HPV (qHPV) vaccine has been shown to reduce anal infection with HPV 16 and HPV 18 in both males and females naïve to those types [42, 43]. Clinical trials to determine the efficacy of vaccination to reduce the incidence of ASIL and penile HPV-associated disease have only been performed with qHPV vaccine. These studies confirmed that qHPV vaccine could reduce the risk of genital warts and anal HSIL in males [43]. In one double-blinded trial, 602 sexually active MSM, age 16–26, were randomized to receive 3 doses of (qHPV vaccine or placebo and evaluated every 6 months by HRA and HPV testing over 3 years. There was significant reduction of anal HSIL associated with any type of HPV (not only those associated with HPV 6, 11, 16 and 18) in those who received the qHPV vaccine compared with those who received the placebo [43]. Wilkin et al. evaluated 112 HIV+ men (ages 27 or older with no evidence of anal HSIL) with the three-dose course of qHPV vaccine and found that all of these HIV+ men seroconverted [44]. Therefore, qHPV vaccine has been demonstrated to be both immunogenic and safe in HIV-infected men. The efficacy of HPV vaccination in prevention of anal HSIL in HIV-infected MSM is being evaluated in an ongoing trial [45]. Deshmukh et al. estimated that qHPV vaccination of HIV-negative MSM age 27 or older treated for anal HSIL would reduce the lifetime risk of anal cancer by 60.77% at an incremental increase of cost-effectiveness ratios (ICER) of $87,240 per quality-adjusted life-year [46]. Their modeling suggests that qHPV vaccination for MSM may decrease their lifetime risk of anal cancer and is a cost-effective strategy because it decreases lifetime costs and increases quality-adjusted life expectancy.

In 2015, the 9-valent (9v) HPV vaccine became available, adding HPV types HPV 31, 33, 45, 52, and 58 to HPV 6, 11 16 and 18 from qHPV. Joura et al. evaluated the safety and efficacy of the 9v HPV vaccine through a double-blind international multicenter trial of 14,215 young women randomized to 9v HPV vaccine or qHPV vaccine. The investigators found that the 9v HPV vaccine prevented infection and disease related to HPV 31, 33, 45, 52, and 58 in a susceptible population and generated an antibody response to HPV 6, 11, 16, and 18 that was non-inferior to that generated by the qHPV vaccine [47]. From these data, it is assumed that the 9v HPV vaccine will provide the same degree of protection as qHPV vaccine from persistent HPV infections and development of anal HSIL (and possibly progression to anal cancer) in patients without evidence of prior vaccine-type HPV infection.

In the long term, HPV vaccination should be an excellent tool for reduction of anal cancer. Vaccination is currently routinely recommended with the 9v vaccine for individuals as young as age 9, with a target of 11–12 years of age. Catch-up vaccination is recommended up to 26 years of age for all women and up to age 21 years for men. However, catch-up vaccination is recommended up to 26 years is recommended for MSM and HIV-infected/immunocompromised individuals. [48]. The qHPV vaccine induces similar antibody titers in two doses as in three doses if the two doses are given 6 months apart in young individuals [49]. In a meta-analysis of seven controlled trials in 11 countries with direct comparisons between two-dose and three-dose HPV vaccine schedules, adolescent girls receiving a two-dose HPV vaccine schedule with a 6-month interval between doses had non-inferior antibody responses to HPV16 and HPV18 (measured as geometric mean concentrations or seropositivity) for at least 2 years after the first dose when compared with post-adolescent women receiving the licensed three-dose schedule [50]. Based on the results of these trials, the CDC currently recommends that children under the age of 15 years receive two doses of HPV vaccine instead of three, whereas the 3-dose regimen is still required for other eligible groups, including HIV-infected individuals [51]. Most HIV-infected individuals currently at risk for anal cancer are older than 26 years and do not qualify for HPV vaccination. Even if the vaccine is made available to them, a high proportion are likely to have already have been exposed to HPV 16 and 18 [52]. Among those who might benefit from vaccination, the impact of the vaccine has been reduced by poor uptake. Uptake of the qHPV vaccine is limited, with only 43% of eligible women and 31% of eligible men receiving all three doses of the vaccine through 2016. The uptake is better for two-dose regimens of qHPV vaccine (55% for women, 43.6% for men) [53]. While herd immunity due to vaccination of females may contribute to protection against HPV even among those who have not been vaccinated, it is likely to be very limited among MSM.

There are several reasons for the poor rates of HPV vaccination in the U.S., including varying levels of access, fear of HPV vaccine side effects, limited understanding of the benefits of HPV vaccination, and fear of vaccination in general. Finally, given the long period of time required for progression from cervical or anal HSIL to invasive cancer, it is expected that it will be decades before any reduction in cancer incidence is realized.

7.4 Secondary Anal Cancer Prevention

Combined with the fact that most HIV-infected men and women are too old for vaccination or were exposed to HPV 16 and 18 before vaccination became available, millions of men and women remain susceptible to HPV 16- and HPV 18-related HSIL and cancer. For these individuals, secondary prevention in the form of identifying and treating HSIL may be the only option to reduce the risk of anal cancer. Determination of the efficacy of HSIL treatment to prevent anal cancer is therefore a current and public health concern for the foreseeable future for this target population.

Secondary prevention of anal cancer consists of detection and treatment of anal HSIL. Anal HSIL can be detected by anal cytology, high-resolution anoscopy (HRA) and/or biopsy. Unlike cervical cancer, United States Preventive Services Task Force guidelines for anal screening are not yet in place (47). This is largely because the efficacy of treating anal HSIL in preventing anal cancer in HIV-infected men and women is not yet known. The Anal Cancer/HSIL Outcomes Research (ANCHOR) Study, supported by the National Cancer Institute and Office of AIDS Research, is an ongoing phase III, randomized, multi-institutional trial to determine whether treating anal HSIL is effective in reducing the incidence of anal cancer in HIV-infected men and women [54]. These results from this major trial should lead to changes in standard-of-care guidelines.

New York State HIV treatment guidelines recommend yearly anal cytology for certain subgroups of HIV-infected individuals [55]. Other guidelines, such as those published by the HIV Medicine Association of the Infectious Diseases Society of America recommend anal cytology screening for anal cancer, but do not specify the frequency of anal cytologies nor the necessity of HRA for follow-up of abnormal anal cytology results [56]. In many regions, resources for anal cytology and the follow-up HRA screening procedures remain limited.

Anal cytology is the test most commonly used to identify individuals who might benefit from HRA. While national organizations such as US Preventive Health Task Force or American Cancer Society do not recommend routine anal cancer screening using anal cytology, we believe that HIV-infected MSM should be considered for screening. Cost-effectiveness analyses have shown that screening MSM regardless of HIV status is justifiable [57]. Other at-risk groups that should be considered for anal cytology include HIV-infected women, HIV-infected men who have sex with women, women with a history of vulvar or cervical cancer, and organ/marrow transplant recipients. Women with a history of vulvar or cervical HSIL may also be considered for screening. Sensitivity of anal cytology is in the range of 50–80%, with sensitivity being higher in the HIV-infected population [58].

Primary HPV testing and cervical cytology have been recommended for primary cervical cancer screening of women between 30 and 65 years of age [59]. HPV testing is also helpful for triage of women with equivocal or low-grade cytologic abnormalities and prediction of the therapeutic outcome after treatment of cervical HSIL [60]. In contrast, the role of HPV testing in anal cytologic specimens is less well-established. Prior studies have shown that molecular tests for the presence of high-risk HPV have high sensitivity but low specificity for anal HSIL [61–63]. HPV testing may be more useful in HIV-uninfected MSM for its negative predictive value [64, 65]. Further research is needed to determine the optimal use of anal HPV testing in screening algorithms for anal HSIL in different at-risk populations.

Individuals with abnormal anal screening cytology are referred for HRA in which the anal canal is examined with a colposcope after the application of 5% acetic acid and/or Lugol's solution and visible lesions are biopsied for histological diagnosis (Fig. 7.3). In many clinical centers patients with histologic results of anal HSIL are recommended for treatment to prevent progression from anal HSIL to invasive cancer, even as the results of the ANCHOR study are awaited, However, unlike the treatment of cervical HSIL where the entire SCJ of the cervix is either ablated or excised, the entire SCJ of the anal canal cannot be surgically treated for concerns of stricture or other complications. Currently, the most commonly used treatment is HRA-directed ablation of apparent anal HSIL lesions. Unfortunately, recurrence rates are very high and frequently additional treatments are needed [66].

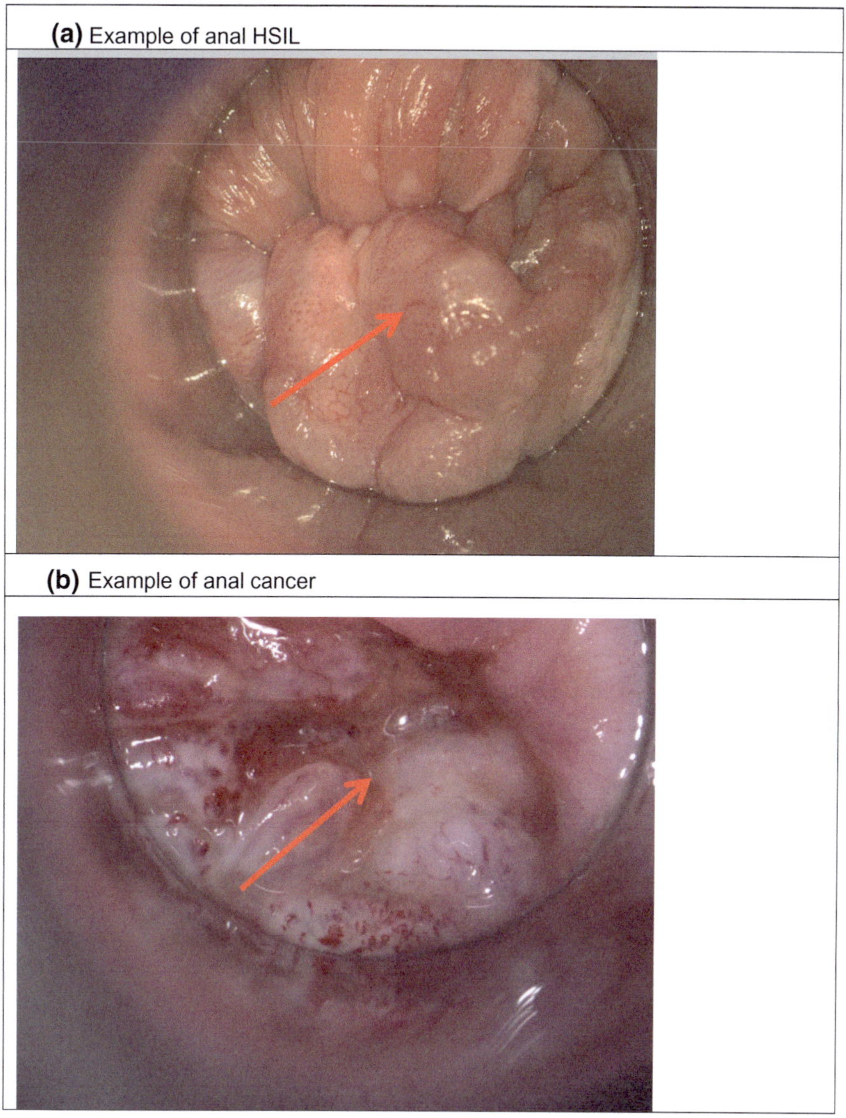

Fig. 7.3 Visualization of Anal HSIL and cancer under high-resolution anoscopy. **a** Anal HSIL, shown in red arrow, is aceto-white with coarse mosaic pattern and punctation. **b** Anal cancer, shown in red arrow, shows atypical vascular changes. It can be friable and ulcerated. A mass should be palpable by digital examination. There may be focal tenderness

7.5 Anal Cancer in HIV-Infected Individuals

The majority of patients with anal cancer present with bleeding, pain or sensation of a mass [67]. Bleeding from a mass at or just above the anal sphincter may be falsely attributed to hemorrhoids and may delay the diagnosis. At initial presentation, most patients have a T1 (tumor 2 cm or smaller) or T2 (tumor more than 2 cm but less than 5 cm) lesion and fewer than 20% are node-positive [68]. The probability of nodal spread is directly related to tumor size and location. It is far more common in cancers that originate in the anal canal than on the perianal skin. Tumor size (T stage) and nodal status (N stage) are the most significant prognostic factors for patients with anal cancer (Table 7 1). In a large series of 270 patients, the 5-year survival by stage was 86% for those with T1-2 disease versus less than 60% for T3-4 disease, and 76% for those with N0 disease versus 54% for those with node-positive disease [69].

Anal cancer is a relatively uncommon cancer in the general population. However, its incidence rate in the US has nearly doubled from the period of 1973–1970 to 1994–2000, and the rate is continuing to rise [70]. According to an analysis of Surveillance, Epidemiology, and End Results (SEER), the incidence of anal cancer increased by 2.9% per year during 1992–2001 [71]. In 2017, there are estimated 8200 new cases of anal cancer (2950 in men and 5250 in women) and 1100 deaths in the US [72].

It is estimated that there are approximately 37 million people worldwide living with HIV/AIDS as of the end of 2015 [73], including about 1.2 million HIV-infected individuals in the United States [74]. Approximately 1% of women and 28% of men with anal cancer are HIV-infected [75]. Prior to the availability of effective ART, the estimated incidence of anal cancer among HIV-infected MSM was nearly 60-fold higher than men in the general population [76]. Since the advent of effective ART, the incidence of malignancies associated with Epstein–Barr virus and Kaposi sarcoma herpesvirus has decreased in HIV-infected individuals. However, the incidence of HPV-associated anal cancer has increased. In a study of 34,189 HIV-infected individuals and 114,260 HIV-uninfected individuals from the North American AIDS Cohort Collaboration on Research and Design (NA-ACCORD) cohort with follow-up between 1996 and 2007, the unadjusted anal cancer incidence rates per 100,000 person-years were 30 for HIV-infected women, 0 for HIV-uninfected women, 131/100,000 for HIV-infected MSM, 46/100,000 for other HIV-infected men, and 2/100,000 for HIV-uninfected men. Therefore, the incidence of anal cancer in HIV-infected MSM is estimated to be 80 times higher than mèn in the general population [6]. This increase in incidence of anal cancer has been shown to be strongly influenced by the HIV epidemic in men [75]. In particular, the HIV prevalence was as high as 84% in anal cancer cases occurring in young African-American men [75]. Low CD4 count was also associated increased incidence rate of anal cancer [77]. The immunosuppression associated with HIV infection reduces the ability to control oncogenic viral processes, which could explain the higher risk of infection-related cancers.

Table 7.1 TNM staging for anal cancer

Anal cancer TNM staging AJCC UICC 2017			
Primary tumor (T)			
T category	**T criteria**		
TX	Primary tumor not assessed		
TO	No evidence of primary tumor		
Tis	High-grade squamous intraepithelial lesion (previously termed carcinoma *in situ,* Bowen disease, anal intraepithelial neoplasia II-III, high-grade anal intraepithelial neoplasia)		
T1	Tumor ≤ 2 cm		
T2	Tumor >2 cm but ≤ 5 cm		
T3	Tumor >5 cm		
T4	Tumor of any size invading adjacent organ(s), such as the vagina, urethra, or bladder		
Regional lymph nodes (N)			
N category	**N criteria**		
NX	Regional lymph nodes cannot be assessed		
ND	No regional lymph node metastasis		
N1	Metastasis in inguinal, mesorectal, internal iliac, or external iliac nodes		
Nla	Metastasis in inguinal, mesorectal, or internal iliac lymph nodes		
N1b	Metastasis in external iliac lymph nodes		
N1c	Metastasis in external iliac with any Nla nodes		
Distant metastasis (M)			
M category	**M criteria**		
MO	No distant metastasis		
Ml	Distant metastasis		
Prognostic stage groups			
When T **is…**	**And N is…**	**And M is…**	**Then the stage group is…**
Tis	NO	MO	0
T1	NO	MO	I
T1	N1	MO	IIIA
T2	NO	MO	IIA
T2	N1	MO	IIIA
T3	NO	MO	IIB
T3	N1	MO	IIIC
T4	NO	MO	nm
T4	N1	MO	IIIC
Any T	Any N	Ml	IV

Used with permission of the American Joint Committee on Cancer (AJCC), Chicago, Illinois. The original and primary source for this information is the AJCC Cancer Staging Manual, Eighth Edition (2017) published by Springer International Publishing

TNM Tumor, node, metastasis; *A* American Joint Committee on Cancer; *UICC* Union for International Cancer Control

A recent analysis from the HIV/AIDS Cancer Match Study, a linkage of population-based state HIV and cancer registries, showed that anal cancer is the third most common cancer occurring in excess in the HIV-infected population. 83% of excess cases of anal cancer occurred among HIV-infected MSM, and 71% among those living five or more years since AIDS onset [15]. As effective ART has greatly prolonged the life expectancy of those with HIV, the proportion of the HIV-infected population in older age groups has increased and will likely continue to increase in the future. A recent publication from NA-ACCORD showed that the annual trend in the cumulative incidence of anal cancer among HIV-infected adults (who were mostly men) was significantly increasing (6% per year) compared to it being stable in HIV-uninfected adults [78]. Yanik et al. used a linkage between data from cancer registries in the SEER program of the National Cancer Institute and Medicare claims (SEER-Medicare) to estimate absolute cancer risk among people age 65 years or alder with an HIV diagnosis and evaluate the association between HIV and cancer in this age group [79]. They also found that HIV-infected individuals aged 65 or older had higher incidence of anal cancer compared with HIV-uninfected elderly persons (adjusted hazard ratio = 34.2) [80]. This highlights a clear need for cancer prevention in this age group and the importance of screening.

Several cohort studies reported that immunodeficiency is associated with anal cancer incidence in HIV-infected individuals. Either nadir CD4 of less than 200 cells/mm^3 or longer duration of time with CD4 less than 200 cells/mm^3 have been shown to be related to higher risk of anal cancer [81, 82], but results from studies evaluating the effect of HIV viral load on incidence of anal cancer have been mixed. Silverberg et al. [83] found that most recent HIV viral load was not associated with anal cancer risk. However, they did not evaluate any cumulative markers of HIV viral load control. Chao et al. performed a retrospective cohort study among male US veterans diagnosed with HIV and followed between 1985 and 2009 using the Veterans Affairs Immunologic Case Registry. They found that individuals with excellent HIV control (between 80 and 100%) during their follow-up time have an approximately half of the risk of anal after adjusting for the effect of CD4 count.

Protease inhibitors (PIs) selectively bind to the catalytic site of HIV protease, interfering with HIV replication. Some PIs (e.g., indinavir, saquinavir, ritonavir, lopinavir, and nelfinavir) at varying concentrations have been shown to be anti-angiogenic and anti-tumorigenic because of their effects on cell invasion and matrix metalloproteinases, as well as because of modulation of the activity of cell proteasome [84, 85]. Nelfinavir also targets the phosphoinositide 3-kinase/AKT pathway, which is thought to play an important role in the development of cancers through multiple mechanisms [86]. There are no in vitro data available regarding the pathways targeted by specific PIs in anal cancer, but anal cancer closely resembles cervical cancer in several ways. In contrast to some of the other studies, a study of HIV-infected male US veterans looked specifically at the use of PIs and their relationship to incidence of anal cancer and showed the contrary. In multivariate analysis, increasing percentage time on PIs was associated with an increased risk of anal cancer [3]. Poor immunologic recovery and virologic control, a history of condylomata acuminata, and case registry enrollment in the late combined ART era were also associated with increased anal cancer risk.

7.6 Treatment and Outcomes of Anal Cancer

Concurrent chemoradiotherapy (CRT) with 5-fluorouracil (5-FU) infusion and mitomycin (or cisplatin) has been established as the standard-of-care regimen for non-metastatic anal cancer [87–92]. In ACCORD 03, induction chemotherapy prior to CRT did not improve response rates, 3-year colostomy-free survival, event-free survival, local control, or overall survival [93]. The UK ACT II trial included over 900 anal cancer patients randomly assigned to radiotherapy with either 5-FU/mitomycin or 5-FU/cisplatin. Patients in each arm were further randomly assigned to maintenance therapy with two cycles of 5-FU/cisplatin or no maintenance. At median follow-up of 5.1 years, no differences were found in complete response rate or progression-free survival [91]. Intensity-modulated radiotherapy (IMRT) has also been shown to reduce acute toxicities compared with conventional three-dimensional radiotherapy [94, 95]. Unfortunately, HIV-infected patients were excluded from these major trials.

When CRT was first applied to HIV-infected patients in the pre-ART era, reduced doses of radiotherapy and/or chemotherapy were administered due to concern for increased hematologic and mucosal toxicity secondary to compromised immunologic status [96, 97]. However, when therapy was applied in standard doses, increased toxicity, requiring treatment breaks or dose reductions, and poorer clinical outcome were reported [98, 99]. In five studies that included 53 HIV-infected patients, the incidence of grade 3–4 skin toxicity was 50–78% [96, 98–101]. Pretreatment CD4 count less than 200 was identified as a factor associated with poorer anal cancer control and increased treatment morbidity in a small retrospective cohort [100].

In the modern ART era, immune restoration with effective suppression of HIV viral load and elevation in CD4 count could be achieved in most HIV-infected patients, with improvement in compliance and reduction in treatment-related side effects. Reports on clinical outcomes of HIV-infected patients with anal cancer treated with standard therapy have been conflicting. Blazy et al. reported that high-dose CRT with radiotherapy doses of 60–70 Gy with concurrent 5-FU and cisplatin is feasible [102]. However, some studies show that HIV-infected patients had comparable disease control and survival to HIV-negative patients [103–109], whereas others suggested that HIV-positive patients may do worse in terms of enhanced treatment-related toxicity and/or an increased risk for local relapse [110–114]. Wexler et al. reported the local failure rate was only 16% in their cohort, but 44% of patients had T1N0 disease [107], which could reflect the fact that many of the referring providers are experienced in caring for HIV-infected individuals and more likely to examine patients for HSIL and anal cancer. Martin et al. reported their single-center experience with standard 5-FU/mitomycin CRT with long-term follow-up. Despite HIV-infected patients having higher nodal stages, the complete response rates after CRT were higher than 80% in both HIV-infected and HIV-uninfected patients [108]. In contrast, one of the largest series of anal cancer patients (total = 107, HIV-infected and HIV-uninfected) showed that HIV-infected patients had significantly worse overall survival and colostomy-free survival compared with a similar cohort of HIV-negative patients, despite having similar treatment approach, patient adherence, and cancer stage [114]. There were also no differences in radiation-related acute toxicity based on HIV status. There are no clear explanations for the differences, or lack of differences, in the outcomes of anal cancer in the HIV-infected versus the HIV-negative population. Almost all of these reports are limited by small patient numbers and the retrospective nature of the data.

Capecitabine, an established treatment alternative to intravenous 5-FU for patients with colorectal cancer, has also been tested in patients with locally advanced anal cancer. Small retrospective comparisons showed capecitabine had similar 3-year locoregional control rates and overall survival compared with 5-FU [115], and the capecitabine group had lower rates of grade 4 hematologic toxicity [116] when combined with mitomycin and radiotherapy. In a small prospective single-arm phase II study, chemoradiotherapy with capecitabine and mitomycin yielded a locoregional control of 86% in 6 months (CI 95% 0.72–0.94) and similar toxicity profile [117]. Despite the limited data, capecitabine has been accepted as an alternative to 5-FU in treatment of non-metastatic anal cancer.

7.7 Novel Therapies for Anal Cancer

Cetuximab is an epidermal growth factor receptor antibody whose activity depends on the presence of wild-type k-ras. It is felt to be a promising agent because k-ras mutations are rare in anal cancer [118]. Cetuximab also prolongs survival when used in combination with radiation therapy in patients with locally advanced squamous cell carcinoma of the oropharynx [119, 120], and enhances the activity of cisplatin in advanced head and neck cancers [121]. The ACCORD 16 anal cancer trial assessed response rates after CRT with 5-FU, cisplatin, and cetuximab. This study was prematurely terminated due to unacceptably high rates of serious adverse events, and 2 of 5 patients who completed planned treatment experienced locoregional recurrences [122, 123]. The AIDS Malignancy Consortium (AMC) and Eastern Cooperative Oncology Group (ECOG) recently completed two trials that were concurrently conducted to determine the effectiveness of cetuximab plus chemoradiation (CRT) in patients with HIV infection (AMC045) and without HIV infection (E3205) [109, 124]. It is important to note that patients with HIV infection had similar clinical outcomes as those who did not have HIV infection, with about 70% being alive and recurrence-free at 3 years. Treatment tolerance and the overall side effect profile were also similar in the two populations. However, the locoregional failure rate of 20% and grade 4 toxicity rate of 26% indicate the continued need for more effective and less toxic treatment in HIV-infected patients with newly diagnosed anal cancer.

Despite the effectiveness of CRT in primary treatment of anal cancer, the locoregional failure rate has been reported as 10–30% [125, 126]. There are 10–20% of anal cancers that present with extrapelvic disease at initial diagnosis [127], and 25% of cases develop distant metastases [128]. For recurrent or metastatic anal cancer, treatment options are quite limited. Patients with biopsy-proven local recurrence of anal cancer can be treated with abdominoperineal resection and colostomy [125]. A small single-arm study of 19 patients showed cisplatin and 5-FU had a response rate of 66% for metastatic anal cancer [129]. However, no consensus exists regarding treatment following progression on first-line therapy for unresectable or metastatic anal cancer.

Given the paucity of treatment options for recurrent or metastatic anal cancer, effective novel therapies are greatly needed. Immune checkpoint therapy is now a new pillar of cancer therapy. Many cancers evade immune surveillance and destruction through upregulation of the immune cell checkpoint molecule programmed death ligand 1 (PD-L1). Program death-1 (PD-1) is an inhibitory receptor expressed by an activated CD4+ and CD8+ T cells. When PD-L1 binds its inhibitory receptor PD-1 on the surface of T cells, T-cell activation is downregulated, which in turn reduces the local anti-tumor immune response [130]. PD-L1 positivity, which had not been previously defined in anal cancer, was found to be high (74% of screened patients) in one study [131]. In addition, the PD-1 pathway may mediate pathogen-specific and cancer-specific CD8+ T-cell dysfunction in chronic HIV infection [132].

Recently, several monoclonal antibodies have been developed that block the binding of PD-1 to PD-L1. Nivolumab is a humanized monoclonal antibody against PD-1 that disrupts this interaction and enables T-cell cytotoxicity. It has activity as a monotherapy in multiple advanced solid cancers, and has been approved by the Food and Drugs Administration for treatment of head and neck cancer, melanoma, non-small-cell lung cancer, and renal cell carcinoma [133–136]. Recently Eng et al. conducted a multicenter phase 2 study of nivolumab for patients with previously treated metastatic anal cancer [137]. Among the 37 patients who received at least one dose of nivolumab, 9 (24%) had either complete or partial response with a side effect profile similar to other trials. In KEYNOTE-028, pembrolizumab had a response rate of 17% (95% CI, 5–37%) and 10 (42%) had confirmed stable disease, for a disease control rate of 58% [131]. The Eastern Cooperative Oncology Group (ECOG) and the American College of Radiology Imaging Network (ACRIN) are planning a trial (ECOG-ACRIN 2165) to determine if there is benefit for adjuvant nivolumab after anal cancer patients have completed standard-of-care CRT. There are also 2 ongoing clinical trials using either nivolumab or pembrolizumab in HIV-infected patients with advanced solid tumors (NCT 02408861 and NCT 02595866). Assuming that HIV-infected patients maintain adequate CD4+ T-cell counts under careful clinical observation with an infectious diseases specialist, HIV-infected patients be considered for participation in future clinical trials with immune checkpoint inhibitors so that the safety and activity of these drugs can be studied further in a larger series.

7.8 HIV-Related Treatment Issues in Treatment of Anal Cancer

For HIV-infected patients with cancer, concurrent treatment with ART and anti-cancer therapy is increasingly common [138]. Extrapolating from treatment studies of HIV-associated lymphomas, combining ART and chemotherapy is tolerable in most cases and is not associated with life-threatening toxic effects, similar to those observed in patients with cancer without HIV infection [139–141]. In HIV-infected patients receiving chemotherapy for cancer, most modern ART regimens can be safely implemented to suppress viral replication to undetectable levels. Less is known about the interaction between ART and new anti-cancer agents such as immune checkpoint inhibitors. Recent guidelines state that integrase strand transfer inhibitor (INSTI)-based regimens may be preferred in cancer patients receiving anti-cancer treatment because of their more favorable drug interaction profile [142]. Zidovudine is often avoided because it commonly causes nausea, anemia, and myelosuppression, which can be potentiated by chemotherapy [143]. Tenofovir may lead to renal dysfunction, particularly in patients receiving other nephrotoxic drugs such as cisplatin. For protease inhibitors (PIs) and non-nucleoside reverse transcriptase inhibitors (NNRTIs), the potential for drug–drug interactions is high because these agents are extensively metabolized by and induce or inhibit the CYP450 system, which mediates the metabolism of more than half of all drugs that undergo hepatic metabolism [144]. PIs also may act as radio-sensitizers by inhibiting proteasome function and causing apoptosis [145], thereby potentially increasing both tumor control and toxicity.

In HIV-infected cancer patients, CD4 count, HIV-1 RNA level, and HAART adherence should be monitored [142]. A low CD4 count might not necessarily reflect suboptimal immunologic response to ART because CD4 counts can be affected by both the malignancies and/or their treatment. Kesselring et al. showed that in the 6 months prior to diagnosis of NADM in HIV-infected patients, a significant CD4 decline could be seen [146]. When administered to immunocompetent individuals, chemotherapy causes a profound decline in CD4 cell counts and a more modest fall in CD8 T cells [147], while the natural killer cell population is relatively spared [148]. The most striking finding from studies in immunocompetent patients is the protracted amount of time needed for recovery of the CD4 cells. In people with HIV there is concern that prolonged CD4 suppression induced by chemotherapy may have a major adverse influence on the course of HIV disease even when suppression of HIV viremia is maintained with ART. Furthermore, although both chemotherapy and radiotherapy lead to decline in CD4 cell count, the effect of radiotherapy on CD4 is more prolonged and significant, whereas chemotherapy does not influence CD4 cell count recovery [149]. Since the major source of bone marrow is radiated, the CD4+ T-cell count may fall severely and may not readily recover to pretreatment values. Scatter of radiation may also affect the intestinal tract, which is also an important compartment for CD4+ T cells [150, 151]. In a single institution study of 60 HIV-infected patients with anal cancer, those who received CRT with effective HAART had higher pretreatment CD4 compared those who received CRT without HAART. However, median CD4 at 3 months after anal cancer diagnosis was more than 50% lower than their pre-treatment value, and their median CD4 at 12 months after diagnosis was only 200 cells/mm^3 [103].

CRT potentiates the neutropenia associated with HIV/AIDS. For anal cancer patients who receive pelvic radiation, myelosuppression may be especially severe. Granulocyte colony-stimulating factors (GCSF) can reduce the effects of chemotherapy-induced neutropenia, and is often liberally used by oncologists when treating cancer in HIV-infected patients. The caveat is that GCSF should not be given concurrently with CRT due to concern for worsening hematologic toxicity [152]. The immunological deterioration following CRT may have an impact on the clinical course of the HIV disease and may be associated with an increased risk of opportunistic infections and diseases. One group reported that 4 patients (11%) developed opportunistic illnesses such as candida esophagitis during long-term follow-up of their anal cancer [153]. Therefore, antibiotic prophylaxis should be implemented to further reduce infectious complications during the treatment of HIV/AIDS-associated anal cancers based on careful assessment of risk.

The guidelines for prophylaxis against opportunistic infections in patients with HIV take into account risk and history of exposure, as well as the status of the immune system, particularly as reflected by the CD4 count, the receipt of and duration of HAART, and the response to HAART [154]. The guidelines for preventing of infections in patients with cancer are centered on the degree and duration of neutropenia, a key risk factor for infection [155]. Both the HIV-related and cancer-related guidelines need to be considered to prevent opportunistic infections in HIV-infected patients with anal cancer.

7.9 Conclusion

As ART helps HIV-infected individuals live longer but without producing a discernible effect on HSIL progression, the impact of HPV on this population can only be expected to increase. HIV-infected patients are often not able to clear HPV infection, and anal HSIL remains common. Guidelines to establish anal cancer screening and management programs as standard of care will await the results of randomized controlled trials of treatment of HSIL to reduce the risk of subsequent cancer. HIV-infected individuals with anal cancer can receive similar treatment as HIV-negative individuals and achieve similar outcomes, but they may require more careful monitoring for toxicities. HIV-infected anal cancer patients should be included in clinical trials of novel cancer drugs such as immune checkpoint inhibitors.

Acknowledgments Dr. Wang has no relevant disclosures.
Dr. Palefsky discloses that he is a member of a Merck scientific advisory board, and receives travel support form Merck. He receives travel and research grant support from Merck. He is a consultant to Ubiome, Agenovir, and Antiva Biosciences.

References

1. Antiretroviral Therapy, Cohort C (2017) Survival of HIV-positive patients starting antiretroviral therapy between 1996 and 2013: a collaborative analysis of cohort studies. Lancet HIV
2. Goncalves PH et al (2016) Cancer prevention in HIV-infected populations. Semin Oncol 43 (1):173–188
3. Shiels MS et al (2011) Cancer burden in the HIV-infected population in the United States. J Natl Cancer Inst 103(9):753–762
4. Bonnet F et al (2009) Changes in cancer mortality among HIV-infected patients: the Mortalite 2005 survey. Clin Infect Dis 48(5):633–639
5. Silverberg MJ, Abrams DI (2007) AIDS-defining and non-AIDS-defining malignancies: cancer occurrence in the antiretroviral therapy era. Curr Opin Oncol 19(5):446–451
6. Silverberg MJ et al (2012) Risk of anal cancer in HIV-infected and HIV-uninfected individuals in North America. Clin Infect Dis 54(7):1026–1034
7. Arbyn M et al (2012) EUROGIN 2011 roadmap on prevention and treatment of HPV-related disease. Int J Cancer 131(9):1969–1982
8. de Martel C et al (2012) Global burden of cancers attributable to infections in 2008: a review and synthetic analysis. Lancet Oncol 13(6):607–615
9. Brickman C, Palefsky JM (2015) Human papillomavirus in the HIV-infected host: epidemiology and pathogenesis in the antiretroviral era. Curr HIV/AIDS Rep 12(1):6–15
10. Hamid NA, Brown C, Gaston K (2009) The regulation of cell proliferation by the papillomavirus early proteins. Cell Mol Life Sci 66(10):1700–1717
11. Munoz N et al (2006) Chapter 1: HPV in the etiology of human cancer. Vaccine 24(Suppl 3):S3/1-10
12. Bosch X, Harper D (2006) Prevention strategies of cervical cancer in the HPV vaccine era. Gynecol Oncol 103(1):21–24
13. Frega A et al (2002) Giant condyloma acuminatum or buschke-Lowenstein tumor: review of the literature and report of three cases treated by CO_2 laser surgery. A long-term follow-up. Anticancer Res 22(2B):1201–1204
14. Machalek DA et al (2012) Anal human papillomavirus infection and associated neoplastic lesions in men who have sex with men: a systematic review and meta-analysis. Lancet Oncol 13(5):487–500
15. Robbins HA et al (2015) Excess cancers among HIV-infected people in the United States. J Natl Cancer Inst 107(4)
16. Ellerbrock TV et al (2000) Incidence of cervical squamous intraepithelial lesions in HIV-infected women. JAMA 283(8):1031–1037
17. Conley L et al (2010) Factors associated with prevalent abnormal anal cytology in a large cohort of HIV-infected adults in the United States. J Infect Dis 202(10):1567–1576
18. Hessol NA et al (2013) Concomitant anal and cervical human papillomavirus infections and intraepithelial neoplasia in HIV-infected and uninfected women. AIDS 27(11):1743–1751
19. Simpson S Jr et al (2016) Front-to-back & dabbing wiping behaviour post-toilet associated with anal neoplasia & HR-HPV carriage in women with previous HPV-mediated gynaecological neoplasia. Cancer Epidemiol 42:124–132
20. Williams AB et al (1994) Anal and cervical human papillomavirus infection and risk of anal and cervical epithelial abnormalities in human immunodeficiency virus-infected women. Obstet Gynecol 83(2):205–211

21. Darragh TM et al (2013) The lower anogenital squamous terminology standardization project for HPV-associated lesions: background and consensus recommendations from the College of American Pathologists and the American Society for Colposcopy and Cervical Pathology. Int J Gynecol Pathol 32(1):76–115
22. Darragh TM (2015) The LAST project and the diagnostic bottom line. Cytopathology 26 (6):343–345
23. D'Souza G et al (2016) Anal cancer screening in men who have sex with men in the multicenter AIDS Cohort study. J Acquir Immune Defic Syndr 71(5):570–576
24. Hernandez AL et al (2014) Incidence of and risk factors for type-specific anal human papillomavirus infection among HIV-positive MSM. AIDS 28(9):1341–1349
25. Palefsky JM et al (1998) Virologic, immunologic, and clinical parameters in the incidence and progression of anal squamous intraepithelial lesions in HIV-positive and HIV-negative homosexual men. J Acquir Immune Defic Syndr Hum Retrovirol 17(4):314–319
26. Libois A et al (2017) Prolonged antiretroviral therapy is associated with fewer anal high-grade squamous intraepithelial lesions in HIV-positive MSM in a cross-sectional study. Sex Transm Infect 93(1):15–17
27. van der Snoek EM et al (2012) Use of highly active antiretroviral therapy is associated with lower prevalence of anal intraepithelial neoplastic lesions and lower prevalence of human papillomavirus in HIV-infected men who have sex with men. Sex Transm Dis 39(7):495–500
28. Hessol NA et al (2009) Anal intraepithelial neoplasia in a multisite study of HIV-infected and high-risk HIV-uninfected women. AIDS 23(1):59–70
29. Stier EALJYLS, Jay N, Berry-Lawhorn JM, Einstein M, Palefsky JM, Wilkin T, Goldstone S, Chiao E (2016) Anal high-grade squamous intraepithelial lesions (HSIL) in HIV-positive (HIV+) women, prevalence and risk factors: AIDS malignancy consortium (AMC) 084. In: International anal neoplasia society (IANS) scientific meeting. San Francisoca, CA
30. Ahdieh L et al (2001) Prevalence, incidence, and type-specific persistence of human papillomavirus in human immunodeficiency virus (HIV)-positive and HIV-negative women. J Infect Dis 184(6):682–690
31. Sun XW et al (1997) Human papillomavirus infection in women infected with the human immunodeficiency virus. N Engl J Med 337(19):1343–1349
32. Lillo FB et al (2001) Human papillomavirus infection and associated cervical disease in human immunodeficiency virus-infected women: effect of highly active antiretroviral therapy. J Infect Dis 184(5):547–551
33. Paramsothy P et al (2009) The effect of highly active antiretroviral therapy on human papillomavirus clearance and cervical cytology. Obstet Gynecol 113(1):26–31
34. Blitz S et al (2013) Evaluation of HIV and highly active antiretroviral therapy on the natural history of human papillomavirus infection and cervical cytopathologic findings in HIV-positive and high-risk HIV-negative women. J Infect Dis 208(3):454–462
35. Konopnicki D et al (2013) Sustained viral suppression and higher CD4+ T-cell count reduces the risk of persistent cervical high-risk human papillomavirus infection in HIV-positive women. J Infect Dis 207(11):1723–1729
36. Minkoff H et al (2010) Influence of adherent and effective antiretroviral therapy use on human papillomavirus infection and squamous intraepithelial lesions in human immunodeficiency virus-positive women. J Infect Dis 201(5):681–690
37. Berry JM et al (2014) Progression of anal high-grade squamous intraepithelial lesions to invasive anal cancer among HIV-infected men who have sex with men. Int J Cancer 134 (5):1147–1155
38. Cachay E, Agmas W, Mathews C (2015) Five-year cumulative incidence of invasive anal cancer among HIV-infected patients according to baseline anal cytology results: an inception cohort analysis. HIV Med 16(3):191–195

39. Mathews WC et al (2014) Natural history of anal dysplasia in an HIV-infected clinical care cohort: estimates using multi-state Markov modeling. PLoS ONE 9(8):e104116
40. Dalla Pria A et al (2014) High-resolution anoscopy screening of HIV-positive MSM: longitudinal results from a pilot study. AIDS 28(6):861–867
41. McCredie MR et al (2008) Natural history of cervical neoplasia and risk of invasive cancer in women with cervical intraepithelial neoplasia 3: a retrospective cohort study. Lancet Oncol 9(5):425–434
42. Kreimer AR et al (2011) Efficacy of a bivalent HPV 16/18 vaccine against anal HPV 16/18 infection among young women: a nested analysis within the costa rica vaccine trial. Lancet Oncol 12(9):862–870
43. Palefsky JM et al (2011) HPV vaccine against anal HPV infection and anal intraepithelial neoplasia. N Engl J Med 365(17):1576–1585
44. Wilkin T et al (2010) Safety and immunogenicity of the quadrivalent human papillomavirus vaccine in HIV-1-infected men. J Infect Dis 202(8):1246–1253
45. AIDS Malignancy Consortium (2016) Vaccine therapy in preventing human papillomavirus infection in young HIV-positive male patients who have sex with males. clinicaltrials.gov [Internet] [cited 2016 March 18]. Available from: https://clinicaltrials.gov/ct2/show/NCT01209325
46. Deshmukh AA et al (2015) Long-term outcomes of adding HPV vaccine to the anal intraepithelial neoplasia treatment regimen in HIV-positive men who have sex with men. Clin Infect Dis 61(10):1527–1535
47. Joura EA et al (2015) A 9-valent HPV vaccine against infection and intraepithelial neoplasia in women. N Engl J Med 372(8):711–723
48. Markowitz LE et al (2014) Human papillomavirus vaccination: recommendations of the advisory committee on immunization practices (ACIP). MMWR Recomm Rep 63 (RR-05):1–30
49. Sankaranarayanan R et al (2016) Immunogenicity and HPV infection after one, two, and three doses of quadrivalent HPV vaccine in girls in India: a multicentre prospective cohort study. Lancet Oncol 17(1):67–77
50. D'Addario M et al (2017) Two-dose schedules for human papillomavirus vaccine: systematic review and meta-analysis. Vaccine 35(22):2892–2901
51. Meites E, Kempe A, Markowitz LE (2016) Use of a 2-dose schedule for human papillomavirus vaccination—updated recommendations of the advisory committee on immunization practices. MMWR Morb Mortal Wkly Rep 65(49):1405–1408
52. Sharma R et al (2013) Prevalence and risk factors for neutralizing antibodies to human papillomavirus types 16 and 18 in HIV-positive men who have sex with men. J Acquir Immune Defic Syndr 64(5):479–487
53. Walker TY et al (2017) National, regional, state, and selected local area vaccination coverage among adolescents aged 13-17 Years—United States, 2016. MMWR Morb Mortal Wkly Rep 66(33):874–882
54. AIDS Malignancy Consortium (2016) Topical or ablative treatment in preventing anal cancer in patients with HIV and anal high-grade squamous intraepithelial lesions. clinicaltrials.gov [Internet] [cited 2016 March 18]. Available from: https://clinicaltrials.gov/show/NCT02135419
55. New York State Department of Health AI (2017) Anal dysplasia and cancer guideline. Accessed 1 June 2017. Available from: http://www.hivguidelines.org/
56. Aberg JA et al (2014) Primary care guidelines for the management of persons infected with HIV: 2013 update by the HIV medicine association of the infectious diseases society of America. Clin Infect Dis 58(1):1–10
57. Goldie SJ et al (1999) The clinical effectiveness and cost-effectiveness of screening for anal squamous intraepithelial lesions in homosexual and bisexual HIV-positive men. JAMA 281 (19):1822–1829
58. Uronis HE, Bendell JC (2007) Anal cancer: an overview. Oncologist 12(5):524–534

59. Sawaya GF, Huchko MJ (2017) Cervical cancer screening. Med Clin North Am 101(4):743–753
60. Arbyn M et al (2012) Evidence regarding human papillomavirus testing in secondary prevention of cervical cancer. Vaccine 30(Suppl 5):F88–F99
61. Burgos J et al (2017) The role of oncogenic HPV determination for diagnosis of high-grade anal intraepithelial neoplasia in HIV-infected MSM. AIDS
62. Hidalgo-Tenorio C et al (2015) The role of polymerase chain reaction of high-risk human papilloma virus in the screening of high-grade squamous intraepithelial lesions in the anal mucosa of human immunodeficiency virus-positive males having sex with males. PLoS One 10(4):e0123590
63. Salit IE et al (2010) The role of cytology (Pap tests) and human papillomavirus testing in anal cancer screening. AIDS 24(9):1307–1313
64. Berry JM et al (2009) Performance characteristics of anal cytology and human papillomavirus testing in patients with high-resolution anoscopy-guided biopsy of high-grade anal intraepithelial neoplasia. Dis Colon Rectum 52(2):239–247
65. Goldstone SE, Enyinna CS, Davis TW (2009) Detection of oncogenic human papillomavirus and other predictors of anal high-grade dysplasia in men who have sex with men with abnormal cytology. Dis Colon Rectum 52(1):31–39
66. Stier EA, Chigurupati NL, Fung L (2016) Prophylactic HPV vaccination and anal cancer. Hum Vaccin Immunother
67. Ryan DP, Compton CC, Mayer RJ (2000) Carcinoma of the anal canal. N Engl J Med 342 (11):792–800
68. Klas JV et al (1999) Malignant tumors of the anal canal: the spectrum of disease, treatment, and outcomes. Cancer 85(8):1686–1693
69. Touboul E et al (1994) Epidermoid carcinoma of the anal canal. Results of curative-intent radiation therapy in a series of 270 patients. Cancer 73(6):1569–1579
70. Jemal A et al (2013) Annual report to the nation on the status of cancer, 1975–2009, featuring the burden and trends in human papillomavirus(HPV)-associated cancers and HPV vaccination coverage levels. J Natl Cancer Inst 105(3):175–201
71. Shiels MS et al (2015) Anal cancer incidence in the United States, 1977–2011: distinct patterns by histology and behavior. Cancer Epidemiol Biomarkers Prev 24(10):1548–1556
72. Siegel RL, Miller KD, Jemal A (2017) Cancer statistics, 2017. CA Cancer J Clin 67(1):7–30
73. UNAIDS, Fact Sheet 2015 (2015) http://www.unaids.org/en/resources/campaigns/HowAIDSchangedeverything/factsheet
74. Control CFD (2015) HIIV in the United States: at a glance. http://www.cdc.gov/hiv/statistics/overview/ataglance.html
75. Shiels MS et al (2012) Impact of the HIV epidemic on the incidence rates of anal cancer in the United States. J Natl Cancer Inst 104(20):1591–1598
76. Frisch M, Biggar RJ, Goedert JJ (2000) Human papillomavirus-associated cancers in patients with human immunodeficiency virus infection and acquired immunodeficiency syndrome. J Natl Cancer Inst 92(18):1500–1510
77. Reekie J et al (2010) Relationship between current level of immunodeficiency and non-acquired immunodeficiency syndrome-defining malignancies. Cancer 116(22):5306–5315
78. Silverberg MJ et al (2015) Cumulative incidence of cancer among persons with HIV in North America: a Cohort study. Ann Intern Med 163(7):507–518
79. Engels EA et al (2011) Use of surveillance, epidemiology, and end results-medicare data to conduct case-control studies of cancer among the US elderly. Am J Epidemiol 174(7):860–870
80. Yanik EL, Katki HA, Engels EA (2016) Cancer risk among the HIV-infected elderly in the United States. AIDS 30(10):1663–1668

81. Guiguet M et al (2009) Effect of immunodeficiency, HIV viral load, and antiretroviral therapy on the risk of individual malignancies (FHDH-ANRS CO4): a prospective cohort study. Lancet Oncol 10(12):1152–1159
82. Powles T et al (2009) Highly active antiretroviral therapy and the incidence of non-AIDS-defining cancers in people with HIV infection. J Clin Oncol 27(6):884–890
83. Silverberg MJ et al (2011) HIV infection, immunodeficiency, viral replication, and the risk of cancer. Cancer Epidemiol Biomarkers Prev 20(12):2551–2559
84. Barillari G et al (2012) Ritonavir or saquinavir impairs the invasion of cervical intraepithelial neoplasia cells via a reduction of MMP expression and activity. AIDS 26(8):909–919
85. Toschi E et al (2011) Human immunodeficiency virus protease inhibitors reduce the growth of human tumors via a proteasome-independent block of angiogenesis and matrix metalloproteinases. Int J Cancer 128(1):82–93
86. Gupta AK et al (2007) The HIV protease inhibitor nelfinavir downregulates Akt phosphorylation by inhibiting proteasomal activity and inducing the unfolded protein response. Neoplasia 9(4):271–278
87. UKCCCR Anal Cancer Trial Working Party (1996) Epidermoid anal cancer: results from the UKCCCR randomised trial of radiotherapy alone versus radiotherapy, 5-fluorouracil, and mitomycin. UK Co-ordinating Committee on Cancer Research. Lancet 348(9034):1049–1054
88. Ajani JA et al (2008) Fluorouracil, mitomycin, and radiotherapy vs fluorouracil, cisplatin, and radiotherapy for carcinoma of the anal canal: a randomized controlled trial. JAMA 299 (16):1914–1921
89. Flam M et al (1996) Role of mitomycin in combination with fluorouracil and radiotherapy, and of salvage chemoradiation in the definitive nonsurgical treatment of epidermoid carcinoma of the anal canal: results of a phase III randomized intergroup study. J Clin Oncol 14(9):2527–2539
90. Gunderson LL et al (2012) Long-term update of US GI intergroup RTOG 98-11 phase III trial for anal carcinoma: survival, relapse, and colostomy failure with concurrent chemoradiation involving fluorouracil/mitomycin versus fluorouracil/cisplatin. J Clin Oncol 30(35):4344–4351
91. James RD et al (2013) Mitomycin or cisplatin chemoradiation with or without maintenance chemotherapy for treatment of squamous-cell carcinoma of the anus (ACT II): a randomised, phase 3, open-label, 2 × 2 factorial trial. Lancet Oncol 14(6):516–524
92. Northover J et al (2010) Chemoradiation for the treatment of epidermoid anal cancer: 13-year follow-up of the first randomised UKCCCR anal cancer trial (ACT I). Br J Cancer 102(7):1123–1128
93. Peiffert D et al (2012) Induction chemotherapy and dose intensification of the radiation boost in locally advanced anal canal carcinoma: final analysis of the randomized UNICANCER ACCORD 03 trial. J Clin Oncol 30(16):1941–1948
94. Chuong MD et al (2013) Intensity-modulated radiation therapy vs. 3D conformal radiation therapy for squamous cell carcinoma of the anal canal. Gastrointest Cancer Res 6(2):39–45
95. Kachnic LA et al (2013) RTOG 0529: a phase 2 evaluation of dose-painted intensity modulated radiation therapy in combination with 5-fluorouracil and mitomycin-C for the reduction of acute morbidity in carcinoma of the anal canal. Int J Radiat Oncol Biol Phys 86 (1):27–33
96. Chadha M et al (1994) Squamous-cell carcinoma of the anus in HIV-positive patients. Dis Colon Rectum 37(9):861–865
97. Peddada AV et al (1997) Chemotherapy and low-dose radiotherapy in the treatment of HIV-infected patients with carcinoma of the anal canal. Int J Radiat Oncol Biol Phys 37 (5):1101–1105
98. Kim JH et al (2001) HIV-positive patients with anal carcinoma have poorer treatment tolerance and outcome than HIV-negative patients. Dis Colon Rectum 44(10):1496–1502

99. Place RJ et al (2001) Outcome analysis of HIV-positive patients with anal squamous cell carcinoma. Dis Colon Rectum 44(4):506–512
100. Hoffman R et al (1999) The significance of pretreatment CD4 count on the outcome and treatment tolerance of HIV-positive patients with anal cancer. Int J Radiat Oncol Biol Phys 44(1):127–131
101. Holland JM, Swift PS (1994) Tolerance of patients with human immunodeficiency virus and anal carcinoma to treatment with combined chemotherapy and radiation therapy. Radiology 193(1):251–254
102. Blazy A et al (2005) Anal carcinomas in HIV-positive patients: high-dose chemoradiotherapy is feasible in the era of highly active antiretroviral therapy. Dis Colon Rectum 48 (6):1176–1181
103. Alfa-Wali M et al (2012) Chemoradiotherapy for anal cancer in HIV patients causes prolonged CD4 cell count suppression. Ann Oncol 23(1):141–147
104. Chiao EY et al (2008) Human immunodeficiency virus-associated squamous cell cancer of the anus: epidemiology and outcomes in the highly active antiretroviral therapy era. J Clin Oncol 26(3):474–479
105. Fraunholz I et al (2011) Concurrent chemoradiotherapy with 5-fluorouracil and mitomycin C for anal carcinoma: are there differences between HIV-positive and HIV-negative patients in the era of highly active antiretroviral therapy? Radiother Oncol 98(1):99–104
106. Seo Y et al (2009) Outcomes of chemoradiotherapy with 5-Fluorouracil and mitomycin C for anal cancer in immunocompetent versus immunodeficient patients. Int J Radiat Oncol Biol Phys 75(1):143–149
107. Wexler A et al (2008) Invasive anal squamous-cell carcinoma in the HIV-positive patient: outcome in the era of highly active antiretroviral therapy. Dis Colon Rectum 51(1):73–81
108. Martin D et al (2017) Are there HIV-specific differences for anal cancer patients treated with standard chemoradiotherapy in the era of combined antiretroviral therapy? Clin Oncol (R Coll Radiol) 29(4):248–255
109. Sparano JA et al (2017) Cetuximab plus chemoradiotherapy for HIV-associated anal carcinoma: a phase II AIDS malignancy consortium trial. J Clin Oncol 35(7):727–733
110. Hogg ME et al (2009) HIV and anal cancer outcomes: a single institution's experience. Dis Colon Rectum 52(5):891–897
111. Munoz-Bongrand N et al (2011) Anal carcinoma in HIV-infected patients in the era of antiretroviral therapy: a comparative study. Dis Colon Rectum 54(6):729–735
112. Oehler-Janne C et al (2008) HIV-specific differences in outcome of squamous cell carcinoma of the anal canal: a multicentric cohort study of HIV-positive patients receiving highly active antiretroviral therapy. J Clin Oncol 26(15):2550–2557
113. Meyer JE et al (2013) HIV positivity but not HPV/p16 status is associated with higher recurrence rate in anal cancer. J Gastrointest Cancer 44(4):450–455
114. Grew D et al (2015) HIV infection is associated with poor outcomes for patients with anal cancer in the highly active antiretroviral therapy era. Dis Colon Rectum 58(12):1130–1136
115. Meulendijks D et al (2014) Chemoradiotherapy with capecitabine for locally advanced anal carcinoma: an alternative treatment option. Br J Cancer 111(9):1726–1733
116. Goodman KA et al (2014) Capecitabine plus mitomycin in patients undergoing definitive chemoradiation for anal squamous cell carcinoma. Int J Radiat Oncol Biol Phys 90:S32–S33
117. Oliveira SC et al (2016) Phase II study of capecitabine in substitution of 5-FU in the chemoradiotherapy regimen for patients with localized squamous cell carcinoma of the anal canal. J Gastrointest Cancer 47(1):75–81
118. Van Damme N et al (2010) Epidermal growth factor receptor and K-RAS status in two cohorts of squamous cell carcinomas. BMC Cancer 10:189
119. Bonner JA et al (2006) Radiotherapy plus cetuximab for squamous-cell carcinoma of the head and neck. N Engl J Med 354(6):567–578

120. Bonner JA et al (2010) Radiotherapy plus cetuximab for locoregionally advanced head and neck cancer: 5-year survival data from a phase 3 randomised trial, and relation between cetuximab-induced rash and survival. Lancet Oncol 11(1):21–28
121. Burtness B et al (2005) Phase III randomized trial of cisplatin plus placebo compared with cisplatin plus cetuximab in metastatic/recurrent head and neck cancer: an Eastern cooperative oncology group study. J Clin Oncol 23(34):8646–8654
122. Deutsch E et al (2013) Unexpected toxicity of cetuximab combined with conventional chemoradiotherapy in patients with locally advanced anal cancer: results of the UNICANCER ACCORD 16 phase II trial. Ann Oncol 24(11):2834–2838
123. Levy A et al (2015) Low response rate after cetuximab combined with conventional chemoradiotherapy in patients with locally advanced anal cancer: long-term results of the UNICANCER ACCORD 16 phase II trial. Radiother Oncol 114(3):415–416
124. Garg M, Lee JY, Kachnic LA, Catalano PJ, Henry DH, Cooley TP, Ratner L, Wachsman W, Aboulafia DM, Benson AB, Palefsky J, Whittington R, Mitsuyasu RT, Sparano JA (2012) Phase II trials of cetuximab (CX) plus cisplatin (CDDP), 5fluorouracil (5FU) and radiation (RT) in immunocompetent (ECOG 3205) and HIVpositive (AMC045) patients with squamous cell carcinoma of the anal canal (SCAC): Safety and preliminary efficacy results. J Clin Oncol 30(suppl; abstr 4030)
125. Mullen JT et al (2007) Results of surgical salvage after failed chemoradiation therapy for epidermoid carcinoma of the anal canal. Ann Surg Oncol 14(2):478–483
126. Schiller DE et al (2007) Outcomes of salvage surgery for squamous cell carcinoma of the anal canal. Ann Surg Oncol 14(10):2780–2789
127. Cummings BJ (2006) Metastatic anal cancer: the search for cure. Onkologie 29(1–2):5–6
128. Eng C (2006) Anal cancer: current and future methodology. Cancer Invest 24(5):535–544
129. Faivre C et al (1999) 5-fluorouracile and cisplatinum combination chemotherapy for metastatic squamous-cell anal cancer. Bull Cancer 86(10):861–865
130. Freeman GJ et al (2000) Engagement of the PD-1 immunoinhibitory receptor by a novel B7 family member leads to negative regulation of lymphocyte activation. J Exp Med 192 (7):1027–1034
131. Ott PA et al (2017) Safety and antitumor activity of the anti-PD-1 antibody pembrolizumab in patients with recurrent carcinoma of the anal canal. Ann Oncol 28(5):1036–1041
132. Day CL et al (2006) PD-1 expression on HIV-specific T cells is associated with T-cell exhaustion and disease progression. Nature 443(7109):350–354
133. Brahmer J et al (2015) Nivolumab versus docetaxel in advanced squamous-cell non-small-cell lung cancer. N Engl J Med 373(2):123–135
134. Ferris RL et al (2016) Nivolumab for recurrent squamous-cell carcinoma of the head and neck. N Engl J Med 375(19):1856–1867
135. Motzer RJ et al (2015) Nivolumab versus everolimus in advanced renal-cell carcinoma. N Engl J Med 373(19):1803–1813
136. Robert C et al (2015) Nivolumab in previously untreated melanoma without BRAF mutation. N Engl J Med 372(4):320–330
137. Morris VK et al (2017) Nivolumab for previously treated unresectable metastatic anal cancer (NCI9673): a multicentre, single-arm, phase 2 study. Lancet Oncol 18(4):446–453
138. Rudek MA, Flexner C, Ambinder RF (2011) Use of antineoplastic agents in patients with cancer who have HIV/AIDS. Lancet Oncol 12(9):905–912
139. Montoto S et al (2012) HIV status does not influence outcome in patients with classical Hodgkin lymphoma treated with chemotherapy using doxorubicin, bleomycin, vinblastine, and dacarbazine in the highly active antiretroviral therapy era. J Clin Oncol 30(33):4111–4116
140. Ratner L et al (2001) Chemotherapy for human immunodeficiency virus-associated non-Hodgkin's lymphoma in combination with highly active antiretroviral therapy. J Clin Oncol 19(8):2171–2178

141. Vaccher E et al (2001) Concomitant cyclophosphamide, doxorubicin, vincristine, and prednisone chemotherapy plus highly active antiretroviral therapy in patients with human immunodeficiency virus-related, non-Hodgkin lymphoma. Cancer 91(1):155–163
142. Gunthard HF et al (2014) Antiretroviral treatment of adult HIV infection: 2014 recommendations of the international antiviral society-USA panel. JAMA 312(4):410–425
143. Margolis AM et al (2014) A review of the toxicity of HIV medications. J Med Toxicol 10 (1):26–39
144. Rudek MA et al (2014) A phase 1/pharmacokinetic study of sunitinib in combination with highly active antiretroviral therapy in human immunodeficiency virus-positive patients with cancer: AIDS malignancy consortium trial AMC 061. Cancer 120(8):1194–1202
145. Pajonk F et al (2002) The human immunodeficiency virus (HIV)-1 protease inhibitor saquinavir inhibits proteasome function and causes apoptosis and radiosensitization in non-HIV-associated human cancer cells. Cancer Res 62(18):5230–5235
146. Kesselring A et al (2011) Immunodeficiency as a risk factor for non-AIDS-defining malignancies in HIV-1-infected patients receiving combination antiretroviral therapy. Clin Infect Dis 52(12):1458–1465
147. Hakim FT et al (1997) Constraints on CD4 recovery postchemotherapy in adults: thymic insufficiency and apoptotic decline of expanded peripheral CD4 cells. Blood 90(9):3789–3798
148. Komada Y et al (1992) Cellular immunosuppression in children with acute lymphoblastic leukemia: effect of consolidation chemotherapy. Cancer Immunol Immunother 35(4):271–276
149. Sankatsing SU et al (2013) Prolonged decrease of CD4+ T lymphocytes in HIV-1-infected patients after radiotherapy for a solid tumor. J Acquir Immune Defic Syndr 62(5):546–549
150. Bouma G, Strober W (2003) The immunological and genetic basis of inflammatory bowel disease. Nat Rev Immunol 3(7):521–533
151. Chun TW et al (1999) Effect of interleukin-2 on the pool of latently infected, resting CD4+ T cells in HIV-1-infected patients receiving highly active anti-retroviral therapy. Nat Med 5(6):651–655
152. Bunn PA Jr et al (1995) Chemoradiotherapy with or without granulocyte-macrophage colony-stimulating factor in the treatment of limited-stage small-cell lung cancer: a prospective phase III randomized study of the Southwest Oncology Group. J Clin Oncol 13 (7):1632–1641
153. Fraunholz IB et al (2014) Long-term effects of chemoradiotherapy for anal cancer in patients with HIV infection: oncological outcomes, immunological status, and the clinical course of the HIV disease. Dis Colon Rectum 57(4):423–431
154. Kaplan JE et al (2009) Guidelines for prevention and treatment of opportunistic infections in HIV-infected adults and adolescents: recommendations from CDC, the National Institutes of Health, and the HIV Medicine Association of the Infectious Diseases Society of America. MMWR Recomm Rep 58(RR-4):1–207; quiz CE1-4
155. Segal BH et al (2008) Prevention and treatment of cancer-related infections. J Natl Compr Canc Netw 6(2):122–174
156. Crosbie EJ et al (2013) Human papillomavirus and cervical cancer. Lancet 382(9895):889–899
157. Glynne-Jones R et al (2014) Anal cancer: ESMO-ESSO-ESTRO clinical practice guidelines for diagnosis, treatment and follow-up. Radiother Oncol 111(3):330–339

Merkel Cell Carcinoma in the HIV-1/AIDS Patient

8

Robert H. Goldstein and James A. DeCaprio

Contents

R. H. Goldstein
Division of Infectious Disease, Department of Medicine, Massachusetts General Hospital and Harvard Medical School, Boston, USA

J. A. DeCaprio (✉)
Department of Medical Oncology, Dana-Farber Cancer Institute, 450 Brookline Avenue, Boston, MA 02215, USA
e-mail: james_decaprio@dfci.harvard.edu

J. A. DeCaprio
Department of Medicine, Brigham and Women's Hospital and Harvard Medical School, Boston, USA

C. Meyers (ed.), *HIV/AIDS-Associated Viral Oncogenesis*, Cancer Treatment and Research 177, https://doi.org/10.1007/978-3-030-03502-0_8

Abstract
Merkel cell carcinoma (MCC) is a highly aggressive, primary neuroendocrine cancer of the skin. The majority of MCC cases are associated with the recently discovered Merkel cell polyomavirus (MCPyV), while the remaining are caused by ultraviolet (UV) light-induced mutations from excessive sunlight exposure. The risk of developing MCC is much higher in the white population relative to all other races. Approximately 10% of all patients with MCC have some form of immunosuppression including HIV-1/AIDS, chronic inflammatory conditions, solid organ transplantation, or hematological malignancies. The age of onset of MCC is lower and the mortality is higher in immunosuppressed individuals than in immune-competent patients. It is plausible that HIV-1/AIDS predisposes to virus-positive MCC, but it should be noted that HIV-1/AIDS increases the risk for developing of UV-induced skin cancers such as cutaneous squamous cell carcinoma and basal cell carcinoma and therefore may also increase the risk for virus-negative MCC. Surgical management is considered standard of care for localized Merkel cell carcinoma with current recommendations advising a wide local excision of the lesion. Most international guidelines support the use of local adjuvant radiotherapy coupled with tumor staging to improve the frequency of cure. For advanced, metastatic, and recurrent MCC, checkpoint blockade inhibitors targeting PD-1 and PD-L1 have shown remarkable activity including durable long-term. MCC in patients living with HIV-1/AIDS are treated with similar modalities as HIV-1 uninfected individuals with MCC.

8.1 Introduction

Merkel cell carcinoma (MCC) is a highly aggressive, primary neuroendocrine cancer of the skin. The majority of MCC cases are associated with the recently discovered Merkel cell polyomavirus (MCPyV), while the remaining are caused by ultraviolet (UV)-light-induced mutations from excessive sunlight exposure. Significantly, immunosuppression from HIV-1 infection, chronic inflammatory conditions, solid-organ transplantation, and hematological malignancies increase the risk of developing MCC. Although MCC is 40 times less common than malignant melanoma, MCC has a dramatically lower survival probability than melanoma, rendering MCC the most lethal skin cancer. Epidemiologic data indicate that there are approximately 2500 new MCC cases per year within the USA, and approximately 1000 of these patients will die from their disease [1]. This high mortality rate is largely due to the fact that until recently none of the standard therapeutic interventions were able to improve the overall survival of patients suffering from the metastatic disease. Since several lines of evidence indicate the outstanding immunogenicity of MCC, immune modulating treatment strategies are particularly attractive.

8.2 Initial Description of Malignancy

MCC was first described by Cyril Toker in 1972 as a trabecular carcinoma of the skin with carcinoid features [2]. Later, using an electron microscope, Toker reported the presence of membrane-bound granules containing dense cores within the tumor cells, a feature common to other tumors of neural crest origin. He also noted that the neurosecretory granules in MCC tumor cells were similar in appearance to those found in normal Merkel cells [3]. The name of the cancer was eventually changed from trabecular carcinoma to Merkel cell carcinoma to reflect the similarity to normal Merkel cells [4, 5].

Normal Merkel cells are the mechano-receptors for a gentle touch and form synapses with afferent nerves [6]. Normal Merkel cells are located in the basal layer of the skin epithelium and in hair follicles. In contrast, nearly all MCC tumors present in the dermal layer of the skin. Immunohistochemistry (IHC) staining was recognized to be useful in distinguishing MCC from other neuroendocrine tumors such as small-cell lung carcinoma (SCLC). In particular, IHC staining for cytokeratin 20 (CK20, KRT20) can distinguish MCC from other skin tumors and can readily detect normal Merkel cells in the basal layer of the skin epidermis and hair follicle [7]. MCC is also frequently positive for additional neuroendocrine markers including neuron cell adhesion molecule 1 (NCAM1; CD56), chromogranin A (CHGA), and synaptophysin (SYP). Staining for TTF-1 (Thyroid Transcription Factor-1, NKX2-1, and NK2 homeobox 1) is rarely positive in MCC and is used to distinguish MCC from SCLC [8]. MCC can present as a pure neuroendocrine tumor or combined neuroendocrine tumors with nonendocrine features such as squamous cell carcinoma [9].

8.3 MCC and Association with Immunosuppression

While it was recognized that individuals with hematologic malignancies that developed MCC had a poor prognosis in the early 1990s [10], it was not until 1997 when a direct link between immunosuppression was postulated [11]. At that time, a correlation was noted between medically induced immunosuppression with azathioprine and cyclosporine and the development and rapid spread of MCC. Early reports highlighted a prolonged period of immunosuppression prior to MCC development.

Around this same time, the incidence of new infections and deaths from the human immunodeficiency virus (HIV-1) were peaking [12]. HIV-1 was noted to be the primary cause of death for men between the ages of 25 and 44 [12]. Most deaths were the result of profound T-cell deficiencies resulting in the acquired immune deficiency syndrome (AIDS), defined by the diagnosis of an opportunistic infection or an AIDS-defining malignancy. The AIDS-defining malignancies included Kaposi's sarcoma, driven by human herpesvirus 8 (HHV-8) also known as Kaposi's sarcoma herpesvirus (KSHV), non-Hodgkin lymphoma, often triggered by

Epstein–Barr virus (EBV), and cervical cancer, resulting from human papillomavirus (HPV). Of note, MCC was never categorized as an AIDS-defining malignancy likely due to the rarity of the malignancy even in individuals with profound immunosuppression.

Despite a dramatically shortened life expectancy, HIV-positive individuals were noted to have MCC as early as 1992 [13]. Using population-based registries of cancer and AIDS, Eric Engel and colleagues were able to determine the relative risk of people with AIDS developing Merkel cell carcinoma to be 13.4 [14]. Additional case reports of HIV-1 associated MCC were described over the next decade [15–21]. Following the introduction of effective antiviral therapy for HIV, it was noted that the prognosis of Merkel cell carcinoma was improved with effective treatment of HIV, suggesting a link with improvement in overall immune function [22].

8.4 Clinical Presentation

An important study assessed the typical presentation of MCC and defined the AEIOU features: **A**symptomatic/lack of tenderness, **E**xpanding rapidly, **I**mmune suppression, **O**lder than 50 years, and **U**ltraviolet-exposed site on a person with fair skin [23]. Most MCC tumors present as asymptomatic pink or red lesion are thought to be benign despite the rapid growth in the prior 3 months. Perhaps reflecting an altered immune state, chronic lymphocytic leukemia (CLL) is highly associated with MCC. This study noted that for several patients with newly diagnosed MCC, the immune-suppressed state (AIDS or CLL) was discovered as part of the MCC workup and recommended that workup for immunosuppression be considered in patients presenting with MCC [23]. The risk of developing MCC is increased in patients with chronic inflammatory disorders such as rheumatoid arthritis or medically induced immunosuppression for solid-organ transplantation [11, 24–26].

Age is a significant risk factor for MCC with 90% of patients over 50 years of age (YOA) and nearly half older than 75 YOA [27]. MCC incidence increases with age, from 0.1 to 1.0 to 9.8 (per 100,000 person-years) between age groups 40 and 44, 60and 64, and 85+ years, respectively [1]. A variety of institutional-based and national cancer registry studies have reported the age at diagnosis for MCC to be 69 years and higher. A report analyzing 6908 MCC cases in the National Cancer Database (NCDB), a national tumor registry for the USA, found the median age at diagnosis was 76 years (range: 20–90 years) [28]. Notably, the age of onset of MCC is lower and the mortality is higher in immunosuppressed individuals than in immune-competent patients [29].

Notably, skin pigmentation seems to protect against MCC, as black, Asian and Hispanic individuals have a considerably lower risk of MCC than white populations. The risk of developing MCC is much higher in the white population relative to all other races. A recent survey of SEER data from 1973 to 2006 identified 3870 cases of MCC. Almost 95% of all MCC cases were identified in the white population while only 1% in the black population [30]. Additional evidence for the risk of UV exposure arises from the frequent occurrence of MCC in elderly patients on the chronically sun-exposed skin, the increased MCC incidence in individuals treated with UVA photochemotherapy and the observation that many patients with MCC have a history of other skin cancers associated with sun exposure. A history of melanoma is also linked with a threefold greater risk of MCC [31]. It should be noted that despite the very high incidence of HIV-1 infection in sub-Saharan Africa, there are few if any reports of MCC. While there could be many reasons for the few reports of MCC, it is possible that dark skin color is highly protective against UV-induced skin damage and the development of MCC even in the presence of profound immunosuppression.

While there are limited data on HIV-infected individuals with Merkel cell carcinoma, the clinical presentation appears different than that seen with HIV-uninfected individuals. In limited case series, HIV-infected individuals are more likely to have Merkel cell carcinomas of non-head and neck skin and to be younger at the age of diagnosis (46 vs. 69 YOA in HIV-uninfected individuals) [16, 19, 21]. Case reports and case series suggest that Merkel cell carcinoma is more aggressive in the context of HIV infection, although there are no large databases that control for HIV viral load, ART, surgical management and chemotherapeutics [15, 16, 19, 21].

8.5 Isolation of Merkel Cell Polyomavirus

Given the increased risk by immunosuppression for developing MCC, Huichen Feng and Masahiro Shuda in the laboratory of Yuan Chang and Patrick Moore began a search for a pathogenic cause for MCC. They performed whole transcriptome sequencing of several MCC tumors and searched for pathogens by first subtracting all human genes from their analysis. In the remaining sequences, novel transcripts distantly related to polyomaviruses were detected in an MCC tumor. Complete sequencing of the viral genome led to the determination that it corresponded to a new human polyomavirus that they called Merkel cell polyomavirus (MCPyV) [32]. They determined that MCPyV DNA was clonally integrated into the genome of MCC tumor cells, when they observed an identical Southern blot integration pattern for a primary tumor and a metastatic tumor involving a lymph node from the same patient. They detected MCPyV by PCR and Southern blotting in 8 of 10 tested MCC tumors, indicating that most but not all MCC tumors contained MCPyV. These results supported the model that MCPyV contributed to MCC in a manner similar to human papillomavirus (HPV) in cancer [33]. It should be noted that Chang and Moore had discovered KSHV by discerning differences in DNA sequences present in AIDS patients and Kaposi's sarcoma tumor tissue compared to normal tissue from non-AIDS patients [34].

When MCPyV was first identified in MCC tumor specimens in 2008, it was only the fifth human polyomavirus to be identified at that time [32]. Its discovery quickly led to the realization that although MCPyV was likely to be causal in MCC, it was a typical polyomavirus, infecting most people at an early age and persisting as a lifelong infection. Although MCPyV can cause the highly aggressive MCC, it normally produces a lifelong, asymptomatic, and innocuous infection in most people. Primary infection with MCPyV does not cause any discernable signs or symptoms [35]. What has come into sharper focus is that although some of the now 14 human polyomaviruses can cause exceptionally catastrophic diseases in immunocompromised patients, MCPyV is the only one clearly associated with cancer [36, 37]. Furthermore, a variety of immunosuppressed conditions including AIDS can significantly increase the risk of developing MCC.

8.6 Polyomaviruses

The first polyomavirus was discovered in 1953, when an infectious agent was reported to cause salivary gland cancer in laboratory mice [38]. The cancer-causing agent was identified as a non-enveloped DNA virus that was named polyomavirus from the Greek poly (many) and oma (tumor). Polyomaviruses are small, non-enveloped, and double-stranded DNA viruses. The circular viral genome is approximately 5200 base pairs and encodes 5–8 viral proteins.

MCPyV is one of the 14 distinct human polyomaviruses species [39, 40]. MCPyV encodes four early genes: Large T antigen (LT); 57kT, an alternatively spliced form of LT; small T antigen (ST); and ALTO (Alternative LT open reading frame) [32, 41]. The late region encodes the major viral capsid protein VP1 and the minor capsid protein VP2 [42].

MCPyV is part of a large group of polyomaviruses, many of which are implicated in human disease. Of these, MCPyV, BK polyomavirus (BKPyV), and JC polyomavirus (JCPyV) are known to have oncogenic potential in cell culture and animal models. BKPyV was first isolated in 1971 from an immunosuppressed renal transplant recipient [43]. Most adults are seropositive (>80%) for BKPyV after asymptomatic childhood infection, but reactivation with complications can be seen in the immunosuppressed population, including those with HIV [44, 45]. In healthy hosts, it is postulated that BKPyV infection occurs via respiratory droplets with infection occurring in tonsillar tissue and spreading to blood mononuclear cells and then disseminating to secondary sites, including the kidney, brain, and lymph nodes [46].

BK polyomavirus (BKPyV) can cause polyomavirus-associated nephropathy in renal transplant recipients and hemorrhagic cystitis in hematopoietic stem cell transplant recipients treated with immunosuppressive therapy [43]. Kidney transplant recipients are most likely to experience reactivation of BKPyV and may develop BKPyV-associated nephropathy and subsequent allograft loss [47], while hematopoietic stem cell transplant recipients more frequently develop post-engraftment hemorrhagic cystitis with BKPyV reactivation [48].

While reactivation of BKPyV in renal transplant and hematopoietic stem cell transplant recipients is well described, reactivation in patients immunosuppressed from HIV is not well understood [49]. BKPyV has been isolated from the urine, blood, and cerebrospinal fluid of HIV-infected individuals and varied manifestations of the disease have been reported, which include hemorrhagic cystitis, renal failure, encephalitis, retinitis, and pulmonary infection [50–52]. BKPyV viruria is seen in a large percentage of HIV-infected individuals compared to HIV-negative controls (57.7% vs. 21.7%) and appears independent of CD4 count or HIV control [51].

JCPyV was also discovered in 1971 after isolation of viral particles from a patient with Hodgkin's disease, who developed progressive multifocal leukoencephalopathy (PML), the most common manifestation of JCPyV pathology [53]. Patients with PML often present with progressive, focal neurologic deficits that may advance to seizures and dementia. Since then, JCPyV has been implicated in granule cell neuronopathy, encephalitis, meningitis, and nephropathy [54] and complications of infection are seen in patients on monoclonal antibodies including natalizumab, efalizumab, and rituximab [55].

Before the introduction of effective antiretroviral therapy (ART) up to 7% of HIV-infected individuals developed PML, but the incidence has substantially decreased since 2000; PML is now more frequently associated with immunosuppressive therapy for multiple sclerosis [56]. Outcomes remain poor for patients with PML and HIV infection, with frequent persistence of neurologic deficits and cognitive decline and, while the introduction of effective ART has decreased the incidence of disease, only about 50% of people with PML and HIV have improvement in outcomes with ART initiation [57].

Human polyomavirus 6 (HPyV6), HPyV7, and Trichodysplasia spinulosa-associated polyomavirus (TSPyV) have been detected on the skin of healthy volunteers [58, 59]. In severely immunocompromised patients, HPyV6 and HPyV7 can cause pruritic dermatoses characterized by hyperproliferation of dyskeratotic (with premature or altered differentiation) keratinocytes that result in brownish skin plaques [60]. TSPyV can cause a hyperkeratotic folliculitis referred to as Trichodysplasia spinulosa in solid-organ transplant recipients [59, 61].

Polyomavirus replication occurs within the cellular nucleus and is dependent on LT. LT forms a double hexamer centered on the viral *origin* of replication. LT functions to melt and unwind the double-stranded viral DNA and recruit cellular DNA polymerases and other host factors to replicate the viral DNA. The replicated viral genome is packaged within the viral capsid comprised of VP1 and VP2. High levels of virus production lead to the lytic destruction of the host cell.

It is not known what cells normally support MCPyV replication since MCPyV LT expression has not yet been detected by immunohistochemistry (IHC) in any normal human tissue. If healthy skin supports MCPyV replication, then cells within hair follicles infected with TSPyV in the Trichodysplasia spinulosa syndrome or in keratinocytes with HPyV6 and HPyV7 in dyskeratotic dermatoses could potentially also support MCPyV replication. Alternatively, a recent report demonstrated that cultures of human dermal fibroblasts could support MCPyV replication [62]. It should be noted that papillomavirus infection is dependent on breaks in the intact epithelium permitting access of the papillomavirus to the basement membrane. A similar mechanism has not been described for MCPyV or any other polyomavirus.

Evidence for persistent infection by a specific polyomavirus is reflected in serum antibodies against the corresponding polyomavirus coat protein VP1. The polyomavirus virion is comprised of 72 pentamers of VP1 together with VP2 on the inner surface of each VP1 pentamer [63]. When expressed in bacteria or yeast, VP1 will spontaneously form pentamers or viruslike particles that generate a useful capture antigen to detect antibodies in serum specific for each human polyomavirus [64, 65].

Based on the VP1 serology assay, it has been inferred that the initial exposure to MCPyV likely occurs in early childhood because the seroprevalence is lower in children and higher in adults. An intriguing study from Cameroon examined serology against the MCPyV VP1 pentamer in 196 children from birth to 5 years of age (YOA) [66]. Significant titers against MCPyV were detected in newborns but these titers gradually decreased to undetectable levels by 16 months of age. Maternal-derived antibodies likely account for the seropositivity in newborns that gradually declined during the first year of life. The maternally derived antibodies were likely to be effective in preventing primary infection during infancy. By 18 months of age when the maternal antibodies were no longer present, children were susceptible to de novo infection and could mount an antibody response of their own. Beginning at 18 months of age, an increasing fraction of children became positive until approximately 80% tested positive by 5 YOA [66]. In a separate cohort from the same study, the strongest correlation of seropositivity was observed between siblings of similar ages suggesting that siblings likely were exposed to MCPyV at the same time and by each other. Similar results were reported from a population study in Australia that investigated the serology of several cutaneous polyomaviruses including MCPyV, HPyV6, HPyV7, and TSPyV as well as BKPyV. Children below 6 months displayed seropositivity rates for all viruses studied comparable to that found in adults with rates decreasing after 6 months of age then starting to increase by 2–3 YOA and continuing to increase with age [67].

Several additional studies support the increasing risk with age for exposure and persistent infection by MCPyV and other polyomaviruses. Seroprevalence of 10 human polyomaviruses was assessed from a population-based skin cancer case–control study conducted in New Hampshire, USA [68]. The overall seropositivity for MCPyV in this study was 70.4%. Of note, all participants were seropositive for at least one polyomavirus and the overall study population had evidence for infection with a mean of 7.3 different polyomaviruses. A study of five polyomaviruses conducted in Italy with participants aged 1–100 YOA found that the seroprevalence for MCPyV rapidly increased with age, from 41.7% in children age 1–4 YOA to 87.6% in 15–19 YOA and remained relatively frequent in adulthood (79.0–96.2%) [69].

Antibodies to MCPyV LT and ST are usually not present in healthy individuals, but can be detected in patients with virus-positive MCC. Antibodies to the common region of MCPyV ST and LT were present in half of the patients with MCC and in less than 1% of healthy individuals [70]. Importantly, antibody titers to MCPyV T antigens decrease upon definitive treatment of the MCC and can be used as a biomarker to follow disease status [70]. Of note, MCC patients often have higher titers of antibodies to VP1 than normal healthy individuals [71].

8.7 Virus-Positive and Virus-Negative MCC

MCC can be distinguished by the presence or absence of integrated MCPyV DNA and viral mRNA and oncoprotein expression. Virus-positive MCC contains integrated copies of the MCPyV DNA. In all cases sequenced to date, the integrated MCPyV DNA has undergone mutations that truncate LT that render it unable to replicate viral DNA. Virus-positive MCC expresses the truncated LT and an intact ST and typically does not express the viral coat proteins VP1 and VP2. Virus-negative MCC does not contain MCPyV DNA and does not express LT or ST. In addition, next-generation DNA sequencing studies of MCC tumors have revealed striking differences in the genomes of virus-positive and virus-negative MCC. Virus-positive MCC typically contains very few somatic mutations and copy number alterations. In contrast, virus-negative MCC shows a very high frequency of DNA mutations that are associated with UV damage with point substitution mutations of cytosine to thymidine (C > T) that occur in the context of dipyrimidines, C[C > T]N and N[C > T]C, typically seen in other sun-exposure-associated skin cancers such as melanoma, basal cell carcinoma, and squamous cell carcinoma [72–77]. While lifelong exposure to UV radiation may be required to introduce all the mutations found in virus-negative tumor DNA, it is less clear why virus-positive MCC also typically occurs in the elderly. Of note, UV exposure could also play a part in viral carcinogenesis by causing local immunosuppression [78].

In a study of 282 cases of MCC, where the presence of virus was established by IHC with two different monoclonal antibodies against the MCPyV Large T antigen as well as PCR detection of viral DNA, the median age at diagnosis was 71 years for both virus-negative and virus-positive MCC [79]. Another study used RNA-fluorescence in situ hybridization (FISH) to detect MCPyV T antigen expression in MCC. This approach yielded a highly accurate determination of whether the MCC was a virus-positive or a virus-negative MCC, detecting MCPyV in 37 of 75 cases (49.3%). They observed that MCC tumors from younger patients and female patients were twice as likely to be virus-positive compared to older male patients [80].

The contrasting mutational profile between virus-positive and virus-negative MCC may provide clues into the oncogenic events necessary to generate the tumor. An important feature of virus-positive MCC is that the tumor maintains expression of LT and ST [32]. In all cases reported to date, the truncated LT preserves the N-terminal J domain and RB-binding (LXCXE) motif but loses the DNA-binding and helicase domains as well as a C-terminal growth inhibitory domain (Fig. 8.1a) [81–83]. Some MCC tumors express a truncated LT that also retains the nuclear

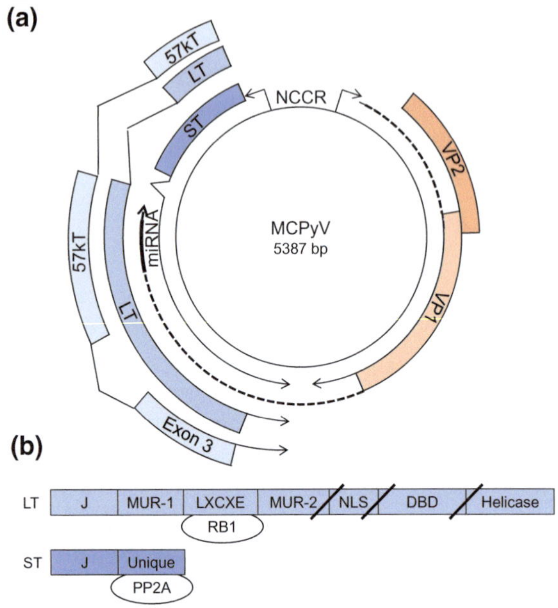

Fig. 8.1 **a** Circular map of MCPyV includes early region genes for LT, ST, and 57 kT and late region for VP1, VP2, and miRNA. The non-coding control region (NCCR) contains a bidirectional promoter and the viral origin of replication. Exon 3 of 57 kT is depicted and is in the same reading frame as LT. ALTO is not depicted. **b** Linear maps of LT and ST. LT and ST share an N-terminal J domain. LT contains the LXCXE or RB-binding motif, MCPyV-unique region (MUR)-1 and -2, nuclear localization signal (NLS), DNA-binding domain (DBD) and helicase domain. In MCC, mutations in LT result in truncations after the LXCXE or NLS and depicted by slashes. ST contains a unique region not shared with LT that binds to protein phosphatase 2A (PP2A)

localization signal (NLS) in addition to the J domain and LXCXE motif [84, 85]. The truncation of LT is probably required for oncogenesis for several reasons. The full-length LT is capable of binding to the viral origin of replication, the first step in replication of the virus that requires the DNA-binding and helicase domains of LT. When full-length MCPyV LT is expressed in virus-positive MCC cell lines, it binds to integrated copies of the MCPyV *origin of replication* and induces in situ replication of the integrated viral DNA leading to a DNA-damage response [81, 83]. This DNA-damage process likely selects against any tumor that expresses full-length LT.

Virus-positive MCC tumors also express MCPyV ST. ST shares the N-terminal J domain with LT and contains a unique region that can bind to protein phosphatase 2A (PP2A) [86] (Fig. 8.1b). MCPyV ST has an additional domain distinct from PP2A binding known as the LT stabilizing domain or LSD. This unique region of MCPyV ST is not well conserved in ST from other polyomaviruses. The LSD motif in ST functions to increase the levels of MCPyV LT and contributes to increased viral replication at least in part by increasing LT levels [87]. Evidence for a role of MCPyV ST in MCPyV replication includes its ability to translocate to viral DNA replication centers within the nucleus in the presence of the viral origin and LT [88]. The LT stabilizing activity may reflect ST's ability to perturb the function of FBXW7, a component of the cullin-RING ligase family of ubiquitin ligases. The ST LSD domain also binds to CDC20 and CDH1, substrate recognition components of the anaphase-promoting complex [89].

MCPyV ST can increase levels of 4EBP1 phosphorylation that, in turn, promotes increased protein translation [90]. Significantly, the PP2A-binding activity of MCPyV ST was not required to increase the levels of phospho-4EBP1. MCPyV ST binding to CDC20 may contribute to increased 4EBP1 phosphorylation [89]. MCPyV ST binds to L-Myc (MYCL), a member of the Myc family of oncogenic transcription factors, and recruits L-Myc to the Tip60/p400 (KAT5/EP400) chromatin remodeling complex [91]. The ST-MYCL-Tip60/p400 complex has potent transcriptional activation activity that likely contributes to the MCPyV oncogenic activity [92]. Expression of ST can promote significant changes in gene expression including induction of pro-glycolytic genes and can induce aerobic glycolysis when expressed in fibroblasts [92]. Whether the ability of ST to induce a Warburg effect in cells is linked to the LSD motif, PP2A binding, or 4EBP1 phosphorylation is not known.

Although its exact molecular functions are not well understood, MCPyV ST has strong oncogenic activity. For example, ST alone can transform Rat-1 fibroblasts [90] and increase cell motility [93]. ST can cooperate with truncated LT to transform human fibroblasts [82]. Combined expression of MCPyV ST with truncated LT in mice keratinocytes led to hyperplasia, hyperkeratosis, and acanthosis of the skin as well as papillomas [94]. ST can induce tumor formation when expressed in mice as a sole transgene [95, 96]. Because of the presence of the integrated MCPyV genome in virus-positive MCC tumors and the oncogenic activities of ST and LT, MCPyV has been classified by the World Health Organization-International Agency for Research on Cancer as probably carcinogenic to humans (Group 2A) [97].

In addition to the expression of MCPyV ST and truncated LT, virus-positive MCC tumors often contain additional mutations in genes that activate the phospho-inositol 3 kinase (PI3 K) pathway such as gain-of-function mutations in PIK3CA or loss-of-function mutations in PTEN and TSC1. Some of these PI3K mutations are also seen in virus-negative tumors. In contrast, most virus-negative MCC contain mutations involving numerous tumor suppressor genes and oncogenes including RB1, TP53, NOTCH, chromatin-modifying enzymes such as KMT2A, KMT2C, KMT2D, ARID1A, ARID1B, SMARCA4, and KAT6A, as well as genes involved in DNA-damage pathways including ATM, MSH2, BRCA1, and BRCA2 [73, 75, 77]. These observations indicate that the MCPyV viral oncogenes contribute the major oncogenic component to virus-positive MCC. In contrast, viral-negative MCC contains a large number of mutations in both oncogenes and tumor suppressor genes.

Despite the significant differences in the tumor mutational burden (TMB) between virus-positive and virus-negative MCC, there are few phenotypic differences in the two types of MCC. Based on histopathological features alone, two subtypes of MCC can be recognized: pure neuroendocrine tumors and combined tumors with neuroendocrine and divergent (mainly squamous) differentiation. Most pure tumors are MCPyV-positive and CK20-positive while combined tumors are uniformly MCPyV-negative and occasionally CK-20 negative [98, 99]. Virus-negative MCC can also present at pure neuroendocrine-type MCC.

What percentage of MCC tumors contain integrated MCPyV is not clear although 80% is a reasonable estimate based on several reports. The original study that identified MCPyV in MCC used Southern blotting with ^{32}P-phosphate labeled viral probe to confirm MCPyV DNA integration into the tumor genome [32]. Very few, if any, studies since then have used radiolabeled Southern blotting to detect integrated viral DNA in tumor DNA. Another approach used in the original study was PCR amplification of viral DNA from tumor DNA. However, PCR amplification of viral DNA is not always reliable for several reasons. The viral DNA has undergone multiple mutations and rearrangements during integration that at a minimum result in truncation of LT and can reduce primer recognition [81]. In addition, the integrated viral DNA may have undergone amplification that could introduce additional mutations to the T antigen genes [76]. There may even be some strain differences in MCPyV common to different parts of the world that could impede detection of integrated virus by PCR [100]. Another challenge to PCR detection arises since most studies use DNA isolated from formalin-fixed paraffin-embedded tumor sections that can result in degradation of DNA. Given the presence of UV-induced DNA damage in virus-negative MCC and integration of MCPyV DNA in virus-positive MCC, it is likely that next-generation sequencing of MCC will serve as the most accurate approach to determine the type of MCC tumor.

While genomic sequencing has revealed that virus-negative MCC has evidence for a high degree of UV damage, this does not exclude a role for UV exposure in the development of virus-positive MCC. The relative lack of UV-damaged DNA in virus-positive MCC indicates that the etiologies are clearly different, suggesting that the precursor to virus-negative MCC was a recipient of lifelong intense UV exposure while the virus-positive MCC were not exposed to the same degree or for as long. However, UV exposure could affect the immune response to virus-negative and virus-positive MCC etiology. The effect of UV radiation in the pathogenesis of MCC has been suggested to be more likely a result of immune modulation than direct effects on DNA itself [101]. It was reported that the early promoter of MCPyV responds to UV exposure and that levels of ST mRNA increased in UV exposed skin from a healthy human volunteer [102].

8.8 Does AIDS Increase the Risk of Virus-Positive or Virus-Negative MCC?

The AIDS-defining cancers of Kaposi's sarcoma, aggressive B-cell lymphomas, and invasive cervical cancer are each associated with human DNA tumor viruses. Similarly, hepatitis B virus (HBV) and hepatitis C virus (HCV) are known to drive hepatocellular carcinoma, which is hastened in the presence of HIV [103, 104].

In contrast, Merkel cell carcinoma caused by MCPyV is not considered an AIDS-defining malignancy. MCPyV DNA loads are significantly higher in HIV-positive men with poorly controlled HIV infection compared to those with well-controlled HIV viral loads [105], but the progression of MCPyV is not known to be directly impacted by the presence of HIV, but rather by the degree of overall immunosuppression [17]. The average time from HIV diagnosis to the diagnosis of Merkel cell carcinoma, 9.5 years, is significantly longer than other AIDS-associated malignancies [105].

Given the association of AIDS-defining malignancies with KHSV, EBV, and HPV, it is a reasonable assumption that AIDS-associated MCC is MCPyV-positive. However, there have not been any molecular studies that have directly determined the virus status in AIDS-associated MCC. Given the association of MCPyV with MCC and the increased risk of HIV-1/AIDS patients developing MCC, it is plausible that MCC is virus-positive. However, it is also plausible that AIDS increases the risk for virus-negative MCC similar to the increased risk for developing of the non-melanoma skin cancers (NMSC) or keratinocyte carcinomas such as cutaneous squamous cell carcinoma and basal cell carcinoma [106–109]. Low CD4 counts and high HIV-1 viral loads have been associated with a twofold increased risk in squamous cell carcinoma [106]. Until more definitive sequencing studies of AIDS-associated MCC, it cannot be stated with certainty that HIV-1 infected individuals have an increased risk of developing virus-positive or virus-negative MCC.

In another example, organ transplant recipients have a lifelong requirement for immunosuppression and are at increased risk for skin cancers. Skin cancers account for 40–50% of all posttransplant malignancies with squamous cell carcinoma (SCC) and basal cell carcinoma (BCC) comprising 90–95% of these skin cancers plus Kaposi's sarcoma, malignant melanoma, and MCC [110]. Importantly, some of the therapeutics used in organ transplantation carry an increased risk for developing skin cancers. The calcineurin inhibitor, cyclosporine, reduces sensitivity to UV-induced apoptosis in keratinocytes. Azathioprine can sensitize cells to UV-induced damage through the incorporation of a metabolite into DNA, which generates reactive oxygen species upon exposure to UV light [111]. In patients with rheumatoid arthritis, methotrexate and anti-TNF drugs were associated with an increased risk of NMSC [112]. The increased risk for skin cancers in organ transplant recipients and rheumatoid arthritis is associated with UV-light-induced mutagenesis for SCC and BCC. This increased risk may also extend to virus-negative MCC.

8.9 Therapy of MCC

Surgical management is considered a standard of care for localized Merkel cell carcinoma with current recommendations advising a wide local excision of the lesion with 1–2 cm of peripheral and deep margins coupled with a sentinel lymph node biopsy given the frequency of metastatic disease at first presentation. Most international guidelines support the use of local adjuvant radiotherapy coupled with tumor staging to improve the frequency of cure [27]. For advanced, metastatic and recurrent MCC, checkpoint blockade inhibitors targeting PD-1 and PD-L1 have shown remarkable activity including durable long-term responses [113, 114]. Importantly, both virus-negative MCC and virus-positive MCC have shown high response rates to checkpoint blockade inhibitors. It is likely that the virus-negative MCC have a high level of neoantigens resulting from the extensive UV-induced mutational rate similar to that observed in melanoma. The high response rate in virus-positive MCC may reflect the presence of the viral tumor antigens although definitive evidence for this model has not been reported.

There are no randomized control trials of Merkel cell carcinoma treatment in the HIV-infected population. In the absence of data to direct therapy, most HIV-infected individuals with Merkel cell carcinoma are treated with similar modalities as HIV-uninfected individuals with Merkel cell carcinoma. These include a combination surgery, radiation therapy, chemotherapy, and immunotherapy.

Most clinical trials involving immunotherapy exclude patients with HIV, immunosuppression, hematological malignancies, and previous organ transplants including a recent trial in MCC [114]. However, current recommendations for the treatment of lymphoma and Hodgkin disease in patients with HIV/AIDS mirror treatment in patients without HIV/AIDS [115, 116]. There have been a few reports of HIV/AIDS patients being treated with checkpoint blockade inhibitors. Reports of patients with HIV infections and advanced melanoma were treated with the PD-1 inhibitor pembrolizumab without significant toxicities [117]. In patients living with HIV/AIDS (PLWHA), non-small-cell lung cancer (NSCLC) is the most common non-AIDS-related malignancy. For PLWHA with NSCLC treated with the PD-1 inhibitor nivolumab, the HIV-related parameters of viral load and CD4 counts were not altered [118]. Another study of seven patients with metastatic NSCLC and HIV infection demonstrated the safety of treatment with PD-1 inhibitors nivolumab or pembrolizumab. All patients received antiretroviral therapy while on anti-PD-1 treatment and none experienced grade 3 or 4 immune-related adverse events or immune reconstitution inflammatory syndrome [119].

Acknowledgements This work was supported in part by U.S. Public Health Service grants R01CA63113, R01CA173023, and P01CA050661 and the DFCI Helen Pappas Merkel Cell Research Fund and the Claudia Adams Barr Program in Cancer Research to J.A.D.

References

1. Paulson KG et al (2017) Merkel cell carcinoma: current United States incidence and projected increases based on changing demographics. J Am Acad Dermatol
2. Toker C (1972) Trabecular carcinoma of the skin. Arch Dermatol 105(1):107–110
3. Tang CK, Toker C (1978) Trabecular carcinoma of the skin: an ultrastructural study. Cancer 42(5):2311–2321

4. Toker C (1982) Trabecular carcinoma of the skin. A question of title. Am J Dermatopathol 4(6):497–500
5. Rywlin AM (1982) Malignant Merkel-cell tumor is a more accurate description than trabecular carcinoma. Am J Dermatopathol 4(6):513–515
6. Maksimovic S et al (2014) Epidermal Merkel cells are mechanosensory cells that tune mammalian touch receptors. Nature 509(7502):617–621
7. Agelli M et al (2010) The etiology and epidemiology of Merkel cell carcinoma. Curr Probl Cancer 34(1):14–37
8. Misch D et al (2015) Value of thyroid transcription factor (TTF)-1 for diagnosis and prognosis of patients with locally advanced or metastatic small cell lung cancer. Diagn Pathol 10:21
9. Pulitzer MP et al (2015) Cutaneous squamous and neuroendocrine carcinoma: genetically and immunohistochemically different from Merkel cell carcinoma. Mod Pathol 28(8): 1023–1032
10. Pitale M, Sessions RB, Husain S (1992) An analysis of prognostic factors in cutaneous neuroendocrine carcinoma. Laryngoscope 102(3):244–249
11. Gooptu C et al (1997) Merkel cell carcinoma arising after therapeutic immunosuppression. Br J Dermatol 137(4):637–641
12. Centers for Disease, Control and Prevention (1997) Update: trends in AIDS incidence, deaths, and prevalence–United States, 1996. MMWR Morb Mortal Wkly Rep 46(8): 165–173
13. Catlett JP, Todd WM, Carr ME Jr (1992) Merkel cell tumor in an HIV-positive patient. Va Med Q 119(4):256–258
14. Engels EA et al (2002) Merkel cell carcinoma and HIV infection. Lancet 359(9305): 497–498
15. Cone LA et al (2006) Merkel cell carcinoma in an HIV-1-infected man. AIDS 20(3): 474–475
16. Manganoni MA et al (2007) Merkel cell carcinoma and HIV infection: a case report and review of the literature. AIDS Patient Care STDS 21(7):447–451
17. Busse PM et al (2008) Case records of the Massachusetts general hospital. Case 19-2008. A 63-year-old HIV-positive man with cutaneous Merkel-cell carcinoma. N Engl J Med 358 (25):2717–2723
18. Ottaviani F et al (2010) Bona fide primary Merkel cell carcinoma of an intraparotid lymph node in a HIV-positive patient. Int J Surg Pathol 18(5):406–408
19. Izikson L et al (2011) Merkel cell carcinoma associated with HIV: review of 14 patients. AIDS 25(1):119–121
20. Wieland U, Kreuter A (2011) Merkel cell polyomavirus infection and Merkel cell carcinoma in HIV-positive individuals. Curr Opin Oncol 23(5):488–493
21. Li M et al (2013) Metastatic Merkel cell carcinoma of the oral cavity in a human immunodeficiency virus-positive patient and the detection of Merkel cell polyomavirus. Oral Surg Oral Med Oral Pathol Oral Radiol 115(5):e66–e71
22. Samarendra P et al (2000) Primary nodal neuroendocrine (Merkel cell) tumor in a patient with HIV infection. South Med J 93(9):920–922
23. Heath M et al (2008) Clinical characteristics of Merkel cell carcinoma at diagnosis in 195 patients: the AEIOU features. J Am Acad Dermatol 58(3):375–381
24. Lanoy E, Engels EA (2010) Skin cancers associated with autoimmune conditions among elderly adults. Br J Cancer 103(1):112–114
25. Clarke CA et al (2015) Risk of Merkel cell carcinoma after solid organ transplantation. J Natl Cancer Inst 107(2)
26. Sahi H et al (2017) History of chronic inflammatory disorders increases the risk of Merkel cell carcinoma, but does not correlate with Merkel cell polyomavirus infection. Br J Cancer 116(2):260–264

27. Bichakjian CK et al (2018) Merkel cell carcinoma, Version 1.2018, NCCN clinical practice guidelines in oncology. J Natl Compr Canc Netw 16(6):742–774
28. Bhatia S et al (2016) Adjuvant radiation therapy and chemotherapy in Merkel cell carcinoma: survival analyses of 6908 cases from the national cancer data base. J Natl Cancer Inst 108(9)
29. Paulson KG et al (2013) Systemic immune suppression predicts diminished Merkel cell carcinoma-specific survival independent of stage. J Invest Dermatol 133(3):642–646
30. Albores-Saavedra J et al (2010) Merkel cell carcinoma demographics, morphology, and survival based on 3870 cases: a population based study. J Cutan Pathol 37(1):20–27
31. Howard RA et al (2006) Merkel cell carcinoma and multiple primary cancers. Cancer Epidemiol Biomark Prev 15(8):1545–1549
32. Feng H et al (2008) Clonal integration of a polyomavirus in human Merkel cell carcinoma. Science 319(5866):1096–1100
33. Durst M et al (1983) A papillomavirus DNA from a cervical carcinoma and its prevalence in cancer biopsy samples from different geographic regions. Proc Natl Acad Sci USA 80(12): 3812–3815
34. Chang Y et al (1994) Identification of herpesvirus-like DNA sequences in AIDS-associated Kaposi's sarcoma. Science 266(5192):1865–1869
35. Tolstov YL et al (2011) Asymptomatic primary Merkel cell polyomavirus infection among adults. Emerg Infect Dis 17(8):1371–1380
36. DeCaprio JA, Garcea RL (2013) A cornucopia of human polyomaviruses. Nat Rev Microbiol 11(4):264–276
37. DeCaprio JA (2017) Merkel cell polyomavirus and Merkel cell carcinoma. Philos Trans R Soc Lond B Biol Sci 372(1732)
38. Gross L (1953) A filterable agent, recovered from Ak leukemic extracts, causing salivary gland carcinomas in C3H mice. Proc Soc Exp Biol Med 83(2):414–421
39. Buck CB et al (2016) The ancient evolutionary history of polyomaviruses. PLoS Pathog 12(4):e1005574
40. Gheit T et al (2017) Isolation and characterization of a novel putative human polyomavirus. Virology 506:45–54
41. Carter JJ et al (2013) Identification of an overprinting gene in Merkel cell polyomavirus provides evolutionary insight into the birth of viral genes. Proc Natl Acad Sci U S A 110(31):12744–12749
42. Schowalter RM, Buck CB (2013) The Merkel cell polyomavirus minor capsid protein. PLoS Pathog 9(8):e1003558
43. Gardner SD et al (1971) New human papovavirus (B.K.) isolated from urine after renal transplantation. Lancet 1(7712):1253–1257
44. Barouch DH et al (2002) BK virus-associated hemorrhagic cystitis in a human immunodeficiency virus-infected patient. Clin Infect Dis 35(3):326–329
45. Knowles WA et al (2003) Population-based study of antibody to the human polyomaviruses BKV and JCV and the simian polyomavirus SV40. J Med Virol 71(1):115–123
46. Ambalathingal GR et al (2017) BK polyomavirus: clinical aspects, immune regulation, and emerging therapies. Clin Microbiol Rev 30(2):503–528
47. van Aalderen MC et al (2012) BK virus infection in transplant recipients: clinical manifestations, treatment options and the immune response. Neth J Med 70(4):172–183
48. Erard V et al (2004) BK virus infection in hematopoietic stem cell transplant recipients: frequency, risk factors, and association with postengraftment hemorrhagic cystitis. Clin Infect Dis 39(12):1861–1865
49. Ramos E et al (2009) The decade of polyomavirus BK-associated nephropathy: state of affairs. Transplantation 87(5):621–630
50. Bratt G et al (1999) BK virus as the cause of meningoencephalitis, retinitis and nephritis in a patient with AIDS. AIDS 13(9):1071–1075

51. Ledesma J et al (2012) BK virus infection in human immunodeficiency virus-infected patients. Eur J Clin Microbiol Infect Dis 31(7):1531–1535
52. Antoniolli L, Borges R, Goldani LZ (2017) BK virus encephalitis in HIV-infected patients: case report and review. Case Rep Med 2017:4307468
53. Padgett BL et al (1971) Cultivation of papova-like virus from human brain with progressive multifocal leucoencephalopathy. Lancet 1(7712):1257–1260
54. Tan CS et al (2012) Increased program cell death-1 expression on T lymphocytes of patients with progressive multifocal leukoencephalopathy. J Acquir Immune Defic Syndr 60(3): 244–248
55. Major EO (2010) Progressive multifocal leukoencephalopathy in patients on immunomodulatory therapies. Annu Rev Med 61:35–47
56. Bowen LN et al (2016) HIV-associated opportunistic CNS infections: pathophysiology, diagnosis and treatment. Nat Rev Neurol 12(11):662–674
57. Cinque P et al (2009) Progressive multifocal leukoencephalopathy in HIV-1 infection. Lancet Infect Dis 9(10):625–636
58. Schowalter RM et al (2010) Merkel cell polyomavirus and two previously unknown polyomaviruses are chronically shed from human skin. Cell Host Microbe 7(6):509–515
59. van der Meijden E et al (2010) Discovery of a new human polyomavirus associated with trichodysplasia spinulosa in an immunocompromized patient. PLoS Pathog 6(7):e1001024
60. Nguyen KD et al (2016) Human polyomavirus 6 and 7 are associated with pruritic and dyskeratotic dermatoses. J Am Acad Dermatol
61. van der Meijden E et al (2016) Primary polyomavirus infection, not reactivation, as the cause of trichodysplasia spinulosa in immunocompromised patients. J Infect Dis
62. Liu W et al (2016) Identifying the target cells and mechanisms of Merkel cell polyomavirus infection. Cell Host Microbe 19(6):775–787
63. Hurdiss DL et al (2016) New structural insights into the genome and minor capsid proteins of BK polyomavirus using cryo-electron microscopy. Structure 24(4):528–536
64. Chromy LR, Pipas JM, Garcea RL (2003) Chaperone-mediated in vitro assembly of polyomavirus capsids. Proc Natl Acad Sci U S A 100(18):10477–10482
65. Norkiene M et al (2015) Production of recombinant VP1-derived virus-like particles from novel human polyomaviruses in yeast. BMC Biotechnol 15:68
66. Martel-Jantin C et al (2013) Merkel cell polyomavirus infection occurs during early childhood and is transmitted between siblings. J Clin Virol 58(1):288–291
67. van der Meijden E et al (2013) Different serologic behavior of MCPyV, TSPyV, HPyV6, HPyV7 and HPyV9 polyomaviruses found on the skin. PLoS ONE 8(11):e81078
68. Gossai A et al (2016) Seroepidemiology of human polyomaviruses in a US population. Am J Epidemiol 183(1):61–69
69. Nicol JT et al (2013) Age-specific seroprevalences of merkel cell polyomavirus, human polyomaviruses 6, 7, and 9, and trichodysplasia spinulosa-associated polyomavirus. Clin Vaccine Immunol 20(3):363–368
70. Paulson KG et al (2017) Viral oncoprotein antibodies as a marker for recurrence of Merkel cell carcinoma: a prospective validation study. Cancer 123(8):1464–1474
71. Pastrana DV et al (2009) Quantitation of human seroresponsiveness to Merkel cell polyomavirus. PLoS Pathog 5(9):e1000578
72. Wong SQ et al (2015) UV-associated mutations underlie the etiology of MCV-negative Merkel cell carcinomas. Cancer Res 75(24):5228–5234
73. Harms PW et al (2015) The distinctive mutational spectra of polyomavirus-negative Merkel cell carcinoma. Cancer Res 75(18):3720–3727
74. Harms PW et al (2016) Next generation sequencing of Cytokeratin 20-negative Merkel cell carcinoma reveals ultraviolet-signature mutations and recurrent TP53 and RB1 inactivation. Mod Pathol 29(3):240–248
75. Goh G et al (2016) Mutational landscape of MCPyV-positive and MCPyV-negative Merkel cell carcinomas with implications for immunotherapy. Oncotarget 7(3):3403–3415

76. Starrett GJ et al (2017) Merkel cell polyomavirus exhibits dominant control of the tumor genome and transcriptome in virus-associated Merkel cell carcinoma. MBio 8(1)
77. Gonzalez-Vela MD et al (2017) Shared oncogenic pathways implicated in both virus-positive and UV-induced Merkel cell carcinomas. J Invest Dermatol 137(1):197–206
78. Popp S et al (2002) UV-B-type mutations and chromosomal imbalances indicate common pathways for the development of Merkel and skin squamous cell carcinomas. Int J Cancer 99(3):352–360
79. Moshiri A et al (2016) Polyomavirus-negative Merkel cell carcinoma: a more aggressive subtype based on analysis of 282 cases using multi-modal tumor virus detection. J Invest Dermatol 137(4):819–827
80. Wang L et al (2017) Age and gender associations of virus positivity in Merkel cell carcinoma characterized using a novel RNA in situ hybridization assay. Clin Cancer Res
81. Shuda M et al (2008) T antigen mutations are a human tumor-specific signature for Merkel cell polyomavirus. Proc Natl Acad Sci U S A 105(42):16272–16277
82. Cheng J et al (2013) Merkel cell polyomavirus large T antigen has growth-promoting and inhibitory activities. J Virol 87(11):6118–6126
83. Li J et al (2013) Merkel cell polyomavirus large T antigen disrupts host genomic integrity and inhibits cellular proliferation. J Virol 87(16):9173–9188
84. Laude HC et al (2010) Distinct Merkel cell polyomavirus molecular features in tumour and non tumour specimens from patients with Merkel cell carcinoma. PLoS Pathog 6(8)
85. Nakamura T et al (2010) Nuclear localization of Merkel cell polyomavirus large T antigen in Merkel cell carcinoma. Virology 398(2):273–279
86. Kwun HJ et al (2015) Restricted protein phosphatase 2A targeting by Merkel cell polyomavirus small T antigen. J Virol 89(8):4191–4200
87. Kwun HJ et al (2013) Merkel cell polyomavirus small T antigen controls viral replication and oncoprotein expression by targeting the cellular ubiquitin ligase SCFFbw7. Cell Host Microbe 14(2):125–135
88. Tsang SH et al (2016) The oncogenic small tumor antigen of Merkel cell polyomavirus is an iron-sulfur cluster protein that enhances viral DNA replication. J Virol 90(3):1544–1556
89. Shuda M et al (2015) CDK1 substitutes for mTOR kinase to activate mitotic cap-dependent protein translation. Proc Natl Acad Sci U S A 112(19):5875–5882
90. Shuda M et al (2011) Human Merkel cell polyomavirus small T antigen is an oncoprotein targeting the 4E-BP1 translation regulator. J Clin Invest 121(9):3623–3634
91. Cheng J et al (2017) Merkel cell polyomavirus recruits MYCL to the EP400 complex to promote oncogenesis. PLoS Pathog 13(10):e1006668
92. Berrios C et al (2016) Merkel cell polyomavirus small T antigen promotes pro-glycolytic metabolic perturbations required for transformation. PLoS Pathog 12(11):e1006020
93. Stakaityte G et al (2018) Merkel cell polyomavirus small T antigen drives cell motility via Rho-GTPase-induced filopodium formation. J Virol 92(2)
94. Spurgeon ME et al (2015) Tumorigenic activity of Merkel cell polyomavirus T antigens expressed in the stratified epithelium of mice. Cancer Res 75(6):1068–1079
95. Shuda M et al (2015) Merkel cell polyomavirus small T antigen induces cancer and embryonic Merkel cell proliferation in a transgenic mouse model. PLoS ONE 10(11): e0142329
96. Verhaegen ME et al (2015) Merkel cell polyomavirus small T antigen is oncogenic in transgenic mice. J Invest Dermatol 135(5):1415–1424
97. Bouvard V et al (2012) Carcinogenicity of malaria and of some polyomaviruses. Lancet Oncol 13(4):339–340
98. Busam KJ et al (2009) Merkel cell polyomavirus expression in merkel cell carcinomas and its absence in combined tumors and pulmonary neuroendocrine carcinomas. Am J Surg Pathol 33(9):1378–1385
99. Kuwamoto S et al (2011) Association of Merkel cell polyomavirus infection with morphologic differences in Merkel cell carcinoma. Hum Pathol 42(5):632–640

100. Martel-Jantin C et al (2014) Molecular epidemiology of merkel cell polyomavirus: evidence for geographically related variant genotypes. J Clin Microbiol 52(5):1687–1690
101. Dowlatshahi M et al (2013) Tumor-specific T cells in human Merkel cell carcinomas: a possible role for Tregs and T-cell exhaustion in reducing T-cell responses. J Invest Dermatol 133(7):1879–1889
102. Mogha A et al (2010) Merkel cell polyomavirus small T antigen mRNA level is increased following in vivo UV-radiation. PLoS ONE 5(7):e11423
103. Engels EA et al (2002) Prevalence of hepatitis C virus infection and risk for hepatocellular carcinoma and non-Hodgkin lymphoma in AIDS. J Acquir Immune Defic Syndr 31(5): 536–541
104. McGovern BH (2007) The epidemiology, natural history and prevention of hepatitis B: implications of HIV coinfection. Antivir Ther 12(Suppl 3):H3–H13
105. Wieland U et al (2011) Merkel cell polyomavirus infection in HIV-positive men. Arch Dermatol 147(4):401–406
106. Asgari MM et al (2017) Association of multiple primary skin cancers with human immunodeficiency virus infection, CD4 count, and viral load. JAMA Dermatol 153(9): 892–896
107. Silverberg MJ et al (2013) HIV infection status, immunodeficiency, and the incidence of non-melanoma skin cancer. J Natl Cancer Inst 105(5):350–360
108. Chang AY, Doiron P, Maurer T (2017) Cutaneous malignancies in HIV. Curr Opin HIV AIDS 12(1):57–62
109. Nehal KS, Bichakjian CK (2018) Update on Keratinocyte Carcinomas. N Engl J Med 379 (4):363–374
110. Garrett GL et al (2017) Incidence of and risk factors for skin cancer in organ transplant recipients in the United States. JAMA Dermatol 153(3):296–303
111. Wheless L et al (2014) Skin cancer in organ transplant recipients: more than the immune system. J Am Acad Dermatol 71(2):359–365
112. Scott FI et al (2016) Risk of nonmelanoma skin cancer associated with the use of immunosuppressant and biologic agents in patients with a history of autoimmune disease and nonmelanoma skin cancer. JAMA Dermatol 152(2):164–172
113. Nghiem PT et al (2016) PD-1 blockade with pembrolizumab in advanced Merkel-cell carcinoma. N Engl J Med 374(26):2542–2552
114. Kaufman HL et al (2016) Avelumab in patients with chemotherapy-refractory metastatic Merkel cell carcinoma: a multicentre, single-group, open-label, phase 2 trial. Lancet Oncol 17(10):1374–1385
115. Montoto S et al (2012) HIV status does not influence outcome in patients with classical Hodgkin lymphoma treated with chemotherapy using doxorubicin, bleomycin, vinblastine, and dacarbazine in the highly active antiretroviral therapy era. J Clin Oncol 30(33):4111–4116
116. Little RF, Dunleavy K (2013) Update on the treatment of HIV-associated hematologic malignancies. Hematol Am Soc Hematol Educ Program 2013:382–388
117. Davar D et al (2015) PD-1 blockade in advanced melanoma in patients with Hepatitis C and/or HIV. Case Rep Oncol Med 2015:737389
118. Lavole A et al (2018) PD-1 blockade in HIV-infected patients with lung cancer: a new challenge or already a strategy? Ann Oncol 29(4):1065–1066
119. Ostios-Garcia L et al (2018) Safety and efficacy of PD-1 inhibitors among HIV-positive patients with non-small cell lung cancer. J Thorac Oncol 13(7):1037–1042

HIV–HBV and HIV–HCV Coinfection and Liver Cancer Development

9

Jianming Hu, Kuancheng Liu and Jun Luo

Contents

Abstract

Liver diseases that are caused by the hepatitis B virus (HBV) and hepatitis C virus (HCV), including cirrhosis and hepatocellular carcinoma (HCC), have

J. Hu (✉) · K. Liu · J. Luo
Department of Microbiology and Immunology, The Pennsylvania State University College of Medicine, 500 University Drive, Hershey, PA 17033, USA
e-mail: juh13@psu.edu

C. Meyers (ed.), *HIV/AIDS-Associated Viral Oncogenesis*, Cancer Treatment and Research 177, https://doi.org/10.1007/978-3-030-03502-0_9

become increasingly important in patients infected with the human immunodeficiency virus (HIV) as their life expectancy is getting longer with successful anti-HIV therapy. Due to their shared transmission routes, dual infection by HIV and HBV or HIV and HCV, and triple infection by all three viruses are fairly common and affect millions of people worldwide. Whereas the immunodeficiency caused by HIV enhances the likelihood of HBV and HCV persistence, hepatotoxicity associated with anti-HIV therapy can worsen the liver diseases associated with HBV or HCV persistence. Evidence suggests HIV infection increases the risk of HBV- or HCV-associated HCC risk although the precise mechanisms of enhanced hepatocarcinogenesis remain to be fully elucidated. Recent success in curing HCV infection, and the availability of therapeutic options effective in long-term suppression of both HIV and HBV replication, bring hope, fortunately, to those who are coinfected but also highlight the need for judicious selection of antiviral therapies.

Keywords

Hepatitis B virus · Hepatitis C virus · Human immunodeficiency virus
HBV · HCV · HIV · Hepatocellular carcinoma · HCC · Coinfection

9.1 Introduction

Before the advent of highly active antiretroviral therapy (HAART), human immunodeficiency virus (HIV)-infected patients were most likely to succumb to opportunistic bacterial or fungal infections, secondary to HIV-induced immune suppression. As anti-retroviral therapies continue to improve, patients infected with HIV are living longer and the health problems that are of primary concern to these patients have been changing, at least for those who have access to HARRT. With longer survival times, liver diseases including chronic viral hepatitis and hepatocellular carcinoma (HCC), have become increasingly important in these patients [1–5]. In fact, approximately 10–15% of mortalities in HIV-infected patients are now due to liver diseases. In patients infected with HIV, most liver diseases are due to chronic viral hepatitis. This is not surprising considering that agents causing viral hepatitis, like the hepatitis B virus (HBV) and hepatitis C virus (HCV), are transmitted through similar routes to HIV. An additional complication involving the liver in HIV-infected patients is the concern that many anti-HIV drugs are hepatotoxic, which can be further exacerbated by viral hepatitis. Fortunate for the millions of HIV-infected patients who are also inflicted with viral hepatitis is the recent development of curative therapies for HCV infection and treatment options that are effective in suppressing both HIV and HBV infection.

9.2 Chronic HBV and HCV Infections

Worldwide, there are ca. 325 million people who are chronically infected with either HBV or HCV, or both [5]. HBV is a small, enveloped DNA virus that belongs to the *Hepadnaviridae* family. HBV is unusual for a DNA virus in that replication of its DNA genome is through reverse transcription of an RNA intermediate [6, 7]. The small, 3.2 kb HBV DNA genome encodes four open reading frames (ORFs) (Fig. 9.1), which are translated to make the viral core (C) protein, the main constituent of the viral nucleocapsid; the reverse transcriptase (RT), the enzyme responsible for DNA replication via reverse transcription; and three envelope glycoproteins. In addition, the HBV X (HBx) protein has a number of pleiotropic effects on viral and cellular gene expression, cell signaling, cell cycle, and apoptosis, although the significance of these in viral replication or pathogenesis remains unresolved [8]. The latest development in this regard is the discovery that HBx triggers the degradation of a host restriction factor that would otherwise suppress HBV transcription [9]. HBV is transmitted by contact with blood or body fluids of an

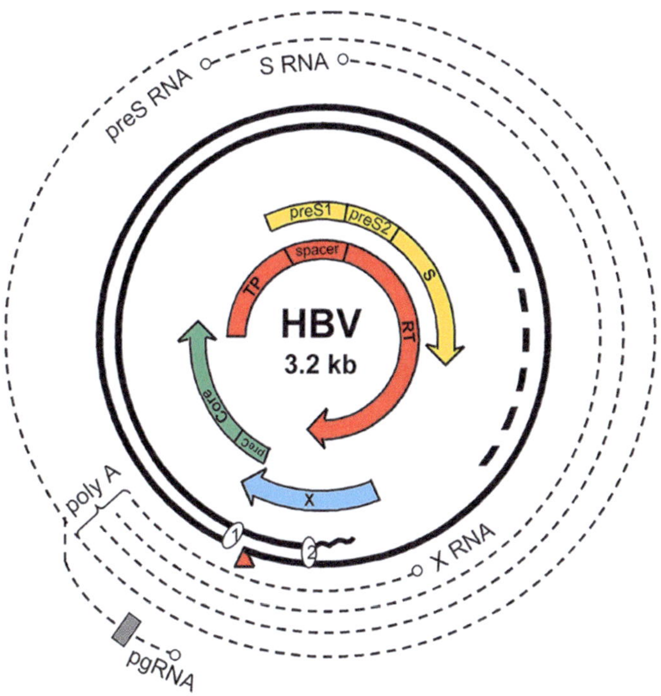

Fig. 9.1 HBV genome organization. Solid lines, the partially double stranded, relaxed circular DNA genome; dotted lines, viral RNA transcripts; solid arrows, encoded proteins. Core, core protein; S, surface protein, RT, reverse transcriptase; X, X protein. Triangle, the RT protein covalently linked to the genome; checked box, the ε RNA packaging signal; ovals, direct repeat 1 and 2, *cis*-acting elements involved in reverse transcription

infected person in the same way as HIV. However, HBV is 50–100 times more infectious than HIV.

Of the 2 billion people who have been infected with HBV, ca. 257 million have chronic, lifelong infections [5, 10–14]. In the US alone, there are 1.25–2 million chronic HBV carriers. In some areas of Asia and Africa, where HBV is endemic, 10–20% of the whole population is chronically infected with HBV (Fig. 9.2). In these parts of the world, HBV infections are acquired mainly perinatally or early in childhood, a high percentage (up to 90%) of which become chronic. In contrast, 5–15% of immune-competent adults who acquire HBV infection will become chronic carriers of the virus. Patients who are chronically infected with HBV are at high risk of premature death from cirrhosis of the liver and HCC, a highly malignant liver cancer [15]. The risk of death from HBV-related liver cancer or cirrhosis is approximately 25% for persons who become chronically infected and are untreated. Together, these diseases kill approximately one million people each year worldwide.

HCV is an enveloped RNA virus belonging to the *Flaviviridae* family [16]. Like other flaviviruses, the 9 kb long, positive-sense, single-stranded RNA genome (Fig. 9.3) of HCV is translated into a polyprotein, which is proteolytically cleaved into the viral structural proteins, including a single capsid (C) protein and two envelope glycoproteins (E1 and E2), and the non-structural proteins required for viral replication, including two proteases (NS2 and NS3) and the viral RNA-dependent RNA polymerase (NS5b). Like HBV, HCV is transmitted through blood and body fluids.

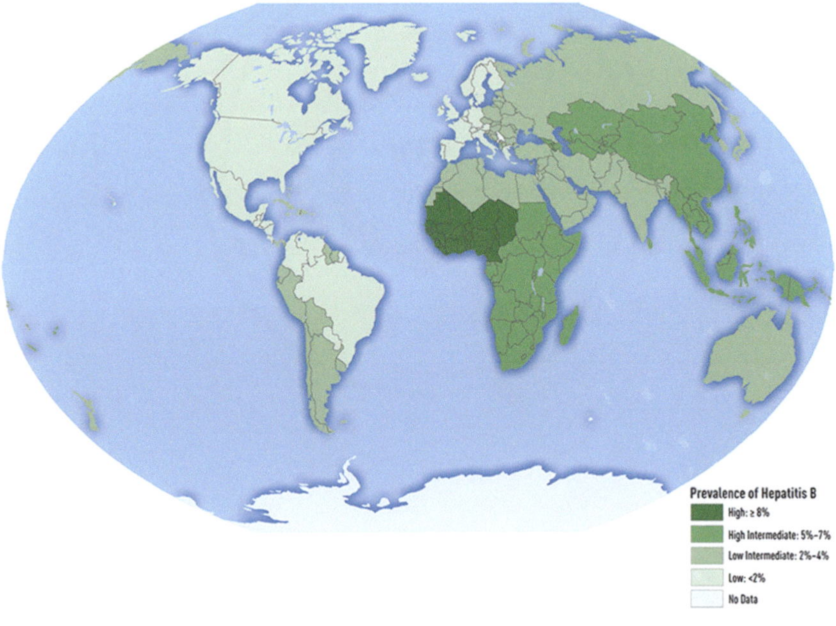

Fig. 9.2 Global prevalence of chronic HBV infection. A map showing the percentage of population chronically infected with HBV in different regions of the world. From U.S. Center for disease control and prevention

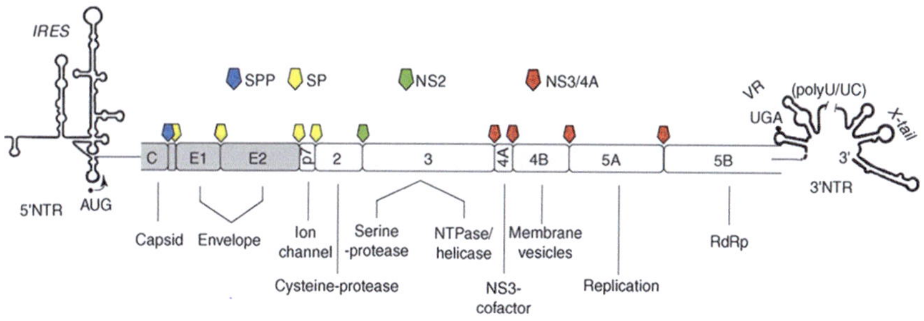

Fig. 9.3 HCV genome organization. The positive-sense, single-stranded HCV RNA genome encodes a long ORF (ca 3000 amino acids). The viral polyprotein is proteolytically cleaved into the structural proteins, C (capsid), E1 and E2 (envelope glycoprotein 1 and 2), as well as the nonstructural (NS) proteins, p7, and NS2 to NS5B, by the host (signal peptidase or SP and signal peptide peptidase or SPP) and viral proteases (NS2 and NS3/4A). The major known function(s) of the viral proteins are indicated below each protein. The structured 5' and 3' non-translated region (NTR) important for viral replication and translation are marked. polyU/UC and X-tail are conserved elements in the 3' NTR important for viral replication. IRES, internal ribosome entry site important for cap-independent translation (note the lack of a 5' cap on the viral RNA); AUG and UGA, the initiation and stop codons for polyprotein translation; VR, variable region; RdRp, RNA-dependent RNA polymerase

HCV frequently causes persistent infection of the liver, although there is no DNA form in its life cycle or latent stage known [17, 18]. In fact, the chance of chronic infection with HCV is approximately 55–85% and this varies little with age, in contrast to HBV infections. An estimated 71 million people worldwide are chronically infected with HCV (Fig. 9.4) [5]. Approximately 3.9 million (1.8%) Americans have been infected with HCV, 2.7 millions of whom are chronically infected. Approximately 5–20% of chronically infected persons develop liver cirrhosis over a period of 20–30 years, and HCC develops in 1–5% of persons with chronic infection.

9.3 Coinfection of HBV or HCV with HIV

Worldwide, ca. 36.7 million people are infected with HIV. Roughly, 1.2 million of these people live in the US. In Europe and the US, approximately 8–16% of HIV patients are also chronically infected with HBV [1, 19–22]. Worldwide, the number of HIV-infected people who are also chronically infected with HBV is estimated at ca. 2.7 million (Fig. 9.5) [5, 23], with a large proportion of these coming from HBV-endemic regions of Asia and Africa [20, 24]. The risk of developing chronic HBV infection is about three- to six-fold higher in HIV-infected patients than in those who are not infected with HIV, likely due to the fact that HIV-induced immune suppression can reduce the patient's ability to clear HBV [25]. Furthermore, HIV-induced immune suppression may also play a role in the reactivation of

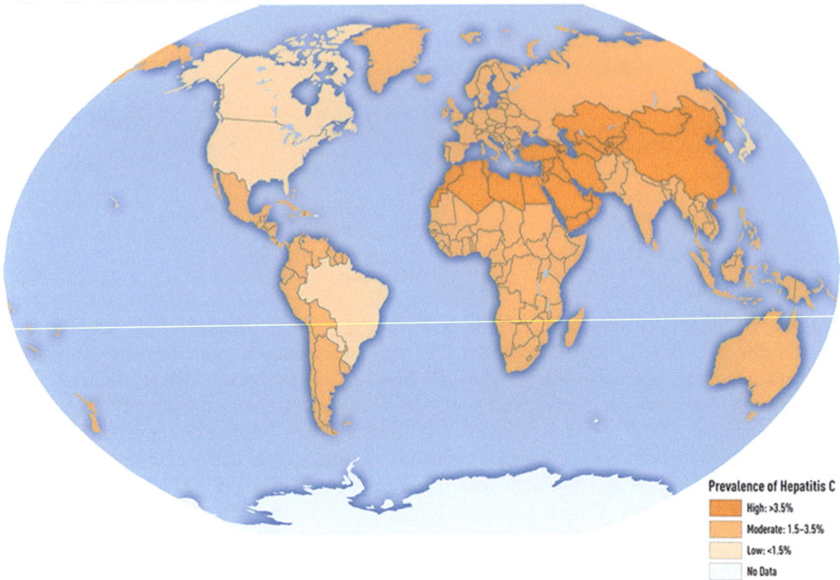

Fig. 9.4 Global prevalence of chronic HCV infection. A map showing the percentage of population chronically infected with HCV in different regions of the world. From U.S. Center for disease control and prevention

Fig. 9.5 Incidences of chronic HBV, HCV, and HIV coinfections. A Venn diagram depicting coinfection of HIV with HBV or HCV, and of HBV with HCV. Shown are estimated numbers (in millions) of patients singly infected with each virus or doubly infected

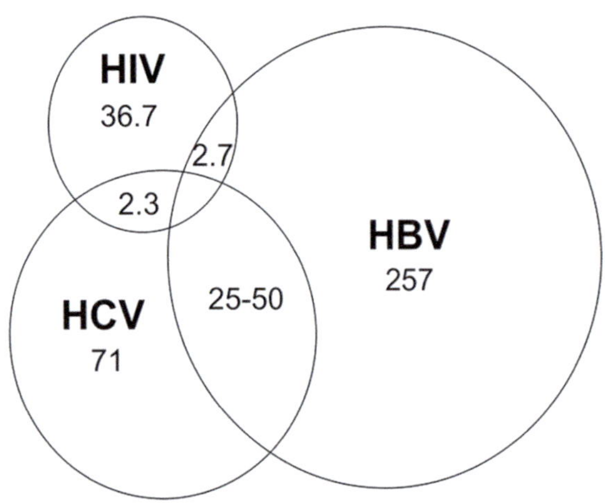

latent HBV infections, which are thought to be under immune control following clinical but not virological resolution [26].

The number of HIV-infected people who are chronically infected with HCV worldwide is approximately 2.3 million [5, 27] (Fig. 9.5). In developed countries, ~25% of HIV patients also have chronic hepatitis C infection [22].

The number of coinfections varies depending on the route of transmission; for example, the incidence of HIV/HCV coinfection is higher in populations of injecting drug users (IUD) than those who were infected by sexual transmission.

9.4 HBV–HCV Dual Infections and Occult HBV Infection

The incidence of HBV–HCV coinfection is not uncommon (Fig. 9.5), as might be predicted from their shared transmission routes [28, 29]. About 10–20% chronic HBV-infected patients are also infected with HCV. In addition, occult HBV infection, defined by the undetectability of HBsAg, the main HBV envelope protein, in the serum but the presence of antibodies against the viral core antigen (HBcAg) or HBV DNA, is fairly common [30]. This may be an important factor in the development of HCC in patients with no serologic evidence of HBV or HCV infection, and additionally, may play a role in the development of HCC in chronically infected HCV patients [31].

9.5 Hepatotropism and Lymphotropism of HBV and HCV

While there is little dispute that HBV and HCV infect the hepatocytes in the human liver, it remains controversial as to whether these viruses also infect other cell types, in particular, lymphoid cells. It has been reported that HBV DNA can be detected in peripheral blood mononuclear cells (PBMC) and some HBV isolates may, in fact, be able to infect PBMC [32], although the question of true infection or passive endocytosis of virus is still being debated. With respect to HCV, there are many reports of infection of the lymphoid cell, particularly B lymphocytes [33] but the level of HCV replication seems to be rather low in general and specific detection is problematic. On the other hand, HIV is lymphotropic and infects only a small percentage of CD4+ T-cells. Therefore, although the possibility exists that HIV and HCV or HBV may infect the same cell types, true dual infection of the same cell seems unlikely.

9.6 Hepatocellular Carcinoma

HCC is the most common primary cancer of the liver, accounting for 60–90% of all hepatic malignancies [34–37]. It is the 6th most common cancer among men and the 11th most common cancer in women worldwide. Particularly, in HBV-endemic regions of sub-Saharan Africa and Eastern Asia, HCC is the most prevalent cancer and incidences can be as high as 50–150 cases per 100,000 population [38]. In North America and Western Europe where the HBV infection rate is relatively low,

HCC incidence is below 10 per 100,000. However, there has been a recent surge of HCC in the developed world, likely due to the prevalence of HCV infections in these areas [39–41]. Together, chronic HBV and HCV infections are responsible for over 80–90% of all HCC on a global scale, and account for 5% of all human cancer burden [42]. Chronic HBV carriers have been shown to be at a 100- to 233-fold increased risk for the development of HCC. HBV infection is responsible for the majority of HCC development in the developing world and accounts for 15–20% of HCC in the US. In contrast, HCV accounts for the majority of HCC in the developed world [43]. In addition, as mentioned earlier, coinfection of both HBV and HCV is also common and can further increase the risk of HCC development.

Despite the clear epidemiological evidence that HBV and HCV are responsible for the vast majority of HCC, the mechanism of viral hepatocarcinogenesis remains incompletely understood (Fig. 9.6) [34–36, 44–47]. There are several possible mechanisms by which liver cancer may develop in patients with chronic viral hepatitis. The first is by an indirect means: chronic viral infection of the liver produces a state of persistent inflammation, in which cancer is a nonspecific side effect of the immune response against the HBV or HCV infection. Thus, HCC may develop as a result of the continuous damage and regeneration of the liver cells in a mutagenic inflammatory environment, which ultimately leads to the aberrant activation of one or more cellular proto-oncogenes or the inactivation of tumor suppressor genes. Evidence is rather strong in support of this nonspecific carcinogenesis mechanism. It is now clear that chronic inflammation and tissue damage over a long period, per se, can be carcinogenic, regardless of the initial trigger events. Thus, not only chronic HBV and HCV infections, but also alcoholic liver damage and metabolic liver damage as a result of genetic mutations including α1-antitrypsin deficiency, Wilson's disease, and hemochromatosis all increase the

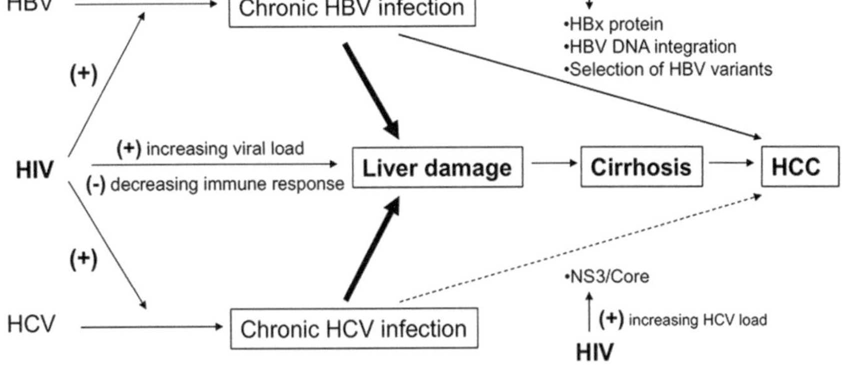

Fig. 9.6 Potential mechanisms HBV- and HCV-induced hepatocarcinogenesis and its enhancement by HIV. Solid arrows, likely mechanisms; dotted arrows, uncertain mechanisms; (+), enhancing effect; (−), inhibitory effect. See text for details

risk of HCC [37]. Indeed, the majority of HCC, regardless of etiology, arises in the background of liver cirrhosis resulting from chronic liver damage.

Another potential mechanism of viral hepatocarcinogenesis involves a more direct role of viral proteins [35, 44, 46, 47]. Thus, although the different causes of chronic liver damage, as outlined above, can all increase the risk of HCC, chronic HBV or HCV infections present a much greater risk of liver cancer than other hepatic inflammatory disorders. In addition, HCC may develop in chronic HBV infections in the absence of cirrhosis. Although the exact viral proteins and their carcinogenic mechanism remain to be elucidated, the HBx protein and the HCV core and NS5A proteins have been reported to have transforming potential in overexpression systems (Fig. 9.6). As HCC generally only develops after decades of HBV or HCV infections, it seems unlikely that these viruses encode any *bona fide* oncogenes. Caution is, therefore, warranted in the interpretation of these overexpression studies out of the context of natural viral infection.

A third molecular mechanism of carcinogenesis, available to HBV but not HCV, is insertional mutagenesis as a result of HBV DNA integration into the host chromosomes [48–50]. It has been known for decades that HBV DNA does integrate into the host genome during infection although integration is not an obligatory step in viral replication, in contrast to retroviruses. In fact, most HBV-related HCCs harbor HBV DNA integration. HBV DNA has been shown to integrate into the host DNA at multiple sites in a seemingly random fashion, although more recent reports suggest that there may be some preferred sites of integration [49, 51, 52]. What is still unclear is the etiological role of DNA integration in HBV carcinogenesis in humans. Elegant studies using the woodchuck hepatitis virus (WHV), a member of the mammalian hepadnaviruses closely related to HBV, have convincingly demonstrated that viral DNA integration, specifically the insertional activation of the cellular *myc* proto-oncogenes, plays a critical role in liver cancer development in chronically WHV-infected woodchucks, virtually all of which develop HCC [53–55]. Reports of potentially similar insertional activation of cellular oncogenes in human liver cancer have appeared [49, 51, 56, 57] but its prevalence and true significance in carcinogenesis remain to be clarified.

9.7 Does HIV Coinfection Increase the Risk of HCC Associated with HBV or HCV Infection?

Epidemiological evidence suggests that coinfection with HIV may increase the risk of HCC development in HBV- or HCV-infected patients, although the data do not seem definitive [19, 28, 58–68]. As mentioned above, HIV coinfection can decrease the rate of HBV or HCV clearance and increase the risk of chronic HBV or HCV infection and the risk of reactivation of latent HBV infections (Fig. 9.6) due to defect in immune-mediated clearance of HBV or HCV [19, 25], including the suppression of anti-HCV activity of NK cells in HIV-infected patients [69]. For both HBV- and HCV-infected patients, HIV coinfection seems to accelerate the progression of liver diseases, leading to increased risk of cirrhosis and possibly liver cancer [19, 70, 71]. In HIV-infected patients, HBV infection carries higher risk of chronicity, higher viral load, lower chance of spontaneous viral clearance, faster progression to cirrhosis, and appears to cause faster and more frequent progression to HCC [63, 72–77]. HIV infection alone has also been reported to increase the risk the HCC, as well as other cancers, presumably as a result of defective immune surveillance against tumorigenesis [68, 78]. Moreover, triple infection with HIV, HBV, and HCV is not uncommon and may present an even greater risk of HCC development than the double infections. No clear data are available in this regard, as it takes years to decades for HCC to develop in HBV- or HCV-infected patients and thus long-term follow-ups are necessary to assess any increased cancer risks in HIV coinfected patients. Before definitive measurements of liver cancer incidences, surrogate markers are sometimes used to predict HCC development. These surrogates include the progression of cirrhosis, which, as mentioned earlier, almost always precedes the development of HCC, and increased viral load, which may predict an increased risk of HCC (Fig. 9.6). In the case of HBV, there is strong evidence now to indicate that HBV viral load is in fact directly correlated with liver disease progression and the risk of HCC development [79–81], and effective antiviral treatment to decrease HBV replication has been shown to decrease the risk of HCC development [82–85]. However, HCV viral load in the blood does not seem to be correlated directly with liver disease progression [86, 87].

A major hurdle in trying to understand the consequences of HIV coinfection in the setting of HBV or HCV chronic infections is the lack of cell culture systems or convenient animal models that can be infected by both HIV and HBV or HCV. Infection of chimpanzees, which are susceptible to all three viruses, is a possibility, but these studies would be very costly and take years to conduct, and are no longer allowed with U.S. government funding. As already alluded to earlier, the chance of HIV coinfecting the same host cell with either HBV or HCV seems to be remote. Any effect of HIV infection on HCC risk associated with HBV or HCV would be unlikely to be exerted at the level of direct virus–virus interactions. Rather, indirect effects of HIV-mediated immune dysfunction on HBV- or HCV-induced hepatocarcinogenesis are more likely (Fig. 9.6).

Mechanistically, the increased chance of chronicity of HBV and HCV infections associated with HIV coinfection, as discussed above, can account for some of the increase in the risk of HCC, as chronic HBV or HCV infection is clearly associated with HCC. The increased HBV or HCV load associated with HIV coinfection could also potentially exacerbate liver damage and thus accelerate disease progression and ultimately cancer development. In this regard, recent studies suggest that under conditions of severe immunodeficiency, the normally non-cytopathic HBV can damage the infected cells directly with uncontrolled high-level replication [88]. Similarly, although HCV is usually considered to be non-cytopathic, HCV replication or HCV proteins may nevertheless directly induce cellular damage such as steatosis [89], which also occurs more frequently in HCV–HIV coinfections [90]. The inflammatory response and cytokines induced by HIV infection may worsen HBV- or HCV-associated liver diseases [77, 91, 92]. In HIV–HCV coinfections, the enteropathy induced by HIV leads to microbial translocation from the intestinal tract to the liver, which is reported to accelerate fibrosis [93] and may thus play a role in enhancing HCV hepatocarcinogenesis. On the other hand, a decreased immune response against HBV or HCV, as a result of HIV-induced immune suppression, might actually reduce liver inflammation and damage [68] and thus, slows down progression to cirrhosis and cancer (Fig. 9.6).

The increased HBV or HCV load in HIV coinfected patients can, in principle, also influence cancer progression by increasing the expression of HBV or HCV proteins that may be more directly involved in cellular dis-regulation and transformation and in the case of HBV, the chance of insertional mutagenesis (Fig. 9.6). Furthermore, some reports suggest that in the HBV/HIV coinfected patients, certain HBV variants that are associated with enhanced carcinogenicity may be selected, which could lead to increased risk of HCC development in these dually infected patients [94, 95].

Although HIV is not known to infect hepatocytes, HIV proteins may nevertheless still be able to influence the HBV- or HCV-infected cells. For example, the HIV transactivating protein, Tat, which can be present systemically during HIV infection, has been reported to enhance the development of liver cancer [96] and may thus influence the development of HCC in HIV coinfected HBV or HCV patients.

9.8 Anti-Retroviral-Therapy-Associated Hepatotoxicity

Anti-retroviral-therapy-induced liver toxicity is an additional concern in HIV coinfected HCV or HBV patients. One of eight patients treated with anti-retroviral drugs show hepatotoxicity, a situation that is more likely to occur in HBV- or HCV-infected patients that further exacerbates liver damage accompanying chronic HBV or HCV infections [1, 19, 76, 97]. Anti-retroviral drugs that have shown hepatotoxicity include certain nucleoside analogs (HIV RT inhibitors) and HIV protease inhibitors. A further complication in the treatment of HIV–HBV coinfected patients is the fact that some nucleoside analogs, such as 3TC (lamivudine), are active against both the HIV and HBV RT, and can select for drug-resistant mutants of both viruses. HIV-infected patients are sometimes treated intermittently, in order to prevent the selection of drug-resistant HIV. However, hepatic flare can result when a patient is taken off anti-retroviral therapy. This is thought to be due to HBV viral rebound upon drug withdrawal and can lead to an increase in liver damage and subsequent progression to cirrhosis and HCC. It is, therefore, important to ensure that the coinfecting HBV infection continues to be treated while the patient is off the HIV treatment, e.g., using nucleoside analogs specific for the HBV RT but inactive against the HIV RT.

9.9 Management of HIV–HBV and HIV–HCV Coinfection

As alluded to above, a number of nucleoside analog inhibitors of the HIV RT (NRTIs) proved to be also active against the HBV RT and indeed have been approved for HBV therapy, including lamivudine, adefovir, and tenofovir [98]. The less potent of these such as lamivudine and adefovir can lead to the rapid selection of drug-resistant HBV mutants, rebound of viral replication, and liver disease progression. Therefore, it is important to select NRTIs that are not only effective against both HIV and HBV but also have a high barrier to resistance, i.e., tenofovir, as no combination therapy is available yet for HBV, in contrast to HIV [73, 76, 99–101]. Some studies suggest that tenofovir may be less effective against HBV in the setting of HBV–HIV coinfection [102], which warrants further studies. As the current treatment for either HBV or HIV is not curative, long-term (and likely lifelong) treatment is required to control either viral infection, careful monitoring of viral resistance as well as drug toxicity is a necessity.

In the case of HCV–HIV coinfection, it is great news that the recently developed anti-HCV drugs that target different HCV proteins (so-called direct-acting antivirals or DAAs) and are highly active against HCV mono-infection remain highly active in HCV–HIV coinfected patients [103–105]. This has raised the hope that effective antiviral therapy against HCV in the HIV-coinfected patients will alleviate the HCV-associated liver diseases. Conversely, effective antiviral therapy against HIV in HIV–HCV coinfected patients can slow liver disease progression induced by HCV infection [106]. On the other hand, drug–drug interactions between anti-HCV and anti-HIV drugs leading to adverse effects in the dually infected and treated patients have been observed and need to be carefully monitored [107, 108].

A concern related to HIV–HBV coinfection is the reported adverse effect of HIV infection on the efficacy of HBV vaccination. A number of studies have indicated that HBV vaccine efficacy, which is normally very high (ca. 95% response rate), can be decreased in HIV-infected patients, especially in those with low CD4 T cell counts [109–114]. As HBV vaccination is an essential part of the global strategy to control HBV infection, this apparent detrimental effect of HIV infection on HBV vaccine efficacy should be closely monitored.

9.10 HGV–HIV Coinfection

An intriguing interaction between HIV and another prevalent human virus, the hepatitis G virus (HGV or GBV-C), may in fact be beneficial to the host [115–117]. Although initially thought to be one of the viruses that can cause hepatitis (hence the name HGV), HGV is not known to cause any human disease but is a relatively common virus that is found worldwide. Like HCV, it is a single-stranded, positive-sense RNA virus that belongs to the family *Flaviviridae*. HGV is transmitted through blood and body fluids, similar to HIV, HBV, or HCV. Different from HCV, HGV primarily infects lymphocytes. Interestingly, in persons coinfected with HIV and HGV, HGV appears to confer protection against the progression to AIDS [118, 119]. This was largely demonstrated before the advent of HAART but has been shown in the post-HAART era as well. Furthermore, HIV replication in vitro was shown to be inhibited by coinfection with HGV, suggesting a mechanism involving direct interaction between the viruses [120]. However, not all studies on the effect of HGV coinfection have reported favorable results and the mechanism that provides this putative protection has yet to be elucidated. For the purposes of this review, HGV does not appear to influence liver disease in HBV or HCV coinfections.

9.11 Summary

Liver diseases caused by chronic HBV or HCV infection, including cirrhosis and HCC, have emerged as an increasingly important problem faced by millions of HIV-infected patients who are coinfected with HBV or HCV. On one hand, HIV-induced immune suppression enhances the risk of chronic viral hepatitis, increases HBV or HCV load, and may hasten the progression to cirrhosis and liver cancer. On the other hand, significant hepatotoxicity is associated with a number of anti-retroviral drugs, further exacerbating liver damage associated with chronic viral hepatitis. The elucidation of the multiple virus–virus and virus–host interactions that underlie viral hepatocarcinogenesis and potential HIV enhancement will be facilitated greatly by the establishment of appropriate in vitro and in vivo model systems. As millions of HIV-infected patients in the developing countries are gaining access to HAART therapy for their HIV infections, endemic HBV and HCV infections and their associated liver diseases will only become more problematic on a global scale.

Fortunately, recent progresses in developing tenofovir-based regimens that are effective against both HIV and HBV infection, and esp. curative therapies for HCV infection, are improving the prognosis for the HIV coinfected patients. Recent focus to find a cure for HBV [121] and ongoing efforts to pursue a cure for HIV infection [122], along with the wider application of the curative HCV therapies, promise to rid the world of these deadly infections and eliminate the vast majority of HCC, a highly malignant cancer for which little treatment options are available. On the other hand, while overall liver-disease-related deaths among HIV–HCV coinfected patients in some countries may be decreasing due to effective surveillance and antiviral treatments, there has been no clear reduction overall in liver disease risk associated with HBV or HCV coinfections in HIV-infected patients and HCC incidence continues to increase [123–125], possibly reflecting the irreversible carcinogenic events that have occurred before effective antiviral treatments and highlighting the need to continue to monitor HCC risks even among the patients undergoing effective antiviral therapy.

References

1. Thomas DL (2006) Growing importance of liver disease in HIV-infected persons. Hepatology 43(2 Suppl 1):S221–S229
2. Cacoub P, Geffray L, Rosenthal E, Perronne C, Veyssier P, Raguin G (2001) Mortality among human immunodeficiency virus-infected patients with cirrhosis or hepatocellular carcinoma due to hepatitis C virus in French departments of internal medicine/Infectious diseases, in 1995 and 1997. Clin Infect Dis 32(8):1207–1214
3. Cooper CL, Mills E, Wabwire BO, Ford N, Olupot-Olupot P (2009) Chronic viral hepatitis may diminish the gains of HIV antiretroviral therapy in sub-Saharan Africa. Int J Infect Dis. 13(3):302–306
4. Koziel MJ, Peters MG (2007) Viral hepatitis in HIV infection. N Engl J Med 356(14): 1445–1454
5. WHO (2017) Global hepatitis report, 2017
6. Hu J (2016) Hepatitis B virus virology and replication. In: Liaw Y-F, Zoulim F (eds) Hepatitis B virus in human diseases. Humana Press, Springer Cham Heidelberg New York Dordrecht London, pp 1–34
7. Ganem D, Prince AM (2004) Hepatitis B virus infection–natural history and clinical consequences. N Engl J Med 350(11):1118–1129
8. Slagle BL, Andrisani OM, Bouchard MJ, Lee CG, Ou JH, Siddiqui A (2015) Technical standards for hepatitis B virus X protein (HBx) research. Hepatology 61(4):1416−1424
9. Decorsiere A, Mueller H, van Breugel PC, Abdul F, Gerossier L, Beran RK et al (2016) Hepatitis B virus X protein identifies the Smc5/6 complex as a host restriction factor. Nature 531(7594):386–389
10. Yim HJ, Lok AS (2006) Natural history of chronic hepatitis B virus infection: what we knew in 1981 and what we know in 2005. Hepatology 43(2 Suppl 1):S173–S181
11. Wright TL (2006) Introduction to chronic hepatitis B infection. Am J Gastroenterol 101(Suppl 1):S1–S6
12. Ott JJ, Stevens GA, Groeger J, Wiersma ST (2012) Global epidemiology of hepatitis B virus infection: new estimates of age-specific HBsAg seroprevalence and endemicity. Vaccine 30(12):2212–2219
13. Schweitzer A, Horn J, Mikolajczyk RT, Krause G, Ott JJ (2015) Estimations of worldwide prevalence of chronic hepatitis B virus infection: a systematic review of data published between 1965 and 2013. Lancet 386(10003):1546–1555
14. Lavanchy D, Kane M (2016) Global epidemiology of hepatitis B virus infection. In: Liaw Y-F, Zoulim F (eds) Hepatitis B virus in human diseases. Humana Press, Springer Cham Heidelberg New York Dordrecht London, pp 187–203
15. Trepo C, Chan HL, Lok A (2014) Hepatitis B virus infection. Lancet 384(9959):2053–2063
16. Lindenbach BD, Rice CM (2005) Unravelling hepatitis C virus replication from genome to function. Nature 436(7053):933–938
17. Chen SL, Morgan TR (2006) The natural history of hepatitis C virus (HCV) infection. Int J Med Sci. 3(2):47–52
18. Seeff LB (1999) Natural history of hepatitis C. Am J Med 107(6B):10S–15S
19. Park JS, Saraf N, Dieterich DT (2006) HBV plus HCV, HCV plus HIV, HBV plus HIV. Curr Gastroenterol Rep 8(1):67–74
20. Alter MJ (2006) Epidemiology of viral hepatitis and HIV co-infection. J Hepatol 44(1 Suppl):S6–S9
21. Spradling PR, Richardson JT, Buchacz K, Moorman AC, Brooks JT, Investigators HIVOS (2010) Prevalence of chronic hepatitis B virus infection among patients in the HIV outpatient study, 1996–2007. J Viral Hepat 17(12):879-86
22. Alexiev I, Alexandrova M, Golkocheva-Markova E, Teoharov P, Gancheva A, Kostadinova A et al (2017) High rate of hepatitis B and C coinfections among people living with HIV-1 in Bulgaria: 2010–2014. AIDS Res Hum Retroviruses 33(3):228–229

23. Martin-Carbonero L, Poveda E (2012) Hepatitis B virus and HIV infection. Semin Liver Dis 32(2):114–119
24. Sulkowski MS (2008) Viral hepatitis and HIV coinfection. J Hepatol 48(2):353–367
25. Lascar RM, Lopes AR, Gilson RJ, Dunn C, Johnstone R, Copas A et al (2005) Effect of HIV infection and antiretroviral therapy on hepatitis B virus (HBV)-specific T cell responses in patients who have resolved HBV infection. J Infect Dis 191(7):1169–1179
26. Yeo W, Lam KC, Zee B, Chan PS, Mo FK, Ho WM et al (2004) Hepatitis B reactivation in patients with hepatocellular carcinoma undergoing systemic chemotherapy. Ann Oncol 15(11):1661–1666
27. Gupta P, Hepatitis C (2013) Virus and HIV type 1 co-infection. Infect Dis Rep. 5(Suppl 1):e7
28. Sterling RK, Sulkowski MS (2004) Hepatitis C virus in the setting of HIV or hepatitis B virus coinfection. Semin Liver Dis 24(Suppl 2):61–68
29. Liu Z, Hou J (2006) Hepatitis B virus (HBV) and hepatitis C virus (HCV) dual infection. Int J Med Sci 3(2):57–62
30. Chemin I, Trepo C (2005) Clinical impact of occult HBV infections. J Clin Virol 34(Suppl 1): S15–S21
31. Tamori A, Nishiguchi S, Shiomi S, Hayashi T, Kobayashi S, Habu D et al (2005) Hepatitis B virus DNA integration in hepatocellular carcinoma after interferon-induced disappearance of hepatitis C virus. Am J Gastroenterol 100(8):1748–1753
32. Michalak TI, Mulrooney PM, Coffin CS (2004) Low doses of hepadnavirus induce infection of the lymphatic system that does not engage the liver. J Virol 78(4):1730–1738
33. Pal S, Sullivan DG, Kim S, Lai KK, Kae J, Cotler SJ et al (2006) Productive replication of hepatitis C virus in perihepatic lymph nodes in vivo: implications of HCV lymphotropism. Gastroenterology 130(4):1107–1116
34. Brechot C (2004) Pathogenesis of hepatitis B virus-related hepatocellular carcinoma: old and new paradigms. Gastroenterology 127(5 Suppl 1):S56–S61
35. Branda M, Wands JR (2006) Signal transduction cascades and hepatitis B and C related hepatocellular carcinoma. Hepatology 43(5):891–902
36. Thorgeirsson SS, Grisham JW (2002) Molecular pathogenesis of human hepatocellular carcinoma. Nat Genet 31(4):339–346
37. Fattovich G, Stroffolini T, Zagni I, Donato F (2004) Hepatocellular carcinoma in cirrhosis: incidence and risk factors. Gastroenterology 127(5 Suppl 1):S35–S50
38. Chen JG, Zhu J, Parkin DM, Zhang YH, Lu JH, Zhu YR, et al (2006) Trends in the incidence of cancer in Qidong, China, 1978–2002. Int J Cancer
39. Davila JA, Morgan RO, Shaib Y, McGlynn KA, El-Serag HB (2004) Hepatitis C infection and the increasing incidence of hepatocellular carcinoma: a population-based study. Gastroenterology 127(5):1372–1380
40. El-Serag HB (2004) Hepatocellular carcinoma: recent trends in the United States. Gastroenterology 127(5 Suppl 1):S27–S34
41. White DL, Thrift AP, Kanwal F, Davila J, El-Serag HB (2017) Incidence of hepatocellular carcinoma in all 50 United States, from 2000 through 2012. Gastroenterology 152(4): 812–20 e5
42. Parkin DM (2006) The global health burden of infection-associated cancers in the year 2002. Int J Cancer 118(12):3030–3044
43. Tradati F, Colombo M, Mannucci PM, Rumi MG, De Fazio C, Gamba G, et al (1998) A prospective multicenter study of hepatocellular carcinoma in italian hemophiliacs with chronic hepatitis C. The study group of the association of Italian hemophilia centers. Blood 91(4):1173–7
44. Liang TJ, Heller T (2004) Pathogenesis of hepatitis C-associated hepatocellular carcinoma. Gastroenterology 127(5 Suppl 1):S62–S71
45. Block TM, Mehta AS, Fimmel CJ, Jordan R (2003) Molecular viral oncology of hepatocellular carcinoma. Oncogene 22(33):5093–5107

46. Anzola M (2004) Hepatocellular carcinoma: role of hepatitis B and hepatitis C viruses proteins in hepatocarcinogenesis. J Viral Hepat 11(5):383–393
47. Guerrieri F, Belloni L, Pediconi N, Levrero M (2016) Pathobiology of Hepatitis B Virus-Induced Hepatocarcingenesis. In: Liaw Y-F, Zoulim F (eds) Hepatitis B virus in Human Diseases. Humana Press, Springer Cham Heidelberg New York Dordrecht London, pp 95–121
48. Cougot D, Neuveut C, Buendia MA (2005) HBV induced carcinogenesis. J Clin Virol 34(Suppl 1):S75–S78
49. Zhao LH, Liu X, Yan HX, Li WY, Zeng X, Yang Y et al (2016) Genomic and oncogenic preference of HBV integration in hepatocellular carcinoma. Nat Commun 7:12992
50. Mason WS, Gill US, Litwin S, Zhou Y, Peri S, Pop O, et al (2016) HBV DNA integration and clonal hepatocyte expansion in chronic hepatitis B patients considered immune tolerant. Gastroenterology 151(5):986–98 e4
51. Murakami Y, Saigo K, Takashima H, Minami M, Okanoue T, Brechot C et al (2005) Large scaled analysis of hepatitis B virus (HBV) DNA integration in HBV related hepatocellular carcinomas. Gut 54(8):1162–1168
52. Minami M, Daimon Y, Mori K, Takashima H, Nakajima T, Itoh Y et al (2005) Hepatitis B virus-related insertional mutagenesis in chronic hepatitis B patients as an early drastic genetic change leading to hepatocarcinogenesis. Oncogene 24(27):4340–4348
53. Hsu T, Moroy T, Etiemble J, Louise A, Trepo C, Tiollais P et al (1988) Activation of c-myc by woodchuck hepatitis virus insertion in hepatocellular carcinoma. Cell 55:627–635
54. Moroy T, Marchio A, Etiemble J, Trepo C, Tiollais P, Buendia M (1986) Rearrangement and enhanced expression of c-myc in hepatocellular carcinoma of hepatitis virus infected woodchucks. Nature 324:276–279
55. Fourel G, Trepo C, Bougueleret L, Henglein B, Ponzetto A, Tiollais P et al (1990) Frequent activation of N-myc genes by hepadnavirus insertion in woodchuck liver tumours. Nature 347(6290):294–298
56. Dejean A, Bougueleret L, Grzeschik KH, Tiollais P (1986) Hepatitis B virus DNA integration in a sequence homologous to v-erb-A and steroid receptor genes in a hepatocellular carcinoma. Nature 322:70–72
57. Wang J, Chenivesse X, Henglein B, Brechot C (1990) Hepatitis B virus integration in a cyclin A gene in a hepatocellular carcinoma. Nature 343(6258):555–557
58. Benhamou Y, Bochet M, Di Martino V, Charlotte F, Azria F, Coutellier A et al (1999) Liver fibrosis progression in human immunodeficiency virus and hepatitis C virus coinfected patients. The multivirc group. Hepatology 30(4):1054–1058
59. Colin JF, Cazals-Hatem D, Loriot MA, Martinot-Peignoux M, Pham BN, Auperin A et al (1999) Influence of human immunodeficiency virus infection on chronic hepatitis B in homosexual men. Hepatology 29(4):1306–1310
60. Weinig M, Hakim JG, Gudza I, Tobaiwa O (1997) Hepatitis C virus and HIV antibodies in patients with hepatocellular carcinoma in Zimbabwe: a pilot study. Trans R Soc Trop Med Hyg 91(5):570–572
61. Tswana SA, Moyo SR (1992) The interrelationship between HBV-markers and HIV antibodies in patients with hepatocellular carcinoma. J Med Virol 37(3):161–164
62. Sherman M (2006) Optimizing management strategies in special patient populations. Am J Gastroenterology 101(Suppl 1):S26–S31
63. Puoti M, Bruno R, Soriano V, Donato F, Gaeta GB, Quinzan GP et al (2004) Hepatocellular carcinoma in HIV-infected patients: epidemiological features, clinical presentation and outcome. AIDS 18(17):2285–2293
64. Kramer JR, Giordano TP, Souchek J, Richardson P, Hwang LY, El-Serag HB (2005) The effect of HIV coinfection on the risk of cirrhosis and hepatocellular carcinoma in U.S. veterans with hepatitis C. Am J Gastroenterol 100(1):56–63

65. Giordano TP, Kramer JR, Souchek J, Richardson P, El-Serag HB (2004) Cirrhosis and hepatocellular carcinoma in HIV-infected veterans with and without the hepatitis C virus: a cohort study, 1992–2001. Arch Intern Med 164(21):2349–2354
66. Garcia-Samaniego J, Rodriguez M, Berenguer J, Rodriguez-Rosado R, Carbo J, Asensi V et al (2001) Hepatocellular carcinoma in HIV-infected patients with chronic hepatitis C. Am J Gastroenterol 96(1):179–183
67. Bruno R, Puoti M, Sacchi P, Filice C, Carosi G, Filice G (2006) Management of hepatocellular carcinoma in human immunodeficiency virus-infected patients. J Hepatol 44(1 Suppl):S146–S150
68. Smukler AJ, Ratner L (2002) Hepatitis viruses and hepatocellular carcinoma in HIV-infected patients. Curr Opin Oncol 14(5):538–542
69. Goeser F, Glassner A, Kokordelis P, Wolter F, Lutz P, Kaczmarek DJ et al (2016) HIV mono-infection is associated with an impaired anti-hepatitis C virus activity of natural killer cells. AIDS 30(3):355–363
70. Hyun CB, Coyle WJ (2004) Hepatocellular carcinoma in a patient with human immunodeficiency virus and hepatitis B virus coinfection: an emerging problem? South Med J 97(4):401–406
71. Chew KW, Bhattacharya D (2016) Virologic and immunologic aspects of HIV-hepatitis C virus coinfection. AIDS 30(16):2395–2404
72. Kew MC, Smuts H, Stewart A (2010) Does HIV infection enhance the hepatocarcinogenic potential of chronic hepatitis B virus infection? J Acquir Immune Defic Syndr 53(3): 413–414
73. Soriano V, Puoti M, Peters M, Benhamou Y, Sulkowski M, Zoulim F et al (2008) Care of HIV patients with chronic hepatitis B: updated recommendations from the HIV-Hepatitis B Virus International Panel. AIDS 22(12):1399–1410
74. Puoti M, Torti C, Bruno R, Filice G, Carosi G (2006) Natural history of chronic hepatitis B in co-infected patients. J Hepatol 44(1 Suppl):S65–S70
75. Brau N, Fox RK, Xiao P, Marks K, Naqvi Z, Taylor LE, et al (2007) Presentation and outcome of hepatocellular carcinoma in HIV-infected patients: a U.S.-Canadian multicenter study. J Hepatol 47(4):527–37
76. Dharel N, Sterling RK (2014) Hepatitis B virus-HIV coinfection: forgotten but not gone. Gastroenterol Hepatol (N Y). 10(12):780–788
77. Iser DM, Lewin SR (2009) The pathogenesis of liver disease in the setting of HIV-hepatitis B virus coinfection. Antivir Ther 14(2):155–164
78. Tanaka T, Imamura A, Masuda G, Ajisawa A, Negishi M, Tanaka S et al (1996) A case of hepatocellular carcinoma in HIV-infected patient. Hepatogastroenterology 43(10):1067–1072
79. Chen CJ, Yang HI, Su J, Jen CL, You SL, Lu SN et al (2006) Risk of hepatocellular carcinoma across a biological gradient of serum hepatitis B virus DNA level. JAMA 295(1):65–73
80. Ohata K, Hamasaki K, Toriyama K, Ishikawa H, Nakao K, Eguchi K (2004) High viral load is a risk factor for hepatocellular carcinoma in patients with chronic hepatitis B virus infection. J Gastroenterol Hepatol 19(6):670–675
81. Hu J, Nguyen D (2004) Therapy for chronic hepatitis B: the earlier, the better? Trends Microbiol 12(10):431–433
82. van Zonneveld M, Honkoop P, Hansen BE, Niesters HG, Murad SD, de Man RA et al (2004) Long-term follow-up of alpha-interferon treatment of patients with chronic hepatitis B. Hepatology 39(3):804–810
83. Akuta N, Suzuki F, Suzuki Y, Sezaki H, Hosaka T, Someya T et al (2005) Favorable efficacy of long-term lamivudine therapy in patients with chronic hepatitis B: an 8-year follow-up study. J Med Virol 75(4):491–498
84. Liaw YF, Sung JJ, Chow WC, Farrell G, Lee CZ, Yuen H et al (2004) Lamivudine for patients with chronic hepatitis B and advanced liver disease. N Engl J Med 351(15): 1521–1531

85. Matsumoto A, Tanaka E, Rokuhara A, Kiyosawa K, Kumada H, Omata M, et al (2005) Efficacy of lamivudine for preventing hepatocellular carcinoma in chronic hepatitis B: a multicenter retrospective study of 2795 patients. Hepatol Res 32(3):173–184
86. Shindo M, Hamada K, Oda Y, Okuno T (2001) Long-term follow-up study of sustained biochemical responders with interferon therapy. Hepatology 33(5):1299–1302
87. Tsuda N, Yuki N, Mochizuki K, Nagaoka T, Yamashiro M, Omura M et al (2004) Long-term clinical and virological outcomes of chronic hepatitis C after successful interferon therapy. J Med Virol 74(3):406–413
88. Meuleman P, Libbrecht L, Wieland S, De Vos R, Habib N, Kramvis A et al (2006) Immune suppression uncovers endogenous cytopathic effects of the hepatitis B virus. J Virol 80(6): 2797–2807
89. Yoon EJ, Hu KQ (2006) Hepatitis C virus (HCV) infection and hepatic steatosis. Int J Med Sci 3(2):53–56
90. Gaslightwala I, Bini EJ (2006) Impact of human immunodeficiency virus infection on the prevalence and severity of steatosis in patients with chronic hepatitis C virus infection. J Hepatol 44(6):1026–1032
91. Shmagel KV, Saidakova EV, Shmagel NG, Korolevskaya LB, Chereshnev VA, Robinson J et al (2016) Systemic inflammation and liver damage in HIV/hepatitis C virus coinfection. HIV Med 17(8):581–589
92. Rivera MM, Soza A, Jazwinski A, Mi L, Kleiner DE, Zhao X et al (2015) HIV through the looking glass: insights derived from Hepatitis B. J Acquir Immune Defic Syndr 68(2):123–127
93. Page EE, Nelson M, Kelleher P (2011) HIV and hepatitis C coinfection: pathogenesis and microbial translocation. Curr Opin HIV AIDS 6(6):472–477
94. Li KW, Kramvis A, Liang S, He X, Chen QY, Wang C et al (2017) Higher prevalence of cancer related mutations 1762T/1764A and PreS deletions in hepatitis B virus (HBV) isolated from HBV/HIV co-infected compared to HBV-mono-infected Chinese adults. Virus Res 227:88–95
95. Sung FY, Lan CY, Huang CJ, Lin CL, Liu CJ, Chen PJ et al (2016) Progressive accumulation of mutations in the hepatitis B virus genome and its impact on time to diagnosis of hepatocellular carcinoma. Hepatology 64(3):720–731
96. Altavilla G, Caputo A, Lanfredi M, Piola C, Barbanti-Brodano G, Corallini A (2000) Enhancement of chemical hepatocarcinogenesis by the HIV-1 tat gene. Am J Pathol 157(4):1081–1089
97. Pol S, Lebray P, Vallet-Pichard A (2004) HIV infection and hepatic enzyme abnormalities: intricacies of the pathogenic mechanisms. Clin Infect Dis 38(Suppl 2):S65–S72
98. Revill P, Locarnini S (2016) The basis for antiviral therapy: drug targets, cross-resistance, and novel small nolecule inhibitors. In: Liaw Y-F, Zoulim F (eds) Hepatitis B virus in human diseases. Humana Press, Springer Cham Heidelberg New York Dordrecht London, pp 303–324
99. Soriano V, de Mendoza C, Pena JM, Barreiro P (2015) Advances in treating drug-resistant hepatitis B virus in HIV-infected patients. Expert Opin Pharmacother 16(2):179–186
100. Soriano V, Labarga P, de Mendoza C, Pena JM, Fernandez-Montero JV, Benitez L et al (2015) Emerging challenges in managing hepatitis B in HIV patients. Curr HIV/AIDS Rep 12(3):344–352
101. Sherman KE (2015) Management of the hepatitis B virus/HIV-coinfected patient. Top Antivir Med 23(3):111–114
102. Audsley J, Bent SJ, Littlejohn M, Avihingsanon A, Matthews G, Bowden S et al (2016) Effects of long-term tenofovir-based combination antiretroviral therapy in HIV-hepatitis B virus coinfection on persistent hepatitis B virus viremia and the role of hepatitis B virus quasispecies diversity. AIDS 30(10):1597–1606
103. Piroth L, Wittkop L, Lacombe K, Rosenthal E, Gilbert C, Miailhes P, et al (2017) Efficacy and safety of direct-acting antiviral regimens in HIV/HCV-coinfected patients—French ANRS CO13 HEPAVIH cohort. J Hepatol 67(1):23–31

104. Meissner EG (2017) Update in HIV-hepatitis C virus coinfection in the direct acting antiviral era. Curr Opin Gastroenterol 33(3):120–127
105. Mandorfer M, Schwabl P, Steiner S, Scheiner B, Chromy D, Bucsics T et al (2016) Interferon-free treatment with sofosbuvir/daclatasvir achieves sustained virologic response in 100% of HIV/hepatitis C virus-coinfected patients with advanced liver disease. AIDS 30(7):1039–1047
106. Woreta TA, Sutcliffe CG, Mehta SH, Brown TT, Higgins Y, Thomas DL et al (2011) Incidence and risk factors for steatosis progression in adults coinfected with HIV and hepatitis C virus. Gastroenterology 140(3):809–817
107. Macias J, Monge P, Mancebo M, Merchante N, Neukam K, Real LM, et al (2016) High frequency of potential interactions between direct-acting antivirals and concomitant therapy in HIV/hepatitis C virus-coinfected patients in clinical practice. HIV Med 18(7):445–451
108. Khatri A, Dutta S, Dunbar M, Podsadecki T, Trinh R, Awni W et al (2016) Evaluation of Drug-Drug Interactions between Direct-Acting Anti-Hepatitis C Virus Combination Regimens and the HIV-1 Antiretroviral Agents Raltegravir, Tenofovir, Emtricitabine, Efavirenz, and Rilpivirine. Antimicrob Agents Chemother 60(5):2965–2971
109. Mehta N, Cunningham CK, Flynn P, Pepe J, Obaro S, Kapogiannis BG et al (2010) Impaired generation of hepatitis B virus-specific memory B cells in HIV infected individuals following vaccination. Vaccine 28(21):3672–3678
110. van den Berg R, van Hoogstraten I, van Agtmael M (2009) Non-responsiveness to hepatitis B vaccination in HIV seropositive patients; possible causes and solutions. AIDS Rev 11(3): 157–164
111. Pollack TM, Trang le TT, Ngo L, Cuong do D, Thuy PT, Colby DJ (2016) Response to hepatitis B vaccination among HIV-infected adults in Vietnam. J Virus Erad 2(2):102–6
112. Njom Nlend AE, Nguwoh PS, Ngounouh CT, Tchidjou HK, Pieme CA, Otele JM et al (2016) HIV-infected or -exposed children exhibit lower immunogenicity to hepatitis B vaccine in Yaounde, Cameroon: an appeal for revised policies in Tropical settings? PLoS ONE 11(9):e0161714
113. Lopes VB, Hassing RJ, de Vries-Sluijs TE, El Barzouhi A, Hansen BE, Schutten M et al (2013) Long-term response rates of successful hepatitis B vaccination in HIV-infected patients. Vaccine 31(7):1040–1044
114. Whitaker JA, Rouphael NG, Edupuganti S, Lai L, Mulligan MJ (2012) Strategies to increase responsiveness to hepatitis B vaccination in adults with HIV-1. Lancet Infect Dis 12(12): 966–976
115. Stapleton JT, Williams CF, Xiang J (2004) GB virus type C: a beneficial infection? J Clin Microbiol 42(9):3915–3919
116. Berzsenyi MD, Bowden DS, Roberts SK (2005) GB virus C: insights into co-infection. J Clin Virol 33(4):257–266
117. Tenckhoff S, Kaiser T, Bredeek F, Donfield S, Menius E, Lail A et al (2012) Role of GB virus C in HIV-1-infected and hepatitis C virus-infected hemophiliac children and adolescents. J Acquir Immune Defic Syndr 61(2):243–248
118. Xiang J, Wunschmann S, Diekema DJ, Klinzman D, Patrick KD, George SL et al (2001) Effect of coinfection with GB virus C on survival among patients with HIV infection. N Engl J Med 345(10):707–714
119. Tillmann HL, Heiken H, Knapik-Botor A, Heringlake S, Ockenga J, Wilber JC et al (2001) Infection with GB virus C and reduced mortality among HIV-infected patients. N Engl J Med 345(10):715–724
120. Xiang J, George SL, Wunschmann S, Chang Q, Klinzman D, Stapleton JT (2004) Inhibition of HIV-1 replication by GB virus C infection through increases in RANTES, MIP-1alpha, MIP-1beta, and SDF-1. Lancet 363(9426):2040–2046
121. Revill P, Testoni B, Locarnini S, Zoulim F (2016) Global strategies are required to cure and eliminate HBV infection. Nat Rev Gastroenterol Hepatol 13(4):239–248

122. Deeks SG, Lewin SR, Ross AL, Ananworanich J, Benkirane M, Cannon P et al (2016) International AIDS Society global scientific strategy: towards an HIV cure 2016. Nat Med 22(8):839–850
123. Gjaerde LI, Shepherd L, Jablonowska E, Lazzarin A, Rougemont M, Darling K et al (2016) Trends in incidences and risk factors for hepatocellular carcinoma and other liver events in HIV and hepatitis C virus-coinfected individuals from 2001 to 2014: a multicohort study. Clin Infect Dis 63(6):821–829
124. Peters L, Klein MB (2015) Epidemiology of hepatitis C virus in HIV-infected patients. Curr Opin HIV AIDS 10(5):297–302
125. Klein MB, Althoff KN, Jing Y, Lau B, Kitahata M, Lo Re V, 3rd, et al (2016) Risk of end-stage liver disease in HIV-viral hepatitis coinfected persons in North America from the early to modern antiretroviral therapy Eras. Clin Infect Dis 63(9):1160–7

MIX
Papier aus verantwortungsvollen Quellen
Paper from responsible sources
FSC® C105338

If you have any concerns about our products,
you can contact us on
ProductSafety@springernature.com

In case Publisher is established outside the EU,
the EU authorized representative is:
Springer Nature Customer Service Center GmbH
Europaplatz 3, 69115 Heidelberg, Germany

Printed by Libri Plureos GmbH
in Hamburg, Germany